POLYMER SCIENCE AND TECHNOLOGY
Volume 12A

ADHESION AND ADSORPTION OF POLYMERS

POLYMER SCIENCE AND TECHNOLOGY

Volume 1 • STRUCTURE AND PROPERTIES OF POLYMER FILMS
Edited by Robert W. Lenz and Richard S. Stein

Volume 2 • WATER-SOLUBLE POLYMERS
Edited by N. M. Bikales

Volume 3 • POLYMERS AND ECOLOGICAL PROBLEMS
Edited by James Guillet

Volume 4 • RECENT ADVANCES IN POLYMER BLENDS, GRAFTS, AND BLOCKS
Edited by L. H. Sperling

Volume 5 • ADVANCES IN POLYMER FRICTION AND WEAR (Parts A and B)
Edited by Lieng-Huang Lee

Volume 6 • PERMEABILITY OF PLASTIC FILMS AND COATINGS
TO GASES, VAPORS, AND LIQUIDS
Edited by Harold B. Hopfenberg

Volume 7 • BIOMEDICAL APPLICATIONS OF POLYMERS
Edited by Harry P. Gregor

Volume 8 • POLYMERS IN MEDICINE AND SURGERY
Edited by Richard L. Kronenthal, Zale Oser, and E. Martin

Volume 9 • ADHESION SCIENCE AND TECHNOLOGY (Parts A and B)
Edited by Lieng-Huang Lee

Volume 10 • POLYMER ALLOYS: Blends, Blocks, Grafts, and Interpenetrating Networks
Edited by Daniel Klempner and Kurt C. Frisch

Volume 11 • POLYMER ALLOYS II: Blends, Blocks, Grafts, and Interpenetrating Networks
Edited by Daniel Klempner and Kurt C. Frisch

Volume 12 • ADHESION AND ADSORPTION OF POLYMERS (Parts A and B)
Edited by Lieng-Huang Lee

A Continuation Order Plan is available for this series. A continuation order will bring
delivery of each new volume immediately upon publication. Volumes are billed only upon
actual shipment. For further information please contact the publisher.

POLYMER SCIENCE AND TECHNOLOGY
Volume 12A

ADHESION AND ADSORPTION OF POLYMERS

Edited by
Lieng-Huang Lee

Xerox Corporation
Rochester, New York

PLENUM PRESS · NEW YORK AND LONDON

Library of Congress Cataloging in Publication Data

International Conference on Adhesion and Adsorption of Polymers, Honolulu, 1979.
 Adhesion and adsorption of polymers.

 (Polymer science and technology; v. 12A-B)
 Includes indexes.
 1. Polymers and polymerization—Congresses. 2. Adhesion—Congresses. 3. Adsorp-
tion—Congresses. I. Lee, Lieng-Huang, 1924-   II. Title. III. Series.
QD380.I53 1979   547.8'4   80-262
 ISBN-13:978-1-4613-3095-0   e-ISBN-13:978-1-4613-3093-6
 DOI: 10.1007/978-1-4613-3093-6

First half of the proceedings of an International Conference on
Adhesion and Adsorption of Polymers, Honolulu, Hawaii, April 2-6, 1979

© 1980 Plenum Press, New York
Softcover reprint of the hardcover 1st edition 1980
A Division of Plenum Publishing Corporation
227 West 17th Street, New York, N.Y. 10011

PREFACE

Honolulu is a most beautiful place, suitable for all occasions. Its choice as the meeting site for the first Joint Chemical Congress between the American Chemical Society and the Chemical Society of Japan was praised by scientists from both sides. During this Congress, the International Conference on Adhesion and Adsorption of Polymers was held at the Hyatt Regency Hotel between April 2 and 5, 1979. We had speakers from ten nations presenting over forty papers related to the subject matter. It was a memorable event.

Unlike our two previous adhesion symposia held in 1971 and 1975, this was the first time in the same conference that we discussed both adhesion and adsorption of polymers simultaneously. These two important phenomena are not only inter-related, but also equally important in adhesive technology as well as biochemical processes. The papers presented to this Conference deal with these two phenomena from both fundamental and practical viewpoints. Furthermore, with the advance of new surface analytical techniques, the actual, microscopic happenings at the interfaces can be pinpointed. Thus, characterization of interface became one of the major focuses of this Conference. As a result, a broad coverage of the subject matter includes statistical thermodynamics, surface physics, surface analysis, fracture mechanics, viscoelasticity, failure analysis, surface modification, adsorption kinetics, biopolymer adsorption, etc. Thanks to the diligence of our contributors, we are now able to publish the final papers in these two volumes.

Fig. 1.   From Left to Right: Dr. T. Hata,
          Mrs. and Dr. L.H. Lee, Dr. K. Nakao.

Fig. 2.   From Left to Right: Dr. and Mrs. A.
          Silberberg, Dr. L.H. Lee.

In these Proceedings we present the revised papers and discussions during the Conference in eight parts:

Part I: Polymer Surface Interactions

Part II: Characterization of Adhesive Interfaces

Part III: Polymeric Structural Adhesives

Part IV: Fracture Strengths of Polymeric Systems

Part V: Modification of Polymer Interfaces

Part VI: Kinetics of Polymer Adsorption

Part VII: Characterization of Adsorbed Interfaces

Part VIII: Adsorption of Biopolymers

Several papers were published elsewhere, so, instead, we have included two "Quickies" and one contribution after the Conference.

I would like to take this opportunity to thank our co-chairmen: Dr. K. Nakao (Fig. 1) and Dr. W.H. Grant. They helped very much in soliciting papers for this Conference. We would like to thank our plenary speakers: Professors I. Prigogine and A. Bellemans, Professor T. Hata (Fig. 1), Dr. S. Tsai, and Professor A. Silberberg (Fig. 2). We are grateful to Dr. Shiro Matsuoka for his talk at our Symposium Banquet on "Energy Conservation Problems for Future Plastics Industries." We also wish to thank our sponsoring organizations, the Division of Organic Coatings and Plastics Chemistry of the American Chemical Society and the Chemical Society of Japan.

Acknowledgement is made to the donors of the Petroleum Research Fund, administered by the American Chemical Society, for partial support of this Symposium.

Last but not least, I would like to thank Ms. Nancy Rickey for her patience in typing the entire Proceedings in camera-ready format.

                                               Lieng-Huang Lee

November, 1979

Contents of Volume 12A

PART ONE: Polymer Surface Interactions

Introductory Remarks   3
  LIENG-HUANG LEE

Statistical Mechanics of Surface Tension and Adsorption   5
  I. PRIGOGINE AND A. BELLEMANS

Surface and Interfacial Tensions of Polymer Melts
and Solutions   15
  TOSHIO HATA AND TOMOYUKI KASEMURA

Donor-Acceptor Interactions at Interfaces   43
  F.M. FOWKES

Surface Tension of Solids: Generalization and Reinterpretation of Critical Surface Tension   53
  SOUHENG WU

A Role of Molecular Forces in Adhesive Interactions of
Polymers   67
  V.A. BELYI, V.A. SMURUGOV AND A.T. SVIRIDYONOK

Discussion   77

PART TWO: Characterization of Adhesive
             Interfaces

Introductory Remarks   85
  LIENG-HUANG LEE

Photoacoustic Spectroscopy for the Study of Adhesion
and Adsorption of Dyes and Polymers   87
  LIENG-HUANG LEE

Adhesive-Adherend Bond Joint Characterization by Auger
Electron Spectroscopy and Photoelectron Spectroscopy   103
  J.S. SOLOMON, D. HANLIN AND N.T. McDEVITT

Techniques for Measuring Adhesive vs. Cohesive Failure   123
  TENNYSON SMITH AND PIERRE SMITH

Failure Characterization of a Structural Adhesive   141
  DAVID W. DWIGHT, EROL SANCAKTAR AND HALBERT F. BRINSON

Dielectric Relaxation Gradients in an Adhesive Bond   165
  JOHN L. CROWLEY AND ARTHUR D. JONATH

New Approach to the Understanding of Adhesive Interface
Phenomena and Bond Strength   175
  ARTHUR D. JONATH

Discussion   195

PART THREE: Polymeric Structural Adhesives

Introductory Remarks   201
  D.H. KAELBLE

Fracture Mechanics and Adherence of Viscoelastic Solids   203
  D. MAUGIS AND M. BARQUINS

The Viscoelastic Shear Behavior of a Structural Adhesive   279
  EROL SANCAKTAR AND H.F. BRINSON

Surface Energetics and Structural Reliability of Adhesive
Bonded Metal Structures   301
  D.H. KAELBLE

Composition and Ageing of a Structural, Epoxy Based Film
Adhesive   313
  C.E.M. MORRIS, A.G. MORITZ AND R.G. DAVIDSON

Viscoelastic Characterization of Structural Adhesive via
Force Oscillation Experiments   321
  *D.L. HUNSTON, W.D. BASCOM, E.E. WELLS, J.D. FAHEY
  AND J.L. BITNER*

Crack Healing in Semicrystalline Polymers, Block Copolymers
and Filled Elastomers   341
  RICHARD P. WOOL

Discussion   363

          PART FOUR: Fracture Strengths in Polymeric
                          Systems

Introductory Remarks   369
  KAZUMUNE NAKAO

Criterion of Interfacial Fracture on Tensile Adhesive Joint   371
  MINEO MASUOKA AND KAZUMUNE NAKAO

Deformation and Shear Strength of FRP Adhesive Joints   393
  HIROO MIYAIRI, HIDEAKI FUKUDA AND
  ATSUYOSHI MURAMATSU

Stress Distribution and Strength in Shear on the Adhesive
Lap Joint Loading Bending Moment   413
  YUKISABURO YAMAGUCHI AND SUSUMU AMANO

Fracture Criteria on Peeling   421
  TAKAHASHI IGARASHI

Superposition of Peel Rate, Temperature and Molecular Weight
for T Peel Strength of Polyisobutylene   439
  TOSHIYA TSUJI, MINEO MASUOKA AND KAZUMUNE NAKAO

Discussion   455

Author Index   xvii

Subject Index   xxxi

Contents of Volume 12B

PART FIVE: Modification of Polymer Interfaces

Introductory Remarks   459
  L.H. SHARPE

Role of Interface in the Strength of Composite Materials   463
  S.W. TSAI AND H.T. HAHN

Surface Modification of Polyethylene by Radiation-Induced
Grafting for Adhesive Bonding   473
  SHINZO YAMAKAWA

Adhesive Tack   497
  R. BATES

Adhesion Measurement of Polymeric Films and Coatings
with Special Reference to Photoresist Materials   503
  K.L. MITTAL AND R.O. LUSSOW

Adhesion of Thin Plasma-Polymerized Films to Metals   521
  PABITRA DATTA AND GRZEGORZ KAGANOWICZ

Adhesion of Filled Polyolefins to Metals and Glass   539
  V.A. BELYI, N.I. EGORENKOV AND A.I. KUZAVKOV

Adhesive Characteristics of Ultraviolet Radiation Curable
Resins   551
  K. NATE AND T. KOBAYASHI

An Evaluation of the Phenomena and Their Final Effects
Resulting from a Corona Discharge on Low Density
Polyethylene   563
  R. VAN DER LINDEN

Discussion   577

PART SIX: Kinetics of Polymer Adsorption

Introductory Remarks   587
  WARREN H. GRANT

Adsorption of Polymers from the Melt and Solutions of
Finite Concentration   591
  A. SILBERBERG

On the Structure of Boundary Layers of Polymers on
Solid Surfaces   601
  YURY S. LIPATOV

Selective Adsorption of Polymers from Solution   629
  RYONG-JOON ROE

Reaction in Models of Bio-Membranes in Connection with
Cell Motion   643
  A. SANFELD, A. STEINCHEN, W. DALLE VEDOVE AND P.M. BISCH

Kinetics of Cell Adhesion to a Hydrophilic-Hydrophobic
Copolymer Model System   667
  JONATHAN J. ROSEN AND MICHAEL B. SCHWAY

The Kinetics of Adsorption of Plasma Proteins to a Series
of Hydrophilic-Hydrophobic Copolymers   677
  THOMAS A. HORBETT

Discussion   683

PART SEVEN: Characterization of Adsorbed
                 Interfaces

Introductory Remarks   689
  D.L. ALLARA

Surface Characterization of Hydrophilic-Hydrophobic Copolymer
Model Systems I.   A Preliminary Study   691
  BUDDY D. RATNER AND ALLAN S. HOFFMAN

Characterization of Adsorption Polymers at the Liquid-Solid
and Liquid-Liquid Interfaces by Interfacial Tension   707
  TSUNETAKA MATSUMOTO AND KATSUHIKO NAKAMAE

Adsorption of Polyelectrolyte Studied by Ellipsometry   729
  AKIRA TAKAHASHI, MASAMI KAWAGUCHI AND TADAYA KATO

Adsorption Studies of Polymers on Metals by Fourier
Transform Reflection Infrared Spectroscopy   751
  D.L. ALLARA

Role of Coupling Agents in Metal-Polymer Adhesion: 1.
The Structure of the Silane Film at Metal-Polymer Interface   757
  C.S. PAIK SUNG, S.H. LEE AND N.H. SUNG

Microstructure, Orientation, and Mechanical Properties
of Polytetrafluoroethylene   775
  THEODORE DAVIDSON AND R.J. GOUNDER

Discussion   791

PART EIGHT: Adsorption of Bipolymers

Introductory Remarks   799
  WARREN H. GRANT

Adsorption of Proteins at Solid Surfaces   801
  WILLEM NORDE

The Adsorption of Human Serum Albumin and γ-Globulin
on Hydrophobic and Hydrophilic Surfaces   827
  WARREN H. GRANT AND RONALD E. DEHL

Adsorption of Soluble Collagen and Adsorption to Adsorbed
Soluble Collagen   837
  A. SILBERBERG

Conformational Changes of Structure of Immunoglobulin (IgG)
and its Fragments F_{ab} and F_c at the Air-Water and Water-Solid
Interfaces   847
  V.V. LAVRENTEV, L. CHASOVNIKOVA AND JU. SOROKIN

Discussion   855

About the Contributors   859

Author Index   867

Subject Index   881

PART ONE

POLYMER SURFACE INTERACTIONS

Introductory Remarks

Lieng-Huang Lee

Wilson Center for Technology

Xerox Corporation

Webster, New York   14580

This is a wonderful opportunity for us to celebrate the first Joint Chemical Congress between the American Chemical Society and the Chemical Society of Japan. The Organizing Committee of this International Conference on Adhesion and Adsorption of Polymers would like to thank you and our speakers for making this meeting possible.

There will be eight sessions of this Conference, dealing first with adhesion and then with adsorption of polymers. The subject of this session is "Polymer Surface Interactions." Some aspects of this subject matter were discussed during the two previous conferences (1,2). Today, as planned in the schedule, the first paper is a plenary lecture on "Statistical Mechanics of Surface Tension and Adsorption" by Professors I. Prigogine and A. Bellemans. Dr. Bellemans is scheduled to deliver this paper; however as a result of the United Airlines strike, he is still staying at the Los Angeles airport and waiting for an opportunity to fly out here. In the meantime, I have brought some extra sets of my own slides and shall briefly discuss the general concepts of surface tension, surface energy and surface free energy.

The second paper is another plenary lecture on "Surface and Interfacial Tension of Polymer Melts and Solutions," to be presented by Professor Toshio Hata, President of Gunma University, Japan. Dr. Hata attended our 1971 Symposium on "Recent Advances on Adhesion" and has been actively publishing papers on adhesion and rheology of polymers. It is indeed fortunate for all of us to hear Dr. Hata again during this Joint Chemical Congress.

The third paper, "Acid-Base Interactions at Interfaces." will be presented by Professor F.M. Fowkes of Lehigh University. Physicists have frequently used a similar term, donor-acceptor interaction. Now, we chemists will be hearing a stronger argument about a more familiar term, acid-base interactions. In this paper, Dr. Fowkes points out that acid-base interactions are also occurring at the polymer-polymer interfaces.

Dr. Souheng Wu will discuss the surface tension of solids in the fourth paper of this morning. Dr. Wu has made experimental measurements of surface and interfacial tensions of polymers. In this paper, he is proposing a theoretical justification for critical surface tension introduced by Dr. W.A. Zisman. Though the original title, "Surface Tension of Solids: An Equation of State Approach," has raised strong comments by reviewers, the data assembled could serve some useful purpose for practioners.

The last paper of this session is "A Role of Molecular Forces in Adhesive Interaction of Polymers," by Professor Belyi et al. Dr. Belyi could not attend this Conference. We understand their situation in the U.S.S.R. This paper is by no means rigorous, but it sums up their view on the effect of dispersion forces on adhesion. Those who are interested in the subject matter should read a new book, Dispersion Forces (3).

REFERENCES

1.   L.H. Lee, Ed., Recent Advances in Adhesion, Gordon and Breach, New York, London and Paris (1973).
2.   L.H. Lee, Ed., Adhesion Science and Technology, Plenum Press, New York (1975).
3.   J. Mahanty and B.W. Ninham, Dispersion Forces, Academic Press, London, New York (1976).

Statistical Mechanics of Surface Tension and Adsorption

I. Prigogine and A. Bellemans

Faculty of Sciences

Université Libre de Bruxelles, 1050 Bruxelles, Belgium

ABSTRACT

This introductory lecture is broadly divided into three parts. In the first one we review some fundamentals of surface phenomena; i.e., the conditions of mechanical equilibrium of an interface, Gibbs thermodynamics of surfaces and the statistical formulation of surface properties. The second part is devoted to a critical discussion of current statistical models of the surface tension of pure liquids and liquid mixtures (cell model, perfect and regular solutions, corresponding state theories, r-mer mixtures, etc...). The last part deals with the adsorption of a chain molecule at an interface, in relation with different physical parameters: interactions, length and concentration.

INTRODUCTION

This lecture will be divided into three parts. In the first one we shall briefly survey some fundamentals of interface phenomena; i.e., the conditions of mechanical equilibrium, Gibbs thermodynamics of surfaces and the general statistical approach of interfaces. The second part will be devoted to a critical presentation of some particular models developed for describing surface properties of pure liquids and liquid mixtures. The third and last part will concern one of the central topics of this Colloquium; i.e., the adsorption of a chain molecule at an interface.

FUNDAMENTALS OF SURFACE PHENOMENA

From the macroscopic point of view the interface between two adjacent fluid phases, a, b appears as a geometrical surface where the local thermodynamic properties of the system change abruptly. As originally stated by Young (1) in 1805, this surface is mechanically equivalent to a non-rigid membrane with zero thickness, isotropically stretched by the so-called interfacial tension γ. Insofar as this membrane may be regarded as massless, its equilibrium shape is determined by the Laplace equation (2)

$$p_a - p_b = \gamma_{ab} (K_1 + K_2) \tag{1}$$

where p_a, p_b are the hydrostatic pressures in phases a, b, and K_1, K_2 are the principal curvatures at the considered locus. Moreover, the equilibrium condition of an edge resulting from the encounter of the interfaces separating three mutually coexisting phases a, b, c, is expressed by the vectorial equation of Neumann

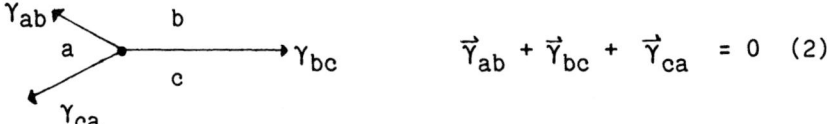

$$\vec{\gamma}_{ab} + \vec{\gamma}_{bc} + \vec{\gamma}_{ca} = 0 \tag{2}$$

(assuming zero edge tension).

Obviously the assumption that two adjacent phases retain their own intensive properties downright to a surface of discontinuity is only a useful way of thinking which, when necessary, must be completed by a detailed picture of what really happens at the molecular level. The interface is actually formed by a thin layer of matter operating the transition between the two bulk phases and both the mechanical and physico-chemical properties of this layer usually differ markedly from those of these phases. In particular, as originally pointed by Bakker (3), the interfacial tension is merely a macroscopic manifestation of the anisotropy of the pressure tensor within the interfacial layer.

The thermodynamics of surface phenomena was formulated by Gibbs (4) in an extremely elegant way. Indeed classical thermodynamics is devised for handling locally homogeneous systems; i.e., the change of any intensive variable is completely negligible over a few molecular diameters. This is clearly not so within the transition layer, the density of which may e.g., drop by a factor of a thousand over perhaps 10 or 20Å when passing from a liquid to a gas phase. Gibbs superseded the difficulty by introducing the concept of dividing surface. This is a geometrical surface located in the transition layer (or at least sensibly close to it) and passing through "points which are similarly situated with respect to the

condition of the adjacent matter", as worded by Gibbs himself (4).
The actual system, consisting of two homogeneous bulk phases separated by a thin inhomogeneous layer, is then replaced by two
homogeneous bulk phases extending on each side of the chosen
dividing surface; this surface itself is treated as a homogeneous
two-dimensional phase and provided therefore with appropriate
mechanical and physico-chemical properties, in order to match the
actual system macroscopically. E.g., its free energy F is split
into three parts, respectively the two bulk phases and the surface
phase s

$$F = F_a + F_b + F_s \qquad\qquad\qquad (3)$$

with the following differentials (5)

$$dF_a = -S_a \, dT - p_a \, dV_a + \Sigma \mu_i \, dN_i^a \qquad\qquad (4a)$$

$$dF_b = -S_b \, dT - p_b \, dV_b + \Sigma \mu_i \, dN_i^b \qquad\qquad (4b)$$

$$dF_s = -S_s \, dT + \gamma \, dA + \Sigma \mu_i \, dN_i^s \qquad\qquad (4c)$$

Note that we limit ourselves to systems in mechanical, thermal and
chemical equilibrium, and that we are essentially interested in
flat interfaces (for which $p_a = p_b$). Although eq. (4c) looks
strictly similar to eqs. (4a)- (4b), one should realize that a
surface phase has no real autonomy in general; its very existence is
conditioned by the presence of the two "parent" bulk phases and its
composition follows from the values of the set of chemical potentials $\{\mu_i\}$.

The variables $\{N_i^s\}$ specify the material content and the composition of the surface phase. In the same way as one defines bulk
concentrations $(c_i^a = N_i^a / V_a,\ c_i^b = N_i^b / V_b)$, it is convenient to
introduce the adsorption

$$\Gamma_i = N_i^s / A, \qquad\qquad\qquad (5)$$

as the mean number of molecules of species i adsorbed on a unit area
of the interface. Similar arguments which lead to the well known
Gibbs-Duhem equation for the bulk, may next be used to derive the
Gibbs adsorption formula

$$d\gamma = -s_s \, dT - \Sigma \, \Gamma_i \, d\mu_i \qquad\qquad (6)$$

where s_s is the surface entropy per unit area. Actually the value
of Γ_i depends on the precise location of the dividing surface
(which for a flat interface is somewhat arbitrary). The relative

adsorption of one component with respect to another is, however, a perfectly well defined quantity, independent of the choice of this surface.

Turning now to statistical mechanics, we see two main routes for developing a molecular expression of surface tension and related properties. One starts from the statistical formulation of the pressure tensor and was fully formulated by Kirkwood and Buff (6). The other route proceeds from the partition (or grand partition) function Z of the system, going to the free energy F and finally to γ by means of the two relationships

$$F = - kT \ln Z \tag{7}$$

$$\gamma = \left(\frac{\partial F}{\partial A}\right)_{T,V,\{N_i\}} \tag{8}$$

In particular, for a fluid of spherical molecules, the final expression of γ is

$$\gamma = (1/2) \int d\vec{R} \int \frac{du}{dR} \frac{X^2 - Z^2}{R} n_2(z,\vec{R}) \, dz \tag{9}$$

where u(R) is the intermolecular pair potential and $n_2(z,\vec{R})$ is the pair density within the transition layer, normally oriented to the z axis; $\vec{R}$ is the vector joining two molecules (with components Y,Y,Z). It is only rather recently that this expression has proven really useful, when reliable information became available on $n_2(z,\vec{R})$. Generalizations of eq. (9) for polyatomic molecules have also recently been established (7).

STATISTICAL MODELS OF INTERFACES

Eq. (9) and its extensions to nonspherical molecules are of limited use because of our rather poor information concerning the pair density n_2. Furthermore one is often interested in complex systems for which equations like (9) would not only be difficult to establish but also completely useless. Hence the deep interest in developing more or less sophisticated models specialized to different kinds of interfaces.

In this respect the Cell Model of the liquid state has been applied rather successfully to describe the liquid-gas interface far from the critical point. Let us recall here that each molecule of the liquid is supposed to be locked in a cage; given w the potential energy at the cage center and ψ the partition function of a molecule in its cage, one gets

$$\gamma = (1/a) \left[(1/2)(w_s - w_L) - kT \ln (\psi_s/\psi_L)\right] \tag{10}$$

where indices L, s respectively refer to "liquid" cells and "surface" cells, a is the area of a surface cell at the interface and the gas phase is treated as a vacuum (8). The results obtained are reasonably good for γ itself but less satisfactory for the surface energy and entropy. A considerable improvement is obtained by allowing a lacunar structure of the surface layer; i.e., by introducing a certain number of empty cells.

Liquid mixtures of different molecular species obviously constitute a more complicated problem as care must now be taken of the preferential adsorption of some of the components at the interface. The so-called model of regular solutions is already a reasonable approach for binary mixtures of spherical molecules of similar size. Given a lattice with each site occupied by either an A or a B molecule and given three energies ε_{AA}, ε_{AB}, ε_{BB} for first neighboring pairs, one gets in the simplest version of the model, i.e., monolayer approximation and random mixing in the bulk and at the surface respectively (9):

$$\gamma = \gamma_A + \frac{kT}{a} \ln \frac{x_A^s}{x_A^L} + \frac{w}{a} \left[(1-2m)x_B^{s^2} - (1-m) x_B^{L^2}\right]$$

$$\tag{11}$$

$$= \gamma_B + \frac{kT}{a} \ln \frac{x_B^s}{x_B^L} + \frac{w}{a} \left[(1-2m)x_A^{s^2} - (1-m) x_A^{L^2}\right]$$

Here x_A, x_B are the mole fractions $(x_A + x_B = 1)$, $w = z(\varepsilon_{AB} - \varepsilon_{AA}/2 - \varepsilon_{BB}/2)$, z is the number of first neighbors of a molecule in the liquid and z (1-m) is the number of first neighbors of a molecule at the interface. In order to get γ, eqs. (11) need to be solved numerically for x_A^s (or x_B^s). Only for w = 0 (i.e., for the so-called perfect solutions) does one obtain an explicit expression for γ in terms of x_A^L, x_B^L

$$e^{-\gamma a/kT} = x_A^L e^{-\gamma_A a/kT} + x_B^L e^{-\gamma_B a/kT} \tag{12}$$

Note that the monolayer model has been shown to present a slight thermodynamic inconsistency to which one can easily remedy by considering a multilayer approach (10).

There are two natural extensions of the model of regular solutions: the first one is to substitute to the lattice a more realistic picture of the liquid, the second one is to generalize the model to molecules occupying more than one lattice site (the so-called r-mer solutions). The first problem has been worked out by means of a combination of the cell model and the theorem of corresponding states (the so-called average potential model or corresponding states theory (11)). The second problem has been extensively studied and will be discussed at length in the third part of this lecture.

The surface tension of electrolyte solutions was initially tackled by Onsager and Samaras (12). In the dilute case image charges arising at the junction of two media of different dielectric constants play the fundamental role. As the concentration in-creases many other factors come into play: the finite size of the ions, their polarizability, the change of the dielectric constant of the solvent itself at the interface and also saturation effects.

Another interesting type of liquids and liquid mixtures for which no serious statistical theory of surface properties seems to exist, are associated substances where hydrogen bonds may extend from molecule to molecule.

ADSORPTION OF CHAIN MOLECULES

Even when presented in a very schematic way this problem is an extremely difficult one: it involves both energetic and entropic effects at the level of a single chain molecule and, if the concentration is not small, which is frequently the case, the situation becomes extremely complex.

The simplest model consists in confining the chain molecules on a semi-infinite lattice, each site being occupied either by a solvent molecule or by a chain segment. Even so there are already three kinds of effects to be taken care of, at infinite dilution:

(a) An entropic effect related to the decrease of the avail-able number of internal configurations of the chain near the surface;

(b) A first energetic effect linked to the energy difference resulting from the replacement of a solvent molecule by a chain segment in the surface layer;

(c) A second energetic effect related to differences in the pair interactions: solvent-solvent, chain segment-chain segment and solvent-chain segment. (This effect vanishes for athermal solutions.)

For finite concentrations, excluded volume effects and energetic effects between different chain molecules come into play. The difficulty of the problem is easy to figure out when realizing that even in the bulk phase the problem of the average conformation of a

single chain can only be solved in an approximate manner, when
retaining the excluded volume effect between distant segments.

Broadly speaking, there seems to exist four kinds of approach
to the adsorption of chains at an interface:

(a) analytical approaches,
(b) mean field theories,
(c) computer simulations
(d) exact enumerations.

Analytical theories are very difficult to develop for long
chains. Even the very special case of dimers is not solved. Edges
effects for a two-dimensional assembly of pure dimers were exactly
treated (13,14) and accurate values were also obtained for pure
trimers (15). Graph theoretical approaches in two and three dimensions have also been developed recently by one of us (A.B.). They
permit in principle to compute the surface tension of monomer-dimer
mixtures at all concentrations by means of an exact series expansion. The number of terms required may, however, become prohibitive
in some cases (16).

Mean field theories of adsorbed chains have been developed by
very many people in different ways. One of us (I.P.) used this kind
of approach for studying short chains (17) (dimers and trimers). All
these works may in some sense be regarded as extensions of the
Flory-Huggins theory of bulk phases, to interfaces; the system is
usually sliced into monolayers and the Flory-Huggins combinatorial
formula is used in each of them. The calculations are very
complicated for long chains. The quantities of interest are usually
the mean number of adsorbed segments of the chain vs. the temperature, the mean length of "trains" of adsorbed segments and the mean
length of "bridges" between two such successive "trains". A recent
account of the situation has been given by Silberberg (18).

Computer simulations have been widely used in the past for
studying properties of chain molecules, either isolated or in the
bulk phase. During the last few years such simulations have been
extended to adsorbed chains. The method used is based on extensions
of the usual Monte Carlo method as it is applied in statistical
mechanics and chains counting 100 segments or more were considered (19). Both the influence of the temperature and of the
presence of other chains were studied in this manner (20).

Let us finally turn to exact enumeration techniques, which
were very successful to analyze the conformation and thermodynamic
properties of chains in homogeneous phases and are presently applied to adsorbed chains (21,22). Because of the direct experience
of one of us (A.B.) in this kind approach, we shall develop it here
in some detail, in the particular case of dilute athermal solutions.

Consider a regular lattice of V sites, A of which are (equivalent) surface sites. Each site is occupied by a solvent molecule (monomer) or by a segment of a chain molecule (n-mer). Let us introduce the following quantities:

Vc_n : the total number of configurations allowed to a single n-mer when boundary effects are neglected (i.e. assuming periodic boundaries and $V^{1/3} \gg n$).

$A\ell_n/2$: the number of configurations lost for a single n-mer on account of the boundaries (these configurations can be figured out as lost when a periodic system is "cut" by a surface).

$Ag_n(\nu)$: the number of configurations allowed to a single n-mer when ν of its segments occupy superficial sites.

The grand partition function of the system is then given by

$$\Xi = 1 + z \left\{ Vc_n + A \left[-\ell_n/2 + \sum_\nu g_n(\nu) (f^\nu - 1) \right] \right\} + O(z^2) \quad (13)$$

where z is the activity of the n-mers and

$$f = \exp(\epsilon/kT), \quad (14)$$

ϵ being the adsorption energy per n-mer segment. It then follows straightforwardly that the modification of the surface tension with respect to that of the pure solvent is

$$\Delta_\gamma = C (kT/a) \left[\ell_n/2 - \sum_\nu g_n(\nu) (f^\nu - 1) \right] /c_n \quad (15)$$

where C is the n-mer concentration and a is the area of a superficial site. The adsorption is similarly given by

$$\Gamma = (C/a) \left[\sum_\nu g_n(\nu) (f^\nu - 1) - \ell_n/2 \right] /c_n \quad (16)$$

and the average number of adsorbed segments of a n-mer touching the surface is

$$\langle \nu \rangle = \frac{\sum_\nu g_n(\nu) \nu f^\nu}{\sum_\nu g_n(\nu) f^\nu} \quad (17)$$

One may also define such quantities as the mean <u>thickness</u> of the adsorbed n-mer, perpendicularly to the surface, and its mean <u>extension</u>, parallel to the surface.

Exact enumerations of c_n, ℓ_n, $g_n(\nu)$ and similar quantities have been made or are presently studied (21,22). Obviously one cannot hope to obtain them for very long chains, a reasonable limit being n = 15 or 16 for a simple cubic lattice. This may appear far too short to match the actual problem but one should realize that one deals here with exact numbers and that powerful extrapolation methods exist wherefrom reliable information can be drawn about the asymptotic limit n $\rightarrow \infty$.

FINAL REMARKS

In spite of an enormous theoretical effort made along different lines of approach and which has, without any doubt, led to a general understanding of the problem, much remains to be done. The lattice model is indeed very crude, but just dropping the lattice and replacing it by a continuum would not do much good as this would negate the solvent structure completely. Future efforts should take care both of the specific nature of the chains and also of the molecular structure of the interface in relation to solvent properties.

REFERENCES

1. T. Young, Phil. Trans. Roy. Soc. (London) 95, 65 (1805).
2. P.S. Laplace, "Mécanique céleste", suppl. 10th Vol. (1806).
3. G. Bakker, "Théorie de la couche capillaire plane des corps purs", Scientia (Paris) 31, 74 (1911).
4. J.W. Gibbs, Collected works, 2 Vols., New York (1928).
5. R. Defay, I. Prigogine, A. Bellemans and D.H. Everett, "Surface Tension and Adsorption", Longmans Green and Co. Ltd., London (1966).
6. J.G. Kirkwood and F.P. Buff, J. Chem. Phys. 17, 338 (1949).
7. H.T. Davis, J. Chem. Phys. 67, 3636 (1977).
8. I. Prigogine and L. Saraga, J. Chim. Phys. 49, 399 (1952).
9. E.A. Guggenheim, Trans. Faraday Soc. 41, 150 (1945).
10. R. Defay and I. Prigogine, Trans. Faraday Soc. 46, 199 (1950).
11. A. Englert-Chowols and I. Prigogine, J. Chim. Phys. 55, 16 (1958).
12. L. Onsager and N.N.T. Samaras, J. Chem. Phys. 2, 528 (1934).
13. M.E. Fisher, Phys. Rev. 124, 1664 (1961).
14. P.W. Kasteleyn, Physica 27, 1209 (1961); J. Math. Phys. 4, 287 (1963).
15. J. Van Craen, J. Chem. Phys. 63, 2591 (1975).
16. A. Bellemans and S. Fuks, Physica 50, 348 (1970); A. Bellemans, Physica 65, 89 (1973); A. Bellemans, J. Coll. Interf. Sci. 58, 521 (1977).
17. I. Prigogine, J. Chim. Phys. 47, 3 (1950); I. Prigogine and L. Saroléa, J. Chim. Phys. 47, 807 (1950).
18. A. Silberberg, Faraday Discussion, Chem. Soc. 59, 203 (1975).

19. M. Lal and R.F.T. Steppo, J. Pol. Sci.: Polymer Symposium 61,
        401 (1977).
20. A.T. Clark and M. Lal, J. Chem. Soc., Faraday Trans. II, 74,
        1857 (1978).
21. A. Bellemans, J. Pol. Soc.: Part C 39, 305 (1972).
22. M.N. Barber, A.J. Guttmann, K.M. Middlemiss, G.M. Torrie and
        S.G. Whittington, J. Phys. A 11, 1833 (1978).

Surface and Interfacial Tensions of Polymer Melts and Solutions

Toshio Hata

Gunma University

3-39-22, Showa-cho, Maebashi-shi, Gunma Prefecture,

Japan, 371

and

Tomoyuki Kasemura

Department of Textile Industry, Faculty of Engineering

Gifu University, Kagamihari-shi, Gifu Prefecture,

Japan, 504

ABSTRACT

Work done in the last ten years at our laboratories is reviewed on the following topics: (1) Methods (sessile bubble and drop methods) for the determination of surface and interfacial tensions; (2) Molecular dependence of surface tension (γ); (3) Molecular structure and γ; (4) Composition dependence of γ in copolymers; (5) Interfacial tension (γ_{12})'s between polymer melts; (6) γ's of polymer solutions and their γ_{12}'s against water; and (7) γ's and γ_{12}'s of polypeptide solutions. Molecular weight dependence of γ is explained on the basis of Sugden's equation and the effects of molecular structure such as methylene sequence in the main chain and the alkyl side chain are pointed out. The sequence effect plays an important role especially in copolymers. Different behaviors of copolymers with different combinations of monomeric units were introduced. The importance of interfacial tension measurements is emphasized for polymer melts and solutions, including polypeptide solutions, to characterize the surface properties of polymeric materials.

INTRODUCTION

Surface and interfacial tensions arising from intermolecular interaction of liquids are also directly related to adsorption, adhesion, wetting, dyeing, printing and other surface processes, which have become increasingly important for polymer systems. Nevertheless, enough systematic studies have not been done to enable one to generalize or characterize the polymer behaviors in surface and interfacial tensions. In the present review, we want to deduce possibly generalized laws or characteristics for polymeric systems from our experimental work, which has been done in the last ten years with our co-workers at the Tokyo Institute of Technology, Gunma, and Gifu University. For other researchers' results, the reviews by Wu (1,2) should be referred to.

1.   METHOD (3)

Surface tension has been measured by the sessile bubble method and interfacial tension by the sessile drop method. Both are thought to be the most adequate for highly viscous substances such as polymer melts and solutions. Figure 1 is the schematic representation of the sessile bubble method. This glass cell is placed in a container filled with deoxidized and dehydrized nitrogen gas. The supporting board can be raised by a lever as high as the horizontal metal plate dips into the liquid sample. A bubble is formed by passing nitrogen gas through a hole at the center of the plate. The entire apparatus is placed in an air bath adjusted to a proper temperature.

The bubble's maximum radius (r) and the distance (h) from the meniscus top to the r plane are measured by a cathetometer. Surface tension (γ) is calculated by Porter's equation (4)

$$\gamma = \Delta\rho g h^2 \{0.5000 - 0.3047(\frac{h}{r}) + 1.219(\frac{h}{r})^3\} \qquad (1)$$

where $\Delta\rho$ is the difference in density of the sample in gaseous and liquid states, and g is the gravitational constant. Density is measured by a proper method according to the sample's state at a normal temperature and with dilatometry at higher temperatures.

In preliminary experiments, we examined the precision of the approximate equations proposed by Porter, Wheeler and Staicopolas, using some standard liquids. As a result we found that the Porter's equation was most reliable in the range of 0.6>h/r>0.4. The error was within 2%.

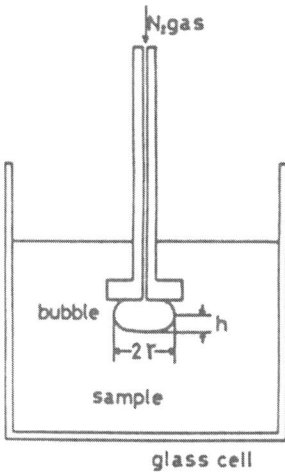

Fig. 1. Scheme of sessile
bubble method.

Interfacial tension can be measured with the same method only by replacing a bubble by a drop of a second liquid. The drop of liquid with density higher than the matrix is placed on a teflon plate at the bottom of the cell.

Here I wish to call attention to the fact that the measurement of interfacial tension is extremely limited for liquid of low molecular weight because of their compatibility or partial miscibility, while, in contrast, polymers are generally incompatible, so we can measure it for almost all combinations of polymer homologs which are miscible in the low molecular wieght region. This fact is of great importance for scientific research in surface chemistry.

2.   MOLECULAR WEIGHT DEPENDENCE OF SURFACE TENSION

2.1 A Formulation and the Experimental Examination

Surface tension of polymer homologs varies with molecular weight generally in the oligomer region. Whether surface tension increases or decreases with molecular weight depends on the nature of repeating unit and end groups, e.g. in n-alkanes it increases, while in alkyl naphthalene and alkyl diphenyls decreases with alkyl chain length as shown by Bascom et al. (5). Examples by us are shown in Fig. 2 for poly (α-methyl styrene) (3) and in Fig. 3 for ethylene and propylene glycol condensates with and without methyl end group (6). (See also Fig. 7 for alkylene glycols.)

We have explained this dependence of surface tension on molecular weight and composite units by using Sugden's equation (8).

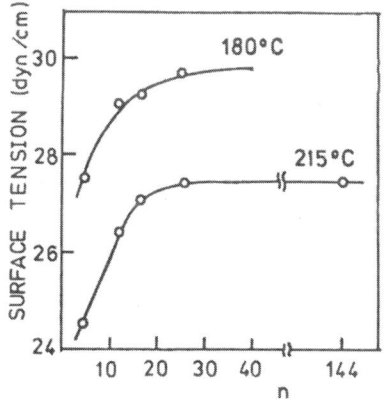

Fig. 2.   Surface tension vs.
          degree of polymeriza-
          tion (n) for poly
          (α-methyl styrene).

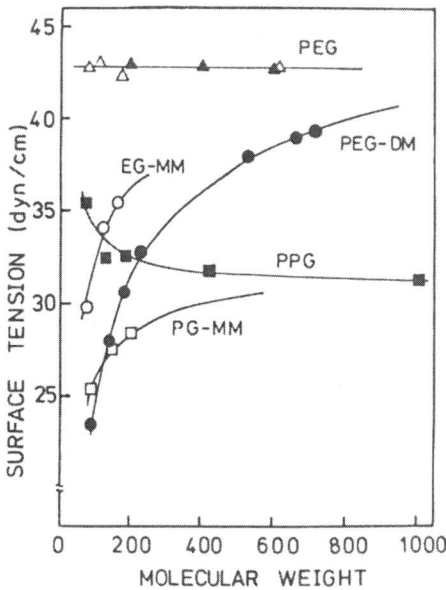

Fig. 3.   Surface tension vs.
          molecular weight at
          25°C.

EG, PEG: Ethylene glycol homologs

PG, PPG: Propylene glycol homo-
         logs

MM: Monomethyl, DM: Dimethyl
(Δ: by A.K. Rastogi and L.E.
   St. Pierre (7))

$$\gamma = \left\{ \frac{(\rho_L - \rho_G)P}{M} \right\}^\beta = \left(\frac{\rho_L P}{M} \right)^\beta = \left(\frac{P}{V} \right)^\beta \tag{2}$$

where ρ_L, ρ_G are density at liquid and gaseous states respectively;
M is molecular weight, and β is Macleod's exponent which was 4 in
Sugden's work. As Parachor P and molecular volume V can be assumed
to obey the additivity law, Eq. (2) is expressed as follows by
dividing P and V into two parts, repeating unit (suffix r) and end
groups (suffix e).

$$\gamma = \left(\frac{mP_r + 2P_e}{mV_r + 2V_e}\right)^\beta \tag{3}$$

where m is the number of repeating unit and at present a homologous series with two same end groups is considered. Equation (3) is transformed as follows and expanded, then an approximate eq. (4) is obtained

$$\gamma = \left(\frac{P_r}{V_r}\right)^\beta \left(1 + \frac{2}{m+2}\frac{P_e - P_r}{P_r}\right)^\beta \left(1 + \frac{2}{m+2}\frac{V_e - V_r}{V_r}\right)^{-\beta}$$

$$\approx \gamma_r \left[1 + \frac{2\beta k}{m+2}\left\{\left(\frac{\gamma_e}{\gamma_r}\right)^{1/\beta} - 1\right\}\right] \tag{4}$$

where $\gamma_r = (P_r/V_r)^\beta$, $\gamma_e = (P_e/V_e)^\beta$, $k = V_e/V_r$, and m+2 can be put n (degree of polymerization) in many cases. Equation (4) shows that, if β would be invariable, γ increases or decreases linearly with $1/(m+2)$ or $1/n$ ($1/M$) according to $\gamma_r > \gamma_e$ or $\gamma_r < \gamma_e$ respectively, and γ keeps constant when $\gamma_r = \gamma_e$. A similar treatment has been applied to glass transition temperature (T_g) by Ueberreiter and Kanig (9), obtaining the linear relationship of $1/T_g$ and $1/M$, or approximately T_g and $1/M$. In the same way for specific volume (v), we have the relation,

$$v = v_r + \frac{2M_e}{M}(v_e - v_r) \tag{5}$$

where $v_r = V_r/M_r$, $v_e = V_e/M_e$ and M_r, M_e are molecular weights of repeating unit and end group respectively. Figure 4 shows that these relations (γ, T_g, $v \propto 1/n$) hold for poly (α-methyl styrene), and Fig. 5 does for γ of polyethers in Fig. 3. Curves in Fig. 5 converge to definite values (γ_∞) of respective polyethers at infinite M, passing through different lines depending on the presence or absence of methyl end group. Strictly speaking, however, β in Eq. (4) varies with M roughly from 5 to 3 in the range of low molecular weight. β in the exponent of (γ_e/γ_r) acts as compensating the change of β in the numerator of the coefficient. This might be a reason why linearity is established in many cases. The effect of β, however, must be taken into consideration especially in low molecular weight region, together with the effect that the higher order terms are omitted in Eq. (4). An exact relation of γ to n (M) is given in the following discussion.

LeGrand and Gaines (10) have proposed the relation

$$\gamma = \gamma_\infty - \frac{A}{M^{2/3}}$$
(6)

where γ_∞ is the surface tension at infinite molecular weight.
Though the authors' discussion on the coefficient A is question-
able, (e.g. according to their discussion, A must be a positive
constant, then Eq. (6) can't explain the decrease of γ with
molecular weight), Eq. (6) seems to represent very well the observed
values of each homologous series. It must be noticed, however, that
the plots of γ vs. $M^{-2/3}$ for poly(ethylene glycol)s and poly-
(propylene glycol)s with and without methyl end group do not
converge to the respective values of γ_∞ at infinite M, where the end
group effect should disappear.

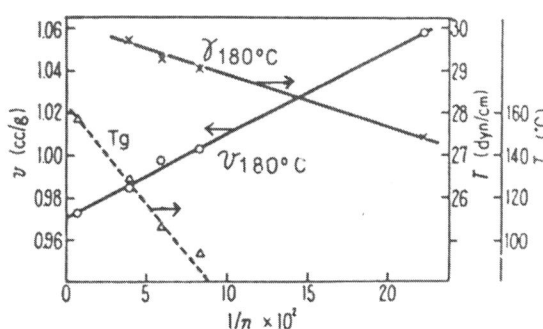

Fig. 4. Specific volume (v),
glass transition tempera-
ture (T_g) and surface
tension (γ) vs. 1/n.

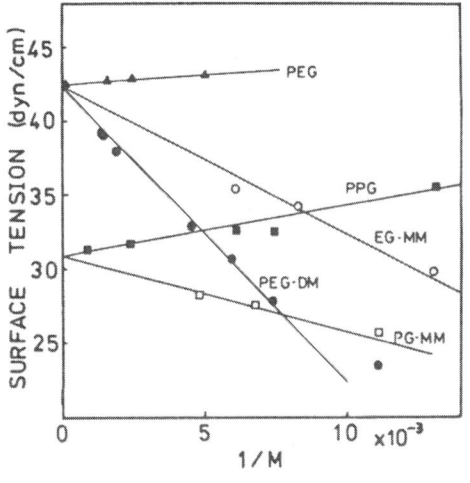

Fig. 5. Surface tension vs.
1/M at 25°C for polyethers
in Fig. 3.

According to Wu the molecular weight dependence of surface tension can be expressed well by the relation

$$\gamma^{-1/4} = \gamma_\infty^{-1/4} + \frac{B}{M_n} \qquad\qquad (7)^{(1)}$$

or

$$\gamma^{1/4} = \gamma_\infty^{1/4} - \frac{C}{M_n} \qquad\qquad (8)^{(2)}$$

where M_n is the number-average molecular weight. If the co-efficients B and C are positive, as pointed out by the author, Eqs. (7) and (8) fail again to represent the decrease of γ with molecular weight. Equations corresponding to Eqs. (7) and (8) can be easily derived from Eq. (3) as

$$\gamma^{1/\beta} = \gamma_r^{1/\beta} + \frac{2k}{m+2k} (\gamma_e^{1/\beta} - \gamma_r^{1/\beta}) \qquad\qquad (9)$$

$$\gamma^{-1/\beta} = \gamma_r^{-1/\beta} + \frac{2k'}{m+2k'} (\gamma_e^{-1/\beta} - \gamma_r^{-1/\beta}) \qquad\qquad (10)$$

where γ_r, γ_e and k are the same as those in Eq. (4) and $k'=P_e/P_r$. (Hata (8) had expressed Eq. (9) as $\gamma =\gamma_r\{1+2/(m+2V_e/V_r).(P_e/P_r-V_e/V_r)\}^4$, putting $\beta=4$.) As γ_r is equal to γ_∞, if we can put $\beta=4$, k=1, k'=1 and m+2=n, Eqs. (9) and (10) reduce to Eqs. (8) and (7) respectively. The derivation of Eqs. (9) and (10) doesn't involve any operation such as expansion and omission of higher order terms; therefore they are more accurate than Eq. (4), at least mathematically. In fact too, the results shown in Figs. 2 and 3 give the respective straight lines by the plots of $\gamma^{1/\beta}$ vs. $1/(m+2k)$ if we take the experimental mean values for β. The line holds also for smaller m, which the approximate Eq. (4) failed to reproduce.

So far we have considered molecules of $X-R_n-X$ or $X-R_n-X'$ type. For molecules of $X-R_n-Y-R_n-X$ type, e.g., dialkyl phthalates; how-ever, it was found that the additivity law above mentioned doesn't hold. The results and discussions for them will be presented in the next chapter. (3.4)

2.2 Estimation of γ_r of Repeating Units and γ_e of End Groups

On the basis of Eq. (4), we can estimate γ_r and γ_e from experimental results on the molecular weight dependence of surface tension. β is obtained by $\log \gamma - \log \rho$ plots, and $k(=V_e/V_r)$ by the molecular weight dependence of specific volume using Eq. (5). The analysis of poly(ethylene glycol)s, poly(propylene glycol)s, poly(isobutylene glycol)s, and their methyl ethers, for instance, gives the values of γ_r and γ_e as listed in Table 1.

The fact that γ of poly(ethylene glycol) is almost independent of the molecular weight arises from $\gamma_r(-CH_2-CH_2-O-) = \gamma_e(-OH)$.

Table 1.   Surface Tension of Repeating Units (γ_r) and End Groups (γ_e) for Some Polyethers at 25°C

Repeating units	γ_r(dyn/cm)	End groups	γ_e(dyn/cm)
$-CH_2-CH_2-O-$	42.7	$-OH$	42 ~ 44
$-CH(CH_3)-CH_2-O-$	31.0	$-CH_3$	11 ~ 13
$-C(CH_3)_2-CH_2-O-$	27.0		

3.   MOLECULAR STRUCTURE AND SURFACE TENSION

3.1 Effect of Tacticity

Temperature dependence of surface tension of isotactic and atactic polypropylene (11) is shown in Fig. 6.   So far as this result indicates, the effect of tacticity on the surface tension is negligibly small.   The flexibility of chain, however, is known to influence various thermodynamic properties of polymer.   Therefore, it may be too early to conclude that the tacticity has no effect on surface tension.   The necessity to consider the chain flexibility will be discussed again in the following section.

3.2 Methyl Group Effect

In general, surface tension decreases by the substitution of a hydrogen for methyl radical, as shown in Figs. 3 and 5 for poly(ethylene, propylene and isobutylene glycol)s.   Further examples are polypropylene (γ = 22.6 dyn/cm) vs. polyethylene (γ = 27.3 dyn/cm for both at 150°C).   This effect of methyl group may arise from the increase of free volume.   For α-hydrogen, however, the methyl   substitution   doesn't   always   reduce γ,   e.g. γ(PE)>γ(PIB)>γ(PP) and γ(PSt)≈γ (P∝MSt).   In the case of poly(isobutylene) it may be said that the flexibility of molecules (intramolecular motion), which is increased by the introduction of α-methyl, acts as hindering the surface orientation of low energy atomic groups, thereby increasing γ.   In poly(isobutylene glycol), this effect may be relatively small, because of the neighboring

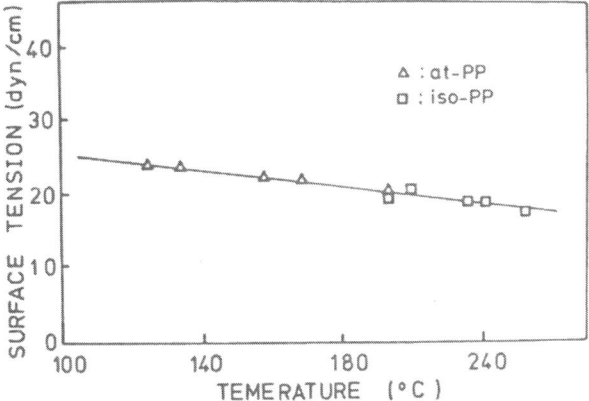

Fig. 6.   Temperature dependence of surface tension
          for isotactic and atactic polypropylenes.

oxygen bond which is able to rotate as well freely as before the
introduction of α-methyl.

Methyl group on the chain end lowers more effectively than the
one on the side chain, if they are compared at the same methyl
content (see original reference) (6).

3.3 Effect of Methylene Sequence in Alkylene-glycols and Their Condensates (12)

Here I discuss the effect of methylene sequence on γ in
alkylene-glycols, $HO-(CH_2)_m-OH$ (AG) and their condensates, poly-
(alkylene glycol)s $HO-[(CH_2)_m-O]_n-H$ (PAG). There has been given
Eq. (4) or Eq. (9) applicable for γ vs. m of AG. For γ vs. m (not n)
of PAG, Eq. (11) can be derived by the same way, putting $n \to \infty$.

$$\gamma = \gamma_{CH_2} \left[1 + \frac{\beta k}{m+1} \left\{ \left(\frac{\gamma_o}{\gamma_{CH_2}}\right)^{1/\beta} - 1\right\}\right] \qquad (11)$$

where γ_{CH_2} and γ_o are surface tension components of CH_2 and O
respectively, and k is ratio of their molecular volumes V_o/V_{CH_2}.
The results are shown in Fig. 7, and plots of γ vs. $1/(m+2)$, for AG
and $1/(m+1)$ for PAG are shown in Figs. 8 and 9, respectively.
Equation (4) holds for AG, while for PAG, experimental curves go
through below the straight line expected from Eq. (11). (The exact
equation corresponding to Eq. (9) gives the same result.) The reason

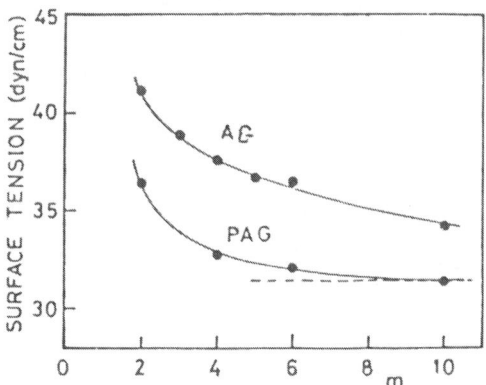

Fig. 7.   Surface tension vs. number of methylene
          groups of alkylene glycols and poly
          (alkylene glycol)s at 90°C.

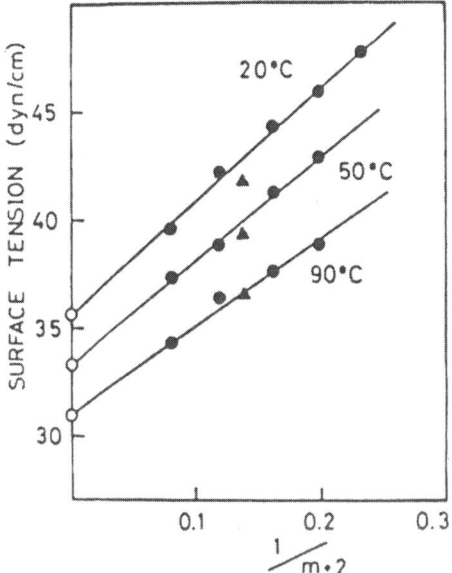

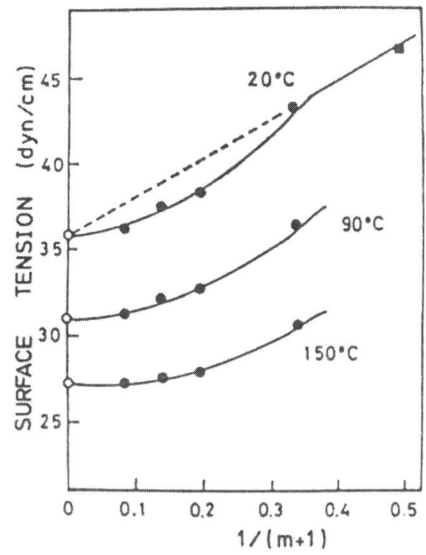

Fig. 8.   Surface Tension vs.
          1/(m+2) for alkylene
          glycols.

Fig. 9.   Surface Tension vs.
          1/(m+1) for poly (alkylene
          glycol)s.

may be attributed to the predominant adsorption of low energy
component, e.g. methylene sequence in this case. The difference of
the effect of methylene sequence between AG and PAG is probably due
to the cooperative adsorption of neighboring sequences in PAG, so
that the length of methylene sequence ficti tiously grows to the
longer one in the adsorbed states as seen in Fig. 9. The sequence
effect is observed more frequently in copolymers as shown in the
next section. (In Fig. 8 we can see that γ of HO-$(CH_2)_5$-OH (Δ) lies
off the curves at $20^\circ C$ and $50^\circ C$. This is presumably cuased by the
ring formation of six members by hydrogen bonding, which disappears
at $90^\circ C$ as shown in the graph.)

3.4 Side Chain Effect

It is well known that properties of vinyl polymers vary with
the length of alkyl side chain. For example, glass transition
temperature decreases rapidly at first and reaches the minimum, at
the chain length of C10 in the case of poly(methacrylate)s. As
surface tension is governed by cohesive energy density similarly to
T_g, it is expected the same behavior to be observed in surface
tension. The dependence of γ on side chain length has been studied
for polyacrylates and poly(vinyl alkylate)s in molten state. In
polyacrylates γ decreased monotonously from 31.0 dyn/cm for
poly(methyl acrylate) to 21.2 dyn/cm for poly(2-ethyl-hexyl
acrylate) at $150^\circ C$ (13). Samples of longer side chain have not been
examined. Here the results for poly (vinyl alkylate)s made up of n-
aliphatic acids (14) are shown in Fig. 10. γ reaches the minimum
value at poly (vinyl caprylate) of C8 side chain.

Similar results are obtained for dialkyl phthalates of various
alkyl chain lengths (15) as shown in Fig. 11. These materials are
ortho compounds, so that their branching structure rather likes
poly(vinyl alkylate)s.

Components of surface tension, γ^d and γ^p in terms of Fowkes
(16), are estimated by the measurements of interfacial tension
(γ_{12}) against polyethylene at $150^\circ C$ in the case of poly(vinyl
alkylate)s, and by the measurements of contact angle (θ) on solid
paraffin in the case of dialkyl phthalates at $20^\circ C$, according to the
equation

$$\gamma_{12} = \gamma_1 + \gamma_2 - 2\sqrt{\gamma_1^d \gamma_2^d} \tag{11}$$

where $\gamma_1 (= \gamma_1^d + \gamma_1^p)$ is taken as the surface tension of the sample,
and $\gamma_2 (= \gamma_2^d)$ is that of polyethylene or paraffin. As shown in Figs.
10 and 11, the first decrease of γ is caused by the decrease of both
γ^d and γ^p in dialkyl phthalates and by γ^p alone in poly(vinyl

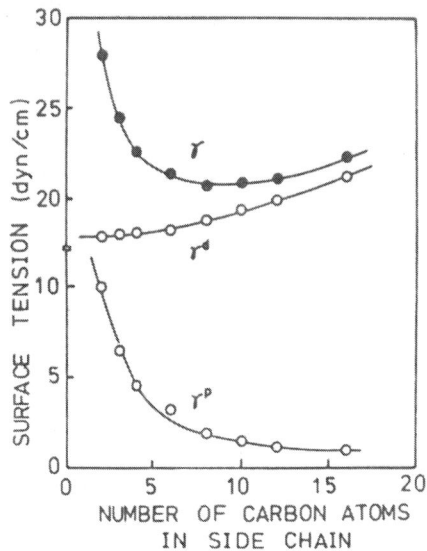

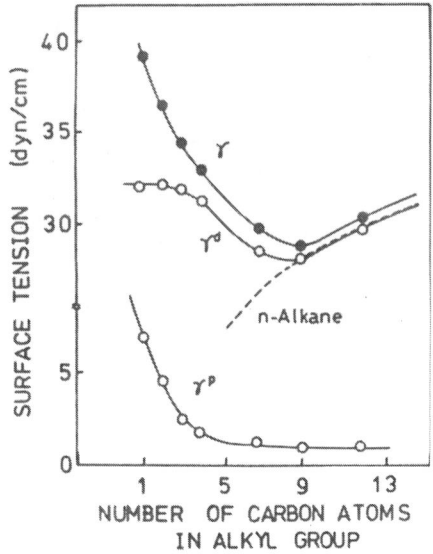

Fig. 10. The effect of side
   chain length of poly(vinyl
   alkylate)s on surface tension
   and its components due to
   dispersion force (γ^d) and
   polar force (γ^p) at 90°C.

Fig. 11. The effect of alkyl
   chain length of dialkyl
   phthalates on surface tension
   and its components, γ^d and
   γ^p, at 20°C.

alkylate)s. Different behaviors of γ^d between the above two series
will be understood by comparing the phthalic and vinyl groups whose
dispersion interactions are far different.

After passing through the minimum, γ increases fairly in
parallel with the increase of γ^d, which shows the predominant
interaction between alkyl side chains. In fact, the change in γ^d in
dialkyl phthalates is essentially the same as that in n-alkanes as
seen in the Figure. For the minimum γ, it can be said that the so-
called HLB (hydrophile-lyophile balance, exactly in this case
"lyophile-lyophob balance") reaches an optimum condition at that
side chain length.

4.   SURFACE TENSION OF COPOLYMERS

4.1. Sequence Effect in Random Copolymers

First we deal with random copolymers. Out of various combinations of monomeric units, we have taken up copolymer samples

classified to each of the following three types of combination:

1)   nonpolar/nonpolar: ethylene/propylene copolymers (E-P series);

2)   polar/polar: ethylene oxide/propylene oxide copolymers (EO-PO series) and tetrahydrofuran/propylene oxide copolymers (THF-PO series); and

3)   nonpolar/polar: ethylene/vinyl acetate copolymers (EVA series).

It is of interest to compare the probably different behaviors of γ for these three types of combination.

In Figs. 12 and 13, results are shown for E-P and THF-PO series respectively (17).   Here x_2 is taken as the mole fraction of low energy component, i.e. propylene in E-P series and propylene oxide in THF-PO series.   As seen in the figures, observed values deviate from the straight line expected from the simple additivity proposed by Rastogi and St. Pierre (18).

If the additivity of P and V in Sugden's equation can be assumed again for monomeric units 1 and 2 in random copolymers, we have

$$\gamma = \gamma_1 \left(\frac{k_2}{k_1}\right)^\beta \left(1 + \frac{k_1 - k_2}{x_2 + k_2}\right)^\beta \qquad (12)$$

where $\gamma_1 = (P_1/V_1)^\beta$, surface tension of homopolymer 1; $k_1 = P_1/(P_2 - P_1)$, $k_2 = V_1/(V_2 - V_1)$ and x_2 is the mole fraction of monomeric unit 2. V_1, V_2 in k_2 are given by specific volume measurements and P_1, P_2 in k_1 are calculated from the relation $P = \gamma^{1/\beta} V$ for each polymer.   Calculated curves are drawn with dotted broken lines (– —— –) in Figs. 12-14, using the mean values of β for each copolymer series.   ($\beta = 3.0$ for E-P series, and 3.5 for THF-PO and EO-PO series.)   As you can see, the observed values can't be represented by Eq. (12), showing the limited applicability of Sugden's equation or the simple additivity of Parachor to copolymers.   Such a discrepancy is probably due to the selective adsorption of low energy component of large sequence length.

In random copolymers, there exists a distribution of sequence length of each component, accordingly do the sequences of low energy component, length of which is larger than the mean value corresponding to x_2.   These sequences will preferentially be adsorbed to surface and reduce γ more than expected from the additivity law.   Mole fraction x_2 larger than 0.4 will be necessary for the sequence to be effective in its length as well as its quantity.   The roles of sequence length and its quantity are clearly demonstrated by block copolymers described in the next section.

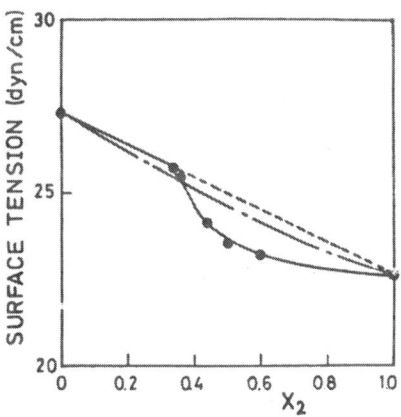

Fig. 12. Composition dependence
of surface tension of ethylen-
propylene copolymers at 150°C.
(- —— - : calculated by
Eq. (12)).

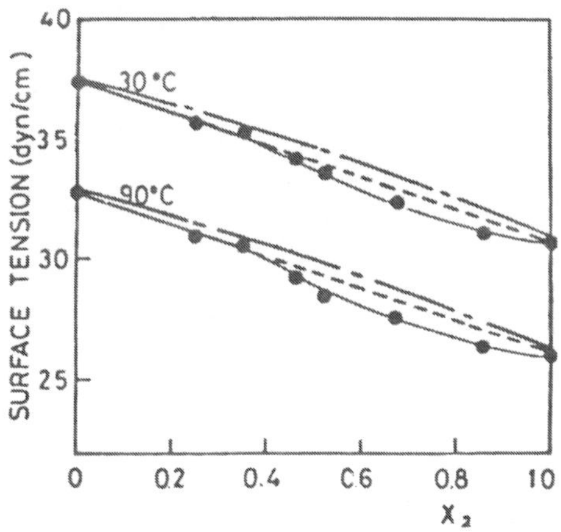

Fig. 13. Composition dependence
of surface tension of tetra-
hydrofuran-propylene oxide
copolymers.

4.2 Comparison Between Random and Block Copolymers

Both random and block copolymers of various compositions can
be synthesized from EO and PO monomers. We (19) compared random EO-
PO copolymers with block ones of the type $(EO)_\ell-(PO)_m-(EO)_\ell$, where
the composition of PO was varied by ℓ, keeping m constant at 16, 30,
and 56. The studies for essentially the same materials had been
done by Rastogi and St. Pierre (18). Nevertheless we tried the
similar experiments by ourselves in order to examine the applicability of the additivity law to random copolymers with additional
experimental values, and further to confirm the sequence effect by
the comparison with block copolymers. The results almost agree with
those of Rastogi et al. under the same conditions. An example in
our observations is shown in Fig. 14. The random copolymers (a in
the figure) behave just like E-P and THF-PO series, while in
contrast the block copolymers lower γ drastically at a few mole
fractions of PO, and the more the sequence length m, the more the
decrease. Each plateau corresponds to each m. It's noteworthy that
the curve of random copolymers intersects and goes below the curve
of block copolymers of m=16 at a mole fraction about 0.8. This
means that PO sequences larger than m=16 appears more frequently in
the random copolymers over this composition. This effect of block
EO-PO copolymers is quite similar to surface-active agents, therefore they are used, in practice, as surfactants. The introduction
of PO block (low energy component) of proper length is of greater
advantage for the reduction of γ than their statistical occurrence

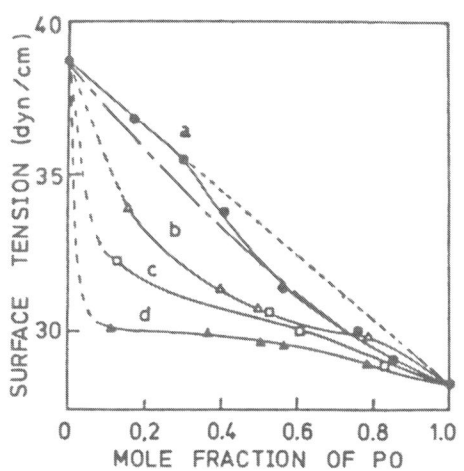

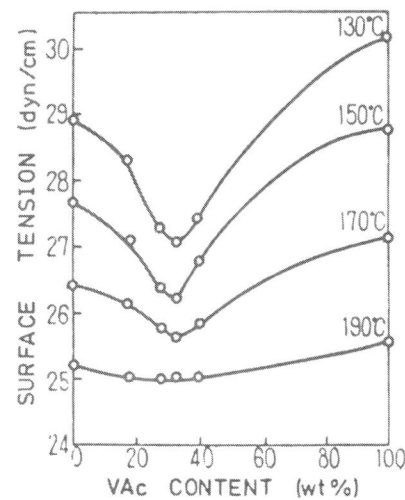

Fig. 14. Composition dependence
of surface tension of ethylene
oxide-propylene oxide copolymers at 60°C.
a: random, b: block (m=16),
c: block (m=30),
d: block (m=56)

Fig. 15. Composition dependence of surface tension of
ethylene-vinyl acetate
copolymers.

at the same composition.

4.3 Characteristics of Ethylene-Vinyl Acetate Copolymers

In preceding sections, it was pointed out that random copolymers composed of nonpolar/nonpolar and polar/polar monomers have
rather common features in γ - x_2 curves. Ethylene-vinyl acetate
copolymers as an example of nonpolar/polar pairs, however, behave
considerably differently as shown in Fig. 15, where γ goes through a
minimum at about 35 wt% of vinyl acetate. A minimum or a maximum
has also been observed in glass transition temperature for some
kinds of copolymers, and it has been attempted to modify the
additivity law to interpret the results. More theoretically,
however, behaviors of polymer surface tension must be founded with
liquid theories of polymers, and an explanation for the results of
EVA above described has been made by our co-worker, T. Nose (20), on
the basis of his hole theory of polymer liquids, applying it to
surface tension of mixtures. Qualitatively speaking, the minimum
appears as a result of the balance between energy and entropy
contribution with the introduction of vinyl acetate into polyethylene.

5.   INTERFACIAL TENSION BETWEEN POLYMER MELTS

Since Good and Girifalco, and Fowkes originally proposed the formulae on interfacial tension (γ_{12}), several modified equations have been reported in the last ten years. The validity of them, however, has not been fully examined because of the lack of enough experimental data especially on γ_{12}. Here I present some results of γ_{12} for rather systematically selected combinations: the one is polyacrylates (PMA, PEA, PnBA, and P2EHA) to each other and to poly(ethylene) (PE) and poly(styrene) (PS); the other is EVA of various vinyl acetate contents to PE, poly(vinyl acetate) (PVA_c) and one of EVA (VAc 25 wt%).

In Figs. 16 and 17, γ_{12} of the former series are shown as a function of alkyl chain length at 150°C (21). For pairs of each polyacrylate, the smaller the difference of alkyl chain length is, the smaller γ_{12} is obtained as expected, and for pairs to PE, the larger the chain length, the smaller γ_{12}. For pairs to PS, γ_{12} is minimized about at poly(propyl acrylate), showing the best affinity of the pair. Results for EVA series (22) are shown in Fig. 18 against vinyl acetate contents, from which a similar conclusion as above mentioned can be drawn.

Using these data as well as γ of each polymer, we examined the proposed equations on γ_{12}. Here the following three equations are checked by the data of polyacrylates series.

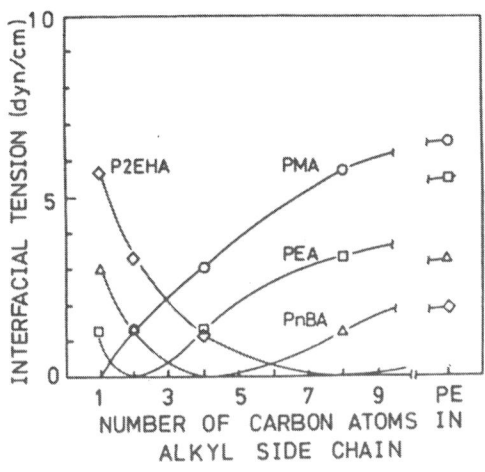

Fig. 16.   Interfacial tension of polyacrylates to each other as a function of side chain length at 150°C.

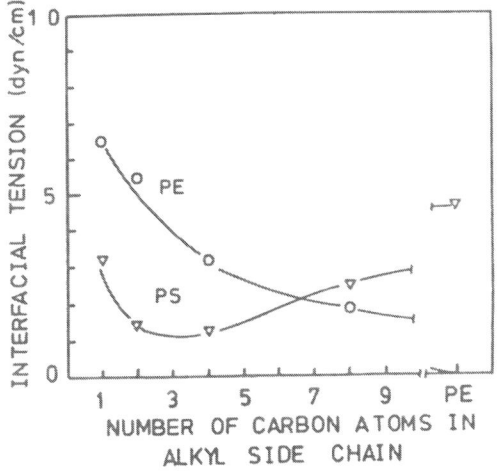

Fig. 17. Interfacial tension of polyacrylates against polyethylene and polystyrene at 150°C.

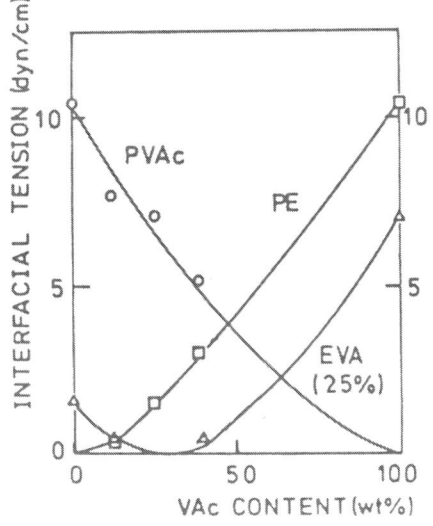

Fig. 18. Interfacial tension of ethylene-vinyl acetate copolymers against poly-ethylene, poly(vinyl acetate) and one of the copolymers (VAc 25 wt%) at 150°C.

D.K. Owens and R.C. Wendt (23), D.H. Kaelble and K.C. Uy (24) and R.K.S. Chan (25)

$$\gamma_{12} = \gamma_1 + \gamma_2 - 2\sqrt{\gamma_1^d \gamma_2^d} - 2\sqrt{\gamma_1^p \gamma_2^p} \qquad (13)$$

T. Hata and Y. Kitazaki (26)

$$\gamma_{12} = \gamma_1 + \gamma_2 - 2\sqrt{\gamma_1^d \gamma_2^d} - 2\sqrt{\gamma_1^a \gamma_2^a} - 2\sqrt{\gamma_1^b \gamma_2^b} \tag{14}$$

S. Wu (27)

$$\gamma_{12} = \gamma_1 + \gamma_2 - \frac{4\gamma_1^d \gamma_2^d}{\gamma_1^d + \gamma_2^d} - \frac{4\gamma_1^p \gamma_2^p}{\gamma_1^p + \gamma_2^p} \tag{15}$$

γ_1^d and γ_1^p in Eqs. (13) and (15) were determined with the observed value of γ_{12} against PE and γ_1 of each polymer, putting $\gamma_2 = \gamma_2^d$ for PE. γ_1^d, γ_1^a and γ_1^b in Eq. (14) were determined with γ_{12} values against PE and PS, assuming $\gamma_2 = \gamma_2^d + \gamma_2^a$ and $\gamma_2^b = 0$ for the latter. The results are summarized in Tables 2 and 3. As you can see in Table 3, the calculated values are much smaller than the observed ones, except for some values calculated with Wu's equation. The terms of work of adhesion seem to be overestimated by these equations.

Table 2.   Surface Tension Components of Polyacrylates
(at 150°C)

Polymers	γ	By Equations (13) and (14)					By Equation (15)		
		γ^d	γ^p	γ^a	γ^b	γ^p/γ (x100)	γ^d	γ^p	γ^p/γ (x100)
PMA	31.0	24.5	6.5	3.4	3.1	21.0	24.5	6.5	21.0
PEA	26.4	21.0	5.4	4.4	1.0	20.0	21.0	5.4	20.5
P$_n$BA	22.8	19.7	3.1	2.6	0.5	13.6	19.9	2.9	12.7
P2EHA	21.1	19.4	1.7	1.1	0.6	8.1	19.5	1.6	7.6
W-PE	23.2	23.2	0	0	0	0	23.2	0	0
PS	31.5	26.9	4.6	4.6	0	14.6	27.1	4.4	14.0

Table 3. Comparison Between Calculated and Observed
Interfacial Tensions (at $150^{\circ}C$)

Polymer Pairs	γ_{12} cal.(dyn/cm)			γ_{12}obs. (dyn/cm)
	Eq. (13)	Eq. (14)	Eq. (15)	
PS-PMA	0.2		0.5	3.2
PS-PEA	0.4		0.9	1.4
PS-PnBA	0.7		1.4	1.4
PS-P2EHA	1.3		2.5	2.5
P2EHA-PMA	1.8	1.9	3.5	5.7
P2EHA-PEA	1.1	1.2	2.1	3.3
P2EHA-PnBA	0.2	0.3	0.4	1.2
PnBA-PMA	0.9	1.4	1.9	3.0
PnBA-PEA	0.3	0.3	0.8	3.0
PEA-PMA	0.2	0.8	0.4	1.3

6. SURFACE TENSION OF POLYMER SOLUTIONS AND THE INTERFACIAL
TENSION AGAINST WATER

Now I want to introduce some examples to show that γ's of
polymer solutions do not always realize the characteristics of
solute polymers, but γ_{12}'s of the solutions against water can show
it clearly.

Figure 19 is surface tensions of toluene solutions of EVA
(γ_{sol}) and their interfacial tensions to water (γ_{sol/H_2O}) plotted
against VAc contents in EVA (28). γ_{sol}'s do not vary over the whole

range of EVA composition while γ_{sol/H_2O} changes drastically with
the composition, passing through a minimum and a maximum. Toluene
is more surface-active than EVA of any composition in this solution,
accordingly it remains at surface keeping γ_{sol} constant at the value
of toluene. On the other hand, at the interface between the toluene
solution and water, EVA molecules are adsorbed more and more with
the increase of hydrophilic vinyl acetate component. The minimum
and the maximum are caused by the competition between   lyophilicity
and the increasing compatibility of EVA with toluene in this region.
In fact, at the maximum γ_{sol/H_2O} both EVA(87.0 wt% VAc) and toluene
have an equal solubility parameter of 8.9.

Figure 20 is the case of toluene solutions of various polyacrylates (29). In this case, too, γ_{sol} is unchanged in spite of
the difference of alkyl side chain, while γ_{sol/H_2O} varies according
to the hydrophilic nature of side chain.

γ_{sol}/H_2O's of toluene solutions of a styrene-maleic anhydride
(3/1) copolymer were measured against water of various pH's (30).
(Fig. 21) The rapid decrease of γ_{sol/H_2O} near at pH 11 suggests the
dissociation of maleic acid at this pH.

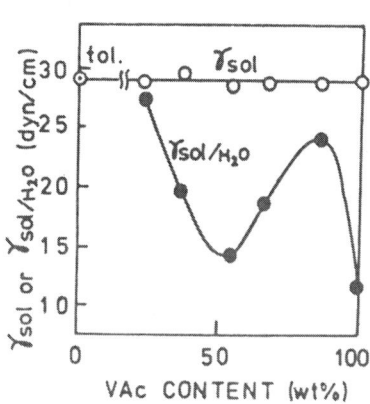

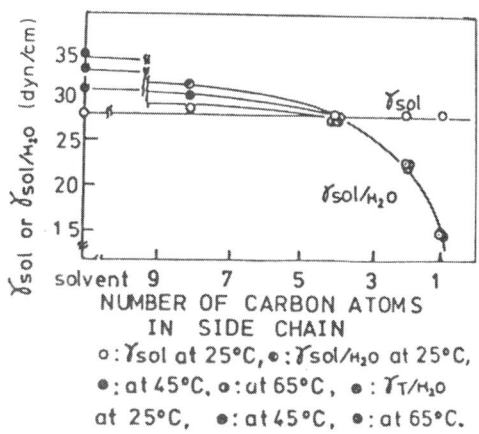

o : γ_{sol} at 25°C, • : γ_{sol/H_2O} at 25°C,
• : at 45°C, ⊙ : at 65°C, • : γ_{T/H_2O}
at 25°C, • : at 45°C, • : at 65°C.

Fig. 19. Surface tension of
 ethylene-vinylacetate copoly-
 mers/toluene solutions (γ_{sol})
 and their interfacial tension
 against water (γ_{sol/H_2O}) as a
 function of vinyl acetate content of the solutes at 20°C,
 C=0.01g/100ml.

Fig. 20. Surface tension of
 polyacrylates/toluene solutions
 (γ_{sol}) and their interfacial
 tension against water (γ_{sol/H_2O})
 as a function of side chain
 length of the solutes at
 C=0.01g/100ml.

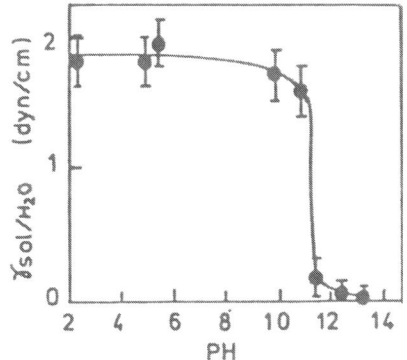

Fig. 21. Interfacial tension
between styrene-maleic
anhydride copolymer/toluene
solutions and water of different pH at 20°C,
C=0.05g/100ml.

    As demonstrated so far, interfacial tensions of polymer solu-
tions against water or other incompatible liquids provide important
information on the characteristics of solute polymers and their
adsorption behavior. It will be applicable as a tool to predict the
polymer adsorption on a hydrophilic solid surface.

7.   SURFACE AND INTERFACIAL TENSIONS OF POLYPEPTIDE SOLUTIONS

    Biological processes seem to take place very often at surface
or interface, where the selective adsorption may play an important
role in specifying the process. So it's useful and interesting for
studies on biological phenomena to make clear the surface-chemical
behavior of biomaterials. Now we want to introduce two topics in
our studies on polypeptide solutions: one is on the surface tension
in relation to the conformational change of polypeptide and another
is on different behaviors between surface and interfacial tensions
of polypeptide solutions.

7.1   A Change of Surface Tension Accompanied with the Conforma-
      tional Transition of Poly-γ-Benzyl-L-Glutamate in Ethylene
      Dichloride Solutions (31)

    We found, for solutions of poly-γ-benzyl-L-glutamate (PBLG) in
ethylene dichloride (EDC), that the conformational transition
occurred with the change of concentration. The measurements of
optical rotatory disperions (ORD) and circular dichroism (CD) have
shown that PBLG takes the α-helical conformation predominantly in
concentrated solutions and the β-structure in diluted solutions.
In the concentration range of their transition, the drastic
decrease of surface tension (γ_{sol}) was observed as shown in Fig. 22,
where helical contents calculated by the Moffitt's equation, using
ORD data, are plotted together. γ_{sol} of further diluted solutions

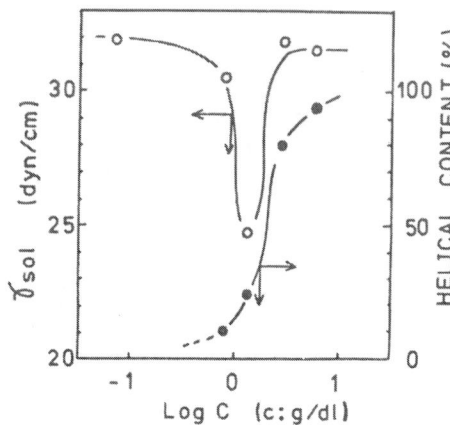

Fig. 22. Surface tension of
poly-γ-benzyl-D-glutamate in
ethylene dichloride and the
content of helical conformation at 25°C plotted against
log C.

(not included in the Figure) scarcely varies keeping the value of
solvent itself.

It is concluded from these facts that PBLG molecules in the β-
form and more or less in associated states may be more compatible
with EDC, accordingly tend to exist in the bulk phase, while in the
helical form they will become more lyophobic, because of the benzyl
side chains oriented towards outside the rodlike molecules. At
20~30% of helical content corresponding to the minimum γ_{sol}, the
lyophobic and lyophilic parts may reach the optimum balance like as
a surface active agent. The following increase of γ_{sol} with
increasing concentration may be attributed to the formation of
liquid crystal (32,33), where helical molecules attract each other
so as to stay in the bulk.

The changes of surface tension accompanied with the conformational transition have been reported by Neumann, Moscarello and
Epand (34) for aqueous solutions of poly-L-lysine, and by Morimoto
and Ueberreiter (35) for dimethylformide and dioxane solutions of
PBLG.

7.2 Surface Tension of Poly-DL-Alanine in Aqueous Solutions of
     Different pH and the Interfacial Tension Against Toluene (30)

The change in γ_{sol} of aqueous solutions of poly-DL-alanine, at
25°C and C=0.02g/dl, are shown in Fig. 23 as a function of pH. γ_{sol}
becomes the maximum at pH 8.0 and the minimum at 4.5. Since poly-
DL-alanine is a random copolymer of D- and L-alanine, its conformation may be thought to be almost random coil and not to vary with
pH. On the basis of the fact that intrinsic viscosities of its

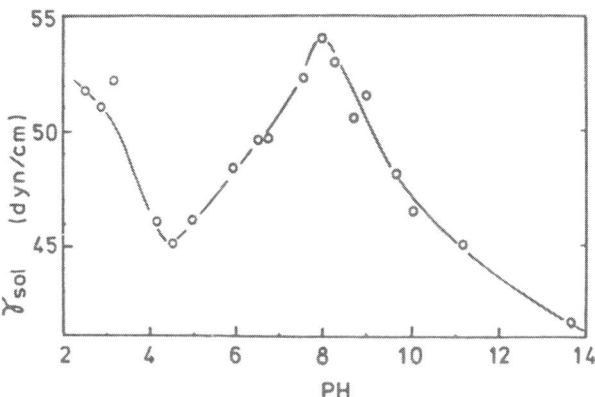

Fig. 23. Surface tension of poly-DL-alanine in
          aqueous solutions of different pH at
          25°C, C=0.01g/100ml.

aqueous solutions reach the minimum at pH 8.0 (Katagai et al. (36)),
this pH can be taken for the isoelectric point of the poly-DL-
alanine. However, poly-L-lysine (34) and egg albumin (37) are
quite different from poly-DL-alanine in that the formers show the
minimum γ_{sol} values near at their isoelectric points respectively.
In these cases, it must be taken into consideration that a change of
molecular structure occurs at the isoelectric point.

    Our results can be explained as follows: Lower values of γ_{sol}
at both sides of the isoelectric point are probably due to the
increasing balance between lyophilic and hydrophilic (electrically
charged) parts in molecules, so that surface adsorption increases
and γ_{sol} decreases. At the isoelectric point, molecules of poly-DL-
alanine decrease their dimension and more or less aggregate with
each other as known by the viscosity minimum. As a result, the
desorption from surface will be accelerated to increase γ_{sol}. Below
pH 4.5, the point of the minimum γ_{sol}, charged molecules tend to
exist in bulk phase and increase γ_{sol} again.

    Interfacial tensions between toluene and aqueous solutions of
poly-DL-alanine ($\gamma_{sol/T}$) behave differently from surface tensions
of the solutions (γ_{sol}). Figure 24 shows $\gamma_{sol/T}$ vs. pH, where we
can see a broad plateau of $\gamma_{sol/T}$ over the range of pH 4~9. The
value of 16~17 dyn/cm at 25°C in this region is much smaller than
that between pure water and toluene (55 dyn/cm). These facts
indicate that the methyl side groups of poly-DL-alanine are more

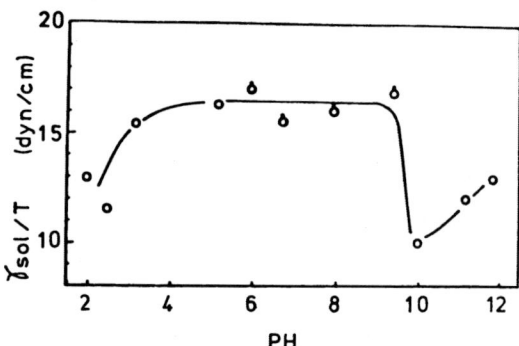

Fig. 24. Interfacial tension between toluene and
          aqueous solutions of poly-DL-alanine as
          a function of pH at 25°C, C≈0.01g/100ml.

strongly adsorbed to the toluene/water interface than to the air/
water interface, and as a result the effect of pH changes on the
adsorbability is screened behind it.

On the behavior of $\gamma_{sol/T}$ in the acidic and alkaline end
regions, discussions should be reserved at present, because $\gamma_{sol/T}$
did not reach equilibrium with time even over 200 hrs. (In the
Figure the values at 100 hrs. are plotted by simple open circles.)
This is probably due to the dissolution of poly-DL-alanine in
toluene in these regions.

Anyhow it can be concluded that comparative studies on surface
and interfacial tensions are quite useful to make clear the important role of interface and contacting liquid and to characterize the
surface behavior of polypeptides in solution.

CONCLUSIONS

So far we have discussed several topics from our studies on
surface and interfacial tensions of polymer melts and solutions. As
you see, the measurements of these quantities provide much useful
information not only for technical purposes but also for basic
researches on polymer physics and surface chemistry. The reason is
that surface or interfacial tension is essentially a thermodynamic
quantity and, in addition, it sensitively reflects the change in
surface or interface. As above-mentioned, to make clear the
selective behavior of polymeric materials at interface becomes
increasingly important. The method is classical but essential and
simple. We hope that surface and interfacial tensions are recognized more rightly as a tool for the studies on polymer science and

technology and experimental results can be further accumulated and their theoretical foundations can be firmly established.

REFERENCES

1. S. Wu,"Interfacial and Surface Tension of Polymers ", J. Macromol. Sci.-Revs., Macromol. Chem. Chem. C10, 1 (1974).
2. S. Wu,"Interfacial Energy, Structure, and Adhesion Between Polymers", (Chap. 6) in Polymer Blends, Vol. 1, D.R. Paul and S. Newman, eds., Academic Press, New York (1978).
3. T. Hata, "Surface Tension of Molten Polymers" (in Japanese), Hyomen (Surface) 6, 281 (1968).
4. A.W. Porter, "The Calculation of Surface Tension from Experiment. Part I, Sessile Drops", Phil. Mag. 15, 163 (1933).
5. W.D. Bascom, R.L. Cottington, C.R. Singleberry, "Dynamic Surface Phenomena in the Spontaneous Spreading of Oils on Solids", in Contact Angle, Wettability and Adhesion (Advances in Chemistry Series, No. 43), p. 355, Am. Chem. Soc., Washington, D.C. (1964).
6. T. Kasemura and T. Hata, "Surface Tension of Poly(ethylene glycol), Poly(propylene glycol) and Their Methyl Ether Homologs--Molecular Weight Dependence and Methyl Group Effect", (in Japanese), Kobunshi Ronbunshu 33, 192 (1976).
7. A.K. Rastogi and L.E. St. Pierre, "Interfacial Phenomena in Macromolecular Systems V. The Surface Free Energies and Surface Entropies of Polyethylene Glycols and Polypropylene Glycols", J. Colloid Interface Sci. 35, 16 (1971).
8. T. Hata, "Surface Chemistry of Polymers", (in Japanese), Kobunshi (Polymers) 17, 594 (1968).
9. K. Ueberreiter and G. Kanig, "Die Kettenlängenabhängigkeit des Volumens, des Ausdehnungs-Koeffizienten und der Einfriertemperatur von fraktionierten Polystyrolen", Z. Naturforsch. 6a, 551 (1951).
10. D.G. LeGrand and G.L. Gaines, Jr., "The Molecular Weight Dependence of Polymer Surface Tension", J. Colloid Interface Sci. 31, 162 (1969).
11. Y. Kitazaki and T. Hata, unpublished data.
12. T. Kasemura, N. Yamashita, K. Suzuki, T. Kondo and T. Hata, "Surface Tension of Alkanediols and Their Condensates, Poly(alkylene glycol)s", (in Japanese), Kobunshi Ronbunshu 35, 215 (1978).
13. H. Tozaki and T. Hata, presented to the 26th Ann. Meeting of Soc. Polym. Sic., Japan (1977).
14. T. Kasemura, F. Uzi, T. Kondo and T. Hata, "Surface Tension of Poly(vinyl alkylate)s", (in Japanese), Kobunshi Ronbunshu 36 (1979), in press.

15. T. Kasemura, N. Yamashita, T. Kondo and T. Hata, "Surface Tension of Dialkyl Phthalates", (in Japanese), Kobunshi Ronbunshu 33, 703 (1976).

16. F.M. Fowkes, "Determination of Interfacial Tensions, Contact Angle, and Dispersion Forces in Surfaces by Assuming Additivity of Intermolecular Interactions in Surfaces", J. Phys. Chem. 66, 382 (1962); "Additivity of Intermolecular Forces at Interface I", ibid. 67, 2538 (1963); "Dispersion Force Contributions to Surface and Interfacial Tensions, Contact Angles, and Heats of Immersion" in Contact Angle, Wettability and Adhesion (Advances in Chemistry Series No. 43), p. 99, Am. Chem. Soc. (1964).

17. T. Kasemura, N. Yamashita, K. Suzuki, T. Kondo and T. Hata, "Composition Dependence of Surface Tension of Copolymers", (in Japanese), Kobunshi Ronbunshu 35, 263 (1978).

18. A.K. Rastogi and L.E. St. Pierre, "Interfacial Phenomena in Macromolecular Systems III, The Surface Free-Energies of Polymers", J. Colloid Interface Sci. 31, 168 (1969).

19. T. Kasemura, K. Suzuki, F. Uzi, T. Kondo and T. Hata, "Composition Dependence of Surface Tension of Ethylene Oxide-Proplyene Oxide Copolymers", (in Japanese), Kobunshi Ronbunshu 35, 779 (1978).

20. T. Nose, "A Hole Theory of Polymer Liquids and Glasses V. Surface Tension of Polymer Liquids", Polymer J. 3, 1 (1972), ibid VI, Mixtures, ibid 3, 196 (1972).

21. H. Tozaki and T. Hata, presented to the 26th Ann. Meeting of Soc. Polym. Sci., Japan (1977).

22. H. Uraki and T. Hata, presented to the 13th Symp.on Adhesion (1975).

23. D.K. Owens and R.C. Wendt, "Estimation of the Surface Energy of Polymers", J. Appl. Polym. Sci. 13, 1741 (1969).

24. D.H. Kaelble and K.C. Uy, "A Reinterpretation of Organic Liquid-Polytetrafluoroethylene Surface Interactions", J. Adhesion 2, 50 (1970).

25. R.K.S. Chan, "Surface Tension of Fluoropolymers I. London Dispersion Term", J. Colloid Interface Sci. 32, 492 (1970), "II. The Polar Attraction Term", ibid. 32, 499 (1970).

26. T. Hata, See Ref. (8); Y. Kitazaki and T. Hata, "Extension of Fowkes' Equation and Estimation of Surface Tension of Polymer Solids", (in Japanese), J. Adhesion Soc., Japan 8, 131 (1972).

27. S. Wu, "Calculation of Interfacial Tension in Polymer Series", J. Polym. Sci. C 34, 19 (1971).

28. M. Sudo, H. Uraki and T. Hata, presented to the 12th Symp. on Adhesion (1974).

29. H. Arakawa, H. Tozaki and T. Hata, presented to the 24th Symp. on Polym. Sci. (1975).

30. H. Tozaki, H. Uraki and T. Hata, presented to the 13th Symp. on Adhesion (1975), see also Ref. 29.

31. H. Arakawa, M. Oya and T. Hata, presented to the 27th Ann. Meeting of Soc. Polym. Sci., Japan (1978).
32. C. Robinson, "Liquid-Crystalline Structures in Solution of a Polypeptide", Trans. Farad. Soc. 52, 571 (1956).
33. K. Kurotsu, K. Kikuchi, A. Tsutsumi and M. Kaneko, "Electric Birefringence of Poly(γ-benzyl-L-glutamate) in Ethylene Dichloride", Polym. J. 6, 571 (1974).
34. A.W. Neumann, M.A. Moscarello and R.M. Epand, "The Application of Surface Tension Measurements to the Study of Conformational Transitions in Aqueous Solutions of Poly-L-Lysine", Biopolym. 12, 1945 (1973).
35. S. Morimoto and K. Ueberreiter, "Surface Tension of Polymer Solutions. II. Poly--Benzyl-L-Glutamate in N, N-Dimethylformamide, Dioxane and Dichloracetic Acid Solutions", Kolloid-Z. u. Z. fur Polymere 253, 1009 (1975).
36. R. Katagai and J. Shinomiya, "The Viscosity Change of Poly-DL-alanine in Aqueous Solution", Makromol. Chem. 127, 282 (1969).
37. E.A. Hauser and L.E. Swearingen, "The Aging of Surfaces of Aqueous Solutions of Egg Albumin", J. Phys. Chem. 45, 644 (1941).

Donor-Acceptor Interactions at Interfaces

F.M. Fowkes

Department of Chemistry

Lehigh University

Bethlehem, PA 18015

ABSTRACT

Nearly all intermolecular interactions in solution and at interfaces can be reduced to two phenomena: London dispersion forces, and electron donor-acceptor (acid-base) interactions. Hydrogen bonds are included in acid-base interactions, and dipole phenomena are usually negligibly small. Earlier popular notions that all "polar" groups can interact with each other are shown untenable; donor-donor and acceptor-acceptor interactions are negligibly small compared to donor-acceptor interactions. Supporting data include surface and interfacial tensions, contact angles, and adsorption of polymers from organic solvents onto inorganic powders.

INTRODUCTION

Intermolecular forces between unlike molecules in solutions or at interfaces determine solution and interfacial properties. There are so many different kinds of intermolecular interactions that for "practical" studies most investigators of interfacial phenomena have tried to lump all interactions into two terms: dispersion force interactions and "polar" interactions. The most widely used equations for "polar" interactions are geometric mean equations suitable for dipole-dipole interactions, but completely unsuitable for acid-base interactions, as will be explained. This unfortunate approach has also been used in solution studies.

R.S. Drago and co-workers have recently correlated enthalpies of interaction of over thirty acids and bases in neutral solvents such as carbon tetrachloride (1). The correlation gives predictions of all hydrogen-bonds, Brønsted acids and bases, and Lewis acids and bases with good accuracy (except for some interactions which are greater than 80 kJoules/mole), and does not need to take into account any dipole-dipole interactions.

One might well distinguish between donor acceptor interactions in which an electron is transferred to the acceptor and acid-base interactions in which protons are transferred to the base. However, for the interactions discussed in this work the electrons are not completely transferred from one molecule to the other, but shifted in distribution; consequently these electron donor-acceptor interactions and acid-base interactions are indistinguishable and both are accurately predicted by the Drago relation. Drago's work suggests that in solution studies one can assume that all intermolecular interactions can be closely approximated by just two phenomena, dispersion force interactions and acid-base interactions, and that dipole-dipole interactions are usually too small to need to consider. This approach is very different from current practice in both principle and in the kinds of equations used.

This paper used the above approach for evaluation of interfacial phenomena (work of adhesion, contact angles, and adsorption) and it is found that many interfacial phenomena can be best predicted by treating interfacial forces as consisting of only dispersion forces and acid-base interactions, and neglecting dipole-dipole interactions.

In 1961 the author proposed that since intermolecular attractions result from several fairly independent phenomena (such as disperion forces (d), dipole interactions (p), hydrogen-bonds (h), etc.), it is reasonable to separate out such terms in the work of adhesion

$$W_A = W_A^d + W_A^p + W_A^h + \ldots \tag{1}$$

and in the surface tension

$$\gamma = \gamma^d + \gamma^p + \gamma^h + \ldots \tag{2}$$

This principle leads to the widely used equation

$$W_a^d = 2\sqrt{\gamma_1^d \, \gamma_2^d} \tag{3}$$

with which interfacial tensions, contact angles, free energies of adsorption, and Hamaker constants were successfully calculated (2,3). In confining the use of this equation to dispersion force interactions the geometric mean expression is correct. There is also some reason to treat dipole-dipole interactions separately with the geometric mean expression

$$W_A^p = 2\sqrt{\gamma_1^p \gamma_2^p} \qquad (4)$$

The interaction energy between two dipoles is $-2\mu_1^2\mu_2^2/3kTr_{12}^6$, (where μ is the dipole moment, k Boltzmann's constant, and T absolute temperature) so if the distance between dipoles r_{12} is the geometric mean of r_{11} and r_{22} of the pure materials, then equation (4) is correct.

The extension of equations (1), (2), and (4) to try to predict hydrogen-bonding with a geometric mean expression is quite incorrect, for hydrogen-bond acceptors such as ethers, esters, or aromatic hydrocarbons cannot themselves form hydrogen-bonds, and therefore γ^h is zero for such materials, even though these materials have a large W_A^h with hydrogen-donors (4). Similarly some hydrogen-donors such as chloroform have zero values of γ^h, but large values of W_A^h (23).

Zisman's extensive series of contact angle measurements of organic liquids on polymer surfaces (5) have tempted several investigators (6-9) to try to solve equation (1) by assuming that all polar and hydrogen-bonding interactions can be predicted by equation (4). This forced fit of data into incorrect equations makes all conclusions at least a little wrong (e.g., finite γ^p values for polyethylene and paraffin).

In solution studies the separation of heats of mixing (ΔH_M) into several terms soon followed. Blanks and Prausnitz (10) used only two terms (polar and nonpolar) while Gardon (11) and Meyer and Wagner (12) included terms for dipole-dipole and for dipole-induced dipole interaction. Although hydrogen-bonding had been recognized much earlier (13,14), it was Hansen (15) who brought forth the widely-used three-dimensional solubility parameter:(δ):

$$\Delta H_M = \Delta H_M^d + \Delta H_M^p + \Delta H_M^h + \cdots \qquad (5)$$

and

$$\delta^2 = \delta_d^2 + \delta_p^2 + \delta_h^2 \qquad (6)$$

The dispersion force term (ΔH^d), or more properly the change in internal energy on mixing (ΔU_M^d), can be correctly evaluated with the geometric mean

$$\Delta H_M^d = \Delta U_M^d = V_m \phi_1 \phi_2 (\delta_1^d - \delta_2^d)^2 \tag{7}$$

and perhaps a dipole-dipole term (ΔU_M^P) can be justified:

$$\Delta U_M + V_M \phi_1 \phi_2 (\delta_1^P - \delta_2^P)^2 \tag{8}$$

but for reasons given in the previous paragraph there is no way that ΔU_m^h can be predicted with a term $(\delta_1^h - \delta_2^h)^2$ and consequently any two sets of correlations are in serious disagreement (16).

The extensive computer-forced correlations of heats of mixing and of works of adhesion (7,15,17,18) are based on treating hydrogen-bonds by the geometric mean, as if the hydrogen-bond were a dipole-dipole interaction, a notion abandoned about twenty years ago (19). A more recent approach has been developed by R.S. Drago (1) who has treated the hydrogen-bond as a Lewis acid-base interaction. Drago has correlated the enthalpy of acid-base interaction ΔH_M^{ab} (in a neutral solvent such as carbon tetrachloride) with two constants (E_B, C_B) for each base and two constants for each acid (E_A, C_A):

$$-\Delta H_M^{ab} = C_A C_B + E_A E_B$$

Drago assumed that ΔH_M^d and ΔH_M^P were negligibly small for the acids and bases, and in his correlation all predicted ΔH_M^{ab} values (up to 80 kJoules/mole) checked measured values within about 5% or less.

Drago's correlation (treating interactions as due only to dispersion forces and acid-base interactions) is much more successful than Kaelble's and Hansen's correlations (treating interactions as due only to dispersion forces and polar interactions predictable from geometric mean equations). It is of special interest that in Drago's correlation the neglect of dipole-dipole interactions gave no problems, suggesting that dipole-dipole interactions are negligibly small compared to acid-base and dispersion force interactions.

We propose that the heat of mixing (ΔH_M) be given by

$$\Delta H_M = P \Delta V_M + V_m \phi_1 \phi_2 (\delta_1^d - \delta_2^d)^2 - x_p (C_A C_B + E_A E_B) + \Delta U_{12}^P \tag{10}$$

where X_p is the mole fraction of acid-base pairs per mole of components present. Similarly, we propose for the work of adhesion:

$$W_A = 2\sqrt{\gamma_A^d \gamma_B^d} - f(C_A C_B + E_A E_B) \times \frac{\text{moles of acid-base pairs}}{\text{unit area}} + W_A^p \quad (11)$$

in which the constant f (near unity) converts enthalpy per unit area into surface free energy, and the last term is usually small.

The Acidity or Basicity of Polymer Surfaces

The literature abounds with contact angle measurements of liquids on the surface of solid polymers (5,20) but few of these measurements are suitable for defining acid-base interactions. Consequently, we have recently concluded a study (21) of well-characterized acidic and basic polymers, and measured the contact angles θ of certain acidic or basic liquids from Drago's tables.

In these studies the work of adhesion is given by:

$$W_A = \gamma_L (1 + \cos \theta) + \pi_e = W_A^d + W_A^{ab} + W_A^p + \ldots \quad (12)$$

where π_e is any reduction of the surface energy of the polymer resulting from adsorption of the vapor of the test liquid. In general π_e is expected to be negligibly small for high energy liquids on low energy solids, and we have measured π_e for several such systems and find it to be indeed zero (22). The experimental approach in this work has been to determine

$$W_A - W_A^d - \pi_e = \gamma_L (1 + \cos \theta) - 2\sqrt{\gamma_S^d \cdot \gamma_L^d} \quad (13)$$

for a series of Drago-characterized acidic or basic liquids on polymers of controlled acidity or basicity.

The findings (Figs. 1 and 2) were that $W_A - W_A^d - \pi_e$ is zero for acidic liquids on acidic surfaces (phenolic liquids on acrylic acid sites) or for basic liquids on basic surfaces (pyridine on vinyl acetate sites), though quite appreciable with acid base-interactions (phenolic liquids on vinyl acetate sites or pyridine on acrylic acid sites).

In general, these two sets of contact angle measurements agree with the general conclusions concerning the Drago correlation: intermolecular forces are dominated by dispersion force interactions and acid-base interactions, while dipole-dipole interactions are negligibly small.

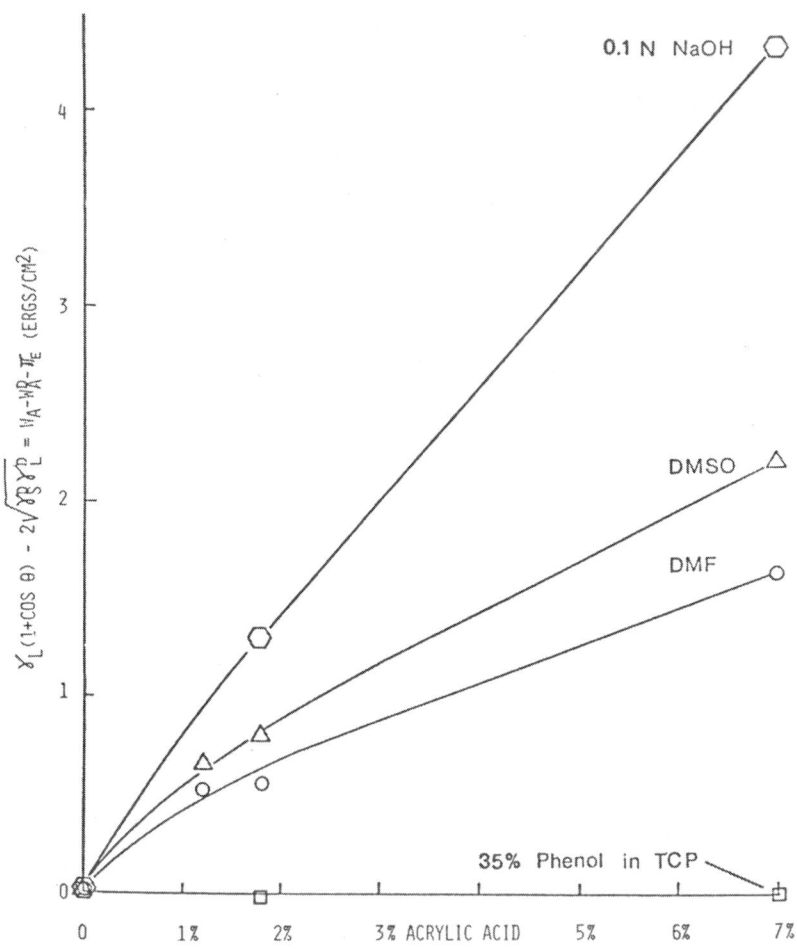

Fig. 1.   Effect of solvent-acrylic copolymer inter-
          action on $W_A - W_A^D - \pi_E$.

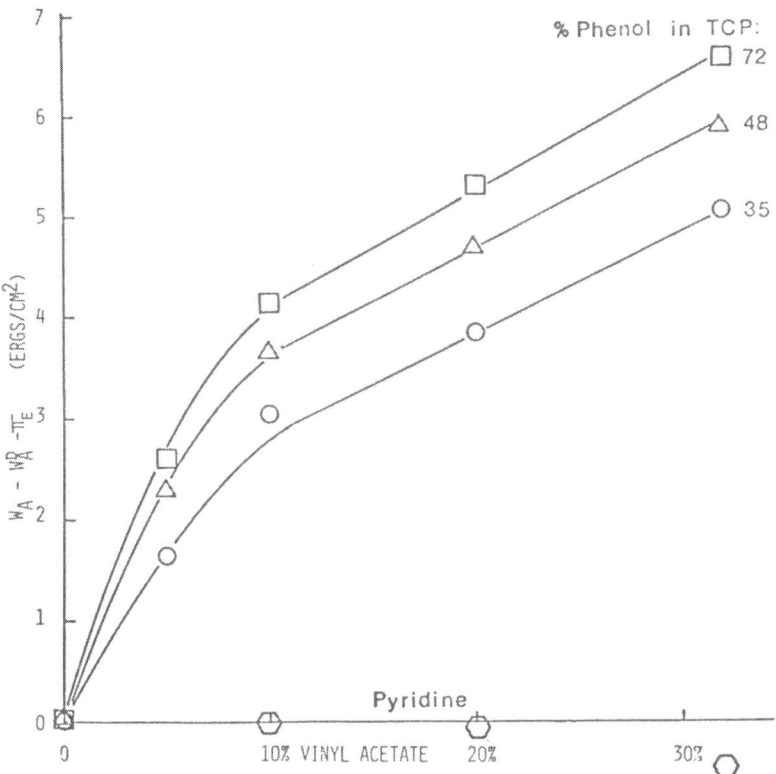

Fig. 2. Effect of solvent-vinylacetate copolymer interaction on $W_A-W_A^D-\pi_E$.

Acid-Base Interactions in Polymer Adsorption

A second route to explore the phenomenon of acid-base inter-actions at interfaces has been to measure the amount of adsorption of acidic or basic polymers onto acidic or basic inorganic powders from solution in acidic or basic organic solvents (23). Polymethyl-methacrylate (PMMA) was the basic polymer and post-chlorinated polyvinylchloride (C1-PVC) was the acidic polymer. The methacryl-ate ester groups are as basic as the acetate groups characterized by Drago and C1-PVC is nearly as acidic as chloroform. The acidic solid was silica; the SiOH silanol groups on the silica surface are well-known to be acidic, and are known to form acid-base bonds with methacrylate groups in the adsorption of methacrylate polymers (24). The basic solid, calcium carbonate, is well-recognized for its basicity.

The basic polymer (PMMA) adsorbed very slightly onto the basic solid $(CaCO_3)$; about 4% of a complete monolayer adsorbed from CCl_4

or benzene, and no adsorption was measurable from the other solvents. In contrast, PMMA adsorbed strongly onto the acidic SiO_2,
especially from the more neutral solvents (Fig. 3).

It appears that the basicity of the methacrylate groups must be
less than the basicity of dioxane, but much greater than the
basicity of benzene. These considerations suggest that the heat of
interaction of PMMA with t-butanol should be less than the 12.3
kJ/mole of dioxane, but much greater than the 5.0 kJ/mole of
benzene. This is confirmed by Drago's 10.5 kJ/mole for the interaction of esters with t-butanol.

The adsorption of the acidic polymer (C1-PVC) onto acidic
silica was so small that it was less than our limits of detection
(10^{-6} g/m^2, or 0.3% of a monolayer); this again illustrates the
unimportance of dipole-dipole interactions between two acids. On
the other hand, the acidic polymer was strongly adsorbed onto the
basic filler ($CaCO_3$) from the more neutral solvents (Fig. 4).

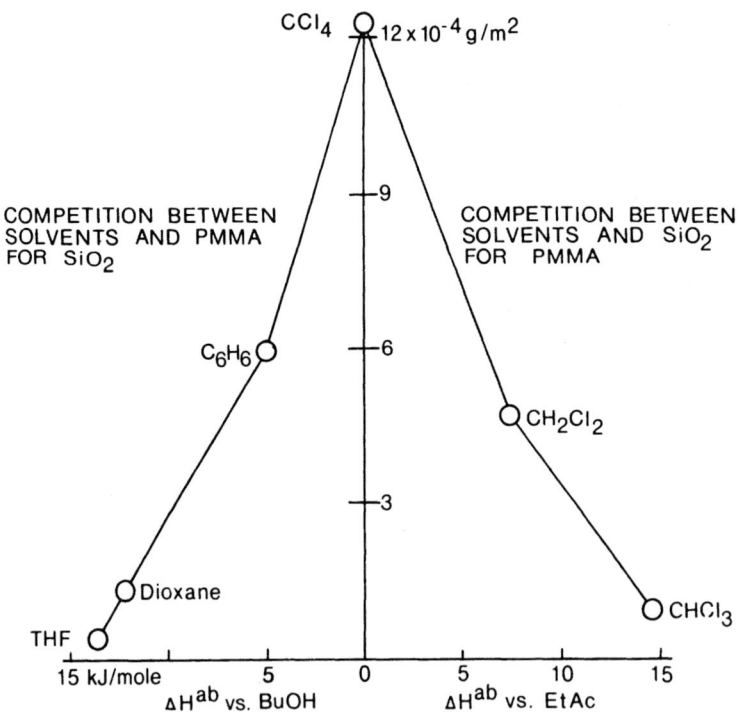

Fig. 3.   Competition between solvents and solid
surfaces during adsorption (PMMA vs. SiO_2).

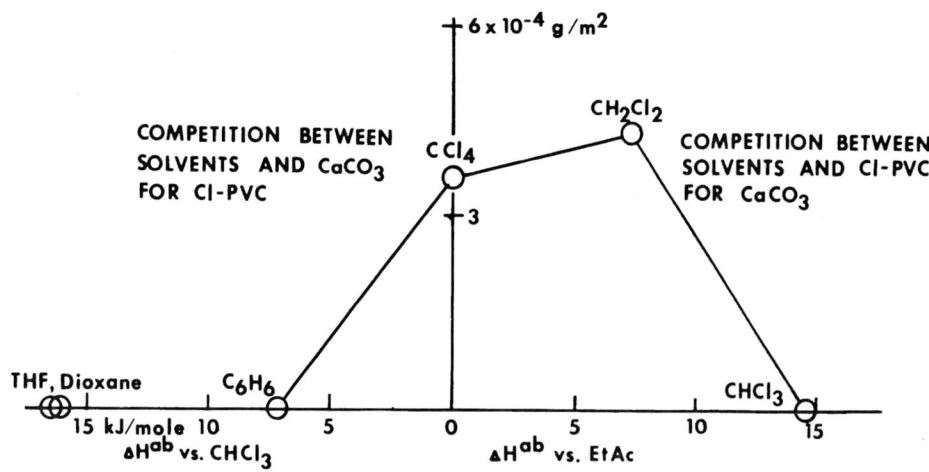

Fig. 4. Competition between solvents and solid surfaces during adsorption (Cl–PVC vs. CaCO$_3$).

In adsorption of polymers from solution, the acidity or basicity of the solvent was found to play an important role; a solvent more basic than the basic polymer is preferentially adsorbed, and a solvent more acidic than the acidic solid is preferentially bound to the basic polymer. Therefore, strong polymer adsorption is observed only from solvents more neutral than the adsorbing polymer or the adsorbent surface sites.

In retrospect we find these acid-base rules for adsorption to govern most of the earlier work. The twenty-year old studies of Koral, Ullman , and Eirich (25) for the adsorption of polyvinylacetate (basic) onto iron powder (acidic) showed strong adsorption from neutral (CCl_4) or weakly acidic ($C_2H_4Cl_2$) solvents but no adsorption from acetonitrile (basic). Similarly, the current studies of G.J. Howard and co-workers (26) show strong adsorption of polyacrylonitrile polymer (basic) on silica (acidic) from weakly acidic solvents (trichloroethylene) but not when proprionitrile is the solvent (equally basic, but more concentrated than the adsorbing polymer).

ACKNOWLEDGEMENTS

The author is deeply indebted to several co-workers; without their work this concept would be only a hazy notion. Special thanks go to S. Maruchi (of Sanyo-Kokusaku Pulp Co., Tokyo), Prof. J.A. Manson, M. Marmo (now at Eastman Kodak), M.A. Mostafa, and G. Sie. Support has come from Ford Motor Company, Sanyo-Kokusaku Pulp Co., and Lehigh University.

REFERENCES

1.   R.S. Drago, G.C. Vogel and T.E. Needham, J. Am. Chem. Soc. 93, 6014 (1971); R.S. Drago, L.B. Parr and C.S. Chamberlain, J. Am. Chem. Soc. 99, 3203 (1977).
2.   F.M. Fowkes, J. Phys. Chem. 66, 382 (1962).
3.   F.M. Fowkes, Ind. Eng. Chem. 12, 40 (1964).
4.   F.M. Fowkes, J. Adhesion 4, 155 (1972).
5.   W.A. Zisman, Advances in Chemistry "Contact Angles, Wettability, and Adhesion", ed. F.M. Fowkes (ACS Special Publication) 43, 1 (1964).
6.   D.K. Owens and R.C. Wendt, J. Appl. Polymer Sci. 13, 1741 (1969).
7.   D.H. Kaelble, J. Adhesion 2, 66 (1970).
8.   S. Wu, J. Adhesion 5, 39 (1973).
9.   J. Kloubek, J. Adhesion 6, 293 (1974).
10. R.F. Blanks and J.M. Prausnitz, Ind. Eng. Chem. 3, 1 (1964).
11. J.L. Gardon, J. Paint Technol. 38, 43 (1966).
12. E.F. Meyer and R.E. Wagner, J. Phys. Chem. 70, 3162; 75, 642 (1971).
13. W. Goody, J. Chem. Phys. 7, 93 (1939); 8, 170 (1949); 9, 204 (1941).
14. P.A. Small, J. Appl. Chem. 3, 71 (1953).
15. C.M. Hansen, J. Paint Technol. 39, 104 and 505 (1967).
16. E.B. Bagley and J.M. Scigliano, Chapter XVI "Solutions and Solubilities", ed. M.R.J. Dack, Vol. VIII of "Techniques of Chemistry", A. Weissberger, ed. Wiley, New York (1975).
17. K.L. Hoy, J. Paint Technol. 42, 118 (1970).
18. C.M. Hansen and A. Beerbower, "Solubility Parameters in "Encyclopedia of Chemical Technology", Supplement Volume, 2nd Edition, Wiley, New York (1971).
19. G.C. Pimental and A.L. McClellan, "The Hydrogen Bond", Freeman, San Francisco (1960).
20. J.R. Dann, J. Colloid Interface Sci. 32, 302 and 321 (1970).
21. F.M. Fowkes and S. Maruchi, Coatings and Plastics Preprints 37, 605 (1977).
22. F.M. Fowkes and D.C. McCarthy, to be published.
23. F.M. Fowkes and M.A. Mostafa, Ind. Eng. Chem. Prod. R&D 17, 3 (1978).
24. B.J. Fontanta and J.R. Thomas, J. Phys. Chem. 65, 480 (1961).
25. J. Koral, R. Ullman and F.R. Eirich, J. Phys. Chem. 62, 541 (1958).
26. G.J. Howard and M.J. McGrath, J. Poly Sci. (Poly. Chem. Ed.) 15, 1721 (1977).

Surface Tension of Solids: Generalization and Reinterpretation of Critical Surface Tension

Souheng Wu

E.I. du Pont de Nemours & Company

Central Research and Development Department

Experimental Station, Wilmington, Delaware   19898

ABSTRACT

The concept of critical surface tension is generalized and reinterpreted in terms of a proposed equation of state. The equation defines a spectrum of critical surface tensions for a given surface, and provides a method by which the surface tension can be accurately determined from the contact angles of a series of testing liquids. The surface tensions obtained for solid polymers, organic solids and monolayers by this method agree remarkably well with those obtained from melt data (temperature dependence), liquid homologs (molecular-weight dependence) and the harmonic-mean equation. In contrast, those obtained by the geometric-mean equation and Zisman's critical surface tensions are often too low. These results also support the validity of the harmonic-mean equation.

CRITIQUE OF SOME EXISTING METHODS FOR DETERMINING THE SURFACE TENSION OF SOLIDS

The surface tension of a solid is difficult, if not impossible, to measure directly, because large reversible deformation of a solid surface cannot be done. A number of indirect methods based on empirical or semi-empirical relations between the wettability and the surface tension have been proposed, including the critical surface tension (1-4), the method based on the harmonic-mean equation (5-9), and the method based on the geometric-mean equation (10-12). Several theoretical treatments are also avail-

53

able (13-20).

Zisman's Critical Surface Tension

This, first proposed by Fox and Zisman (1), has been widely used. However, its value is quite variable, depending on the nature of the testing liquids used. For instance, the critical surface tension for poly(ethylene terephthalate) varies from 27 to 46 dyne/cm at 20°C (21); the value for perfluorodecanoic acid monolayer varies from 11 to 27.5 at 20°C (22). Moreover, the critical surface tensions of liquids have been found to be significantly lower than their surface tensions which can be measured directly and accurately. For instance, the critical surface tension of water is about 20 dyne/cm, compared with its surface tension of 72.8 dyne/cm at 20°C (23,24). Zisman (22) has cautioned against equating the critical surface tension to the surface tension.

Geometric-Mean Method and Harmonic-Mean Method

Recently, semi-empirical equations relating the wettability of the surface tension based on the theory of fractional polarity (6-8) have been proposed, providing methods for determining the surface tension and the polarity of solid surfaces from contact angles of two testing liquids. These are, namely, the geometric-mean method (10-12) and the harmonic-mean method (5-9). We will show that the harmonic-mean method gives accurate results, whereas the geometric-mean method does not.

The geometric-mean equation is an empirical extension of the Fowkes' equation (25). The experimental proofs of the Fowkes' equation consist mainly of the following. (a) The dispersion-force component γ^d of water as calculated from the interfacial tensions between water and a series of n-alkanes is rather constant (25). (b) The equation correctly predicts the interfacial tension between water and mercury (25). However, when the equation is applied to water/organic liquids or mercury/organic liquids systems, large deviations are found. These deviations are attributed to polar interactions which are, however, not accounted for in the Fowkes' equation.

Moreover, the surface tension γ and its dispersion and polar components, γ^d and γ^p, respectively, for a given solid surface as calculated by the geometric-mean equation vary widely, depending on the nature of the testing liquids used. For instance, for nylon 66 at 20°C, γ varies from 35.7 to 45.8 dyne/cm, γ^d from 20.7 to 40.2 dyne/cm, and γ^p from 3.0 to 14.9 dyne/cm (12). Such wide variations show that the geometric-mean equation is only a rough approximation.

On the other hand, Wu (5-8) has accurately measured both the surface and interfacial tensions of many incompatible pairs of polymer liquids and melts.   In such systems, stable equilibrium interfaces can be formed so that the surface and interfacial tensions can be measured accurately and directly.   Such data provide an excellent opportunity for testing and verifying the equations experimentally.   By using the data for fifteen pairs of polymer liquids and melts, Wu (5-8) showed the validity of the harmonic-mean equation, whereas the geometric-mean equation was found to be in serious error in many cases (5-8).

THE EQUATION OF STATE METHOD

Here, we would like to propose a novel equation of state, generalize and reinterpret the critical surface tension, and provide a method by which the surface tension of solids can be accurately determined from wettability data.   We will also show that the surface tensions of solids obtained by the present method agree remarkably well with those obtained from the melt data, the liquid homologs and the harmonic-mean method.   In contrast, the values obtained by the geometric-mean method and the Zisman's critical surface tensions are often too low.

The Proposed Equation of State

Combining the Young's equation (26) with the equation of Good and Girifalco (27) gives

$$\cos \theta = 2\phi (\gamma_S/\gamma_{LV})^{1/2} - 1 - (\pi_e/\gamma_{LV}) \tag{1}$$

where θ is the equilibrium contact angle, ϕ the interaction parameter, γ_S the surface tension of the solid in vacuum, γ_{LV} the surface tension of the liquid in equilibrium with its saturated vapor, and π_e the equilibrium spreading pressure.   The ϕ has been explicitly expressed in terms of molecular parameters (27), or in terms of macroscopic polarities of the two phases. (6-8)   The critical surface tension may be defined as

$$\gamma_c = \lim_{\theta \to 0} \gamma_{LV} \tag{2}$$

Thus, combining eqs. (1) and (2) gives

$$\gamma_c = \phi^2 \gamma_S - \pi_e + (\pi_e^2/4\phi^2\gamma_S) + \ldots \tag{3}$$

which converges rapidly, and therefore, can be truncated to

$$\gamma_{c,\phi} = \phi^2\gamma_S - \pi_e \tag{4}$$

where we have used $\gamma_{c,\phi}$ to replace γ_c, i.e., $\gamma_{c,\phi} = \gamma_c$, to emphasize the fact that γ_c is a function of ϕ. Using eq. (4) in eq. (1) gives

$$\cos\theta = 2(\gamma_{c,\phi}/\gamma_{LV})^{1/2} - 1 \tag{5}$$

which is strikingly simple, and will be termed the equation of state.

Application of the Equation of State

Eq. (5) can be rearranged to

$$\gamma_{c,\phi} = (1/4)(1 + \cos\theta)^2\gamma_{LV} \tag{6}$$

which can be used to calculate $\gamma_{c,\phi}$ from the contact angle of any one testing liquid. When a series of testing liquids are used, a spectrum of $\gamma_{c,\phi}$ values will be obtained. The maximum $\gamma_{c,\phi}$ value is equal to γ_S. This is because $\phi_{max} = 1$ occurring when the polarities of the two phases are identical (6-8,27), and if π_e is small, then from eq. (4),

$$\gamma_{c,\phi,max} = \gamma_S - \pi_e \stackrel{\sim}{=} \gamma_S \tag{7}$$

Thus, if the spectrum of $\gamma_{c,\phi}$ is plotted vs. the polarity of the testing liquids, the data points will fall into a smooth curve with the maximum $\gamma_{c,\phi,max}$ equal to γ_S.

However, since the polarity of many testing liquids are unknown, the $\gamma_{c,\phi}$ spectrum may instead be plotted vs. γ_{LV}. In this case, the data points will be somewhat scattered, since $\gamma_{c,\phi}$ is a function of not only γ_{LV}, but also ϕ (or the polarity). However, a smooth curve can be drawn just to envelop all the data points below, as illustrated in Figs. 1-4 for some polymers, organic solid and monolayers. Table 1 gives the $\gamma_{c,\phi}$ spectra for polyethylene and polytetrafluoroethylene as calculated by eq. (6) as examples. All contact angle data are from Refs. (1-4,11). It can be seen that $\gamma_{c,\phi,max}$ occurs at large θ values, so that we may safely let $\pi_e = 0$. Thus, $\gamma_{c,\phi,max} = \gamma_S$.

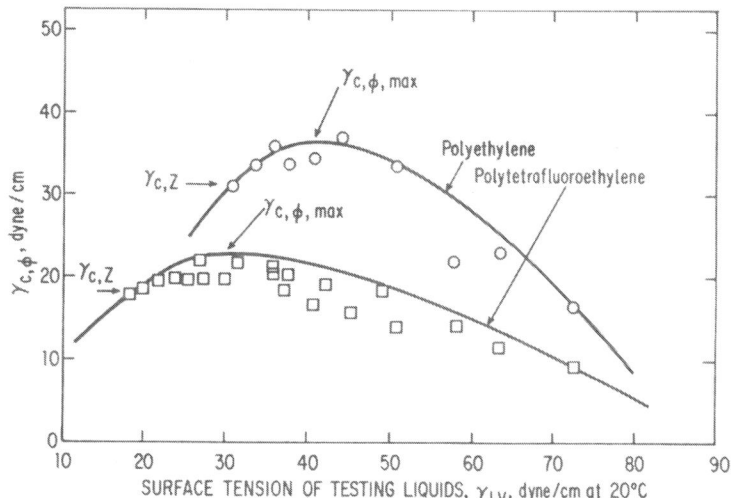

Fig. 1.   Equation of State Plots for Polyethylene
          and Polytetrafluoroethylene.

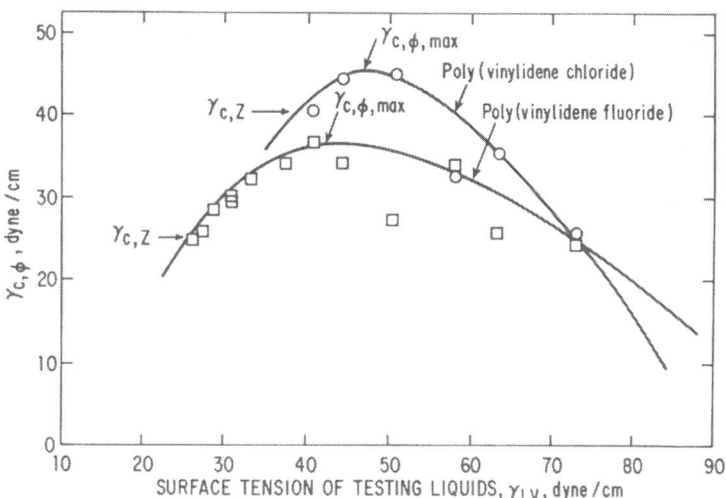

Fig. 2.   Equation of State Plots for Poly(vinylidene
          chloride) and Poly(vinylidene fluoride).

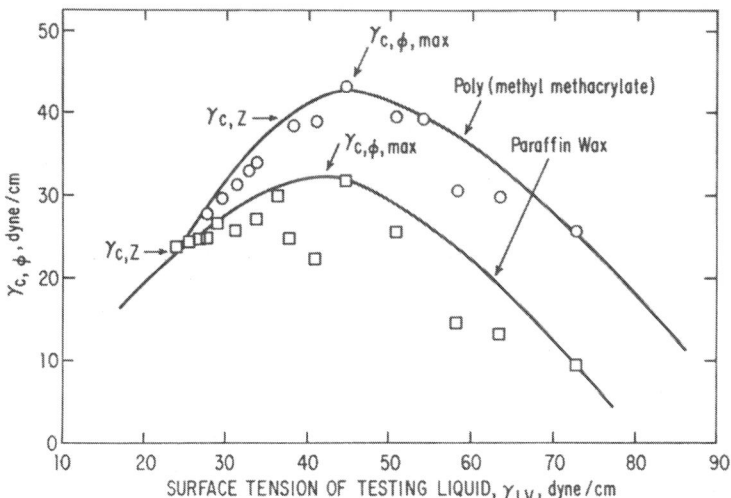

Fig. 3.   Equation of State Plots for Poly(methyl-
          methacrylate) and Paraffin Wax.

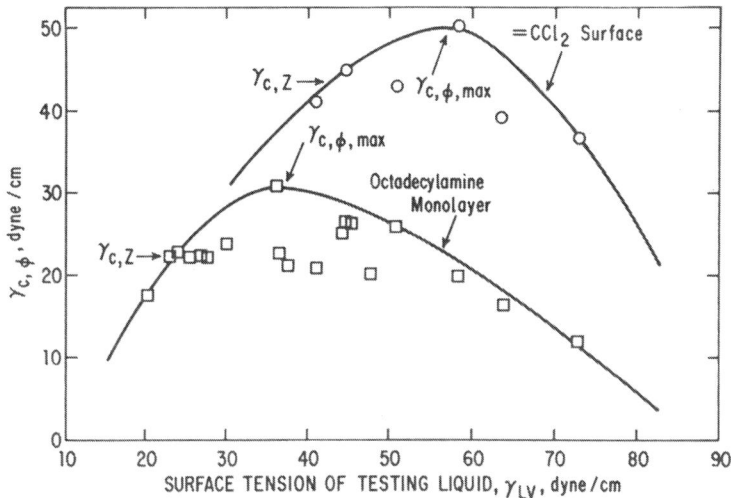

Fig. 4.   Equation of State Plots for $=CCl_2$ Surface
          (i.e., Adsorbed Monolayer of Perchloro-2,4-
          pentadienoic Acid) and $-CH_3$ Surface (i.e.,
          Adsorbed Monolayer of Octadecylamine).

Zisman's critical surface tension, designated as $\gamma_{c,Z}$, can be defined as $\gamma_{c,Z} = |\gamma_{c,\phi}|_{\theta = 0}$. It can be seen in Figures 1-4 that $\gamma_{c,Z}$ is determined only by the few low-contact-angle data. In fact, $\gamma_{c,Z}$ is little affected by all the higher contact-angle data in a Zisman plot.

Comparison Among Various Methods

Table 2 lists the γ_S values obtained by the present equation of state method for various solid polymers, organic solids and monolayers. Also included for comparison are those obtained from molecular-weight dependence of liquid homologs (7,8,28), temperature dependence of melts (5-10), the harmonic-mean method (5,6,9), the geometric-mean method (5,11,12) and Zisman's critical surface tension (4). The results obtained from the melt data and the liquid homolog data are accurate within 0.1 to 0.5 dyne/cm (8). It can be seen that the results obtained by the equation of state method, the melt data, the liquid homolog data are generally in excellent agreement with one another, whereas those obtained by the geometric-mean method and Zisman's critical surface tensions are often too low.

Remarks on Some Materials

Polyethylene: The γ_S at $20°C$ is 35.9 dyne/cm by the equation of state method, 35.7 dyne/cm from melt data (29), 34.7 dyne/cm from liquid homologs (8), 36.1 dyne/cm by the harmonic-mean method (5), in contrast to 32.4 dyne/cm by the geometric-mean method (11,12) and 31 dyne/cm for Zisman's critical surface tension (4).

Polytetrafluoroethylene: The γ_S at $20°C$ is 22.6 dyne/cm by the equation of state method, 23.9 dyne/cm from liquid homologs (8), 22.5 dyne/cm by the harmonic-mean method (5), in contrast to 15.6-19.1 dyne/cm by the geometric-mean method (11,12) and 18.5 dyne/cm for Zisman's critical surface tension (4).

The surface tension of a fluorocarbon oil $C_{21}F_{44}$ has been directly measured to be 21.5 dyn/cm at $20°C$ by Dettre and Johnson (30). As the molecular weight increases, the surface tension should increase. It is thus noteworthy that the γ_S values for solid polytetrafluoroethylene at $20°C$ obtained by the equation of state method, the liquid homologs and the harmonic-mean method are all in the range of 22 to 24 dyn/cm, consistent with expectation. In contrast, the geometric-mean method and the Zisman's critical surface tension are too low.

Table 1. CALCULATION OF CRITICAL SURFACE TENSION SPECTRA FOR POLYETHYLENE AND POLYTETRAFLUOROETHYLENE

Liquids	γ_{LV} dyne/cm at 20°C	Polyethylene		Polytetrafluoroethylene	
		θ degree	$\gamma_{c,\phi}$ dyne/cm	θ degree	$\gamma_{c,\phi}$ dyne/cm
n-Alkanes					
Hexane	18.4	–	–	12	18.0
Heptane	20.3	–	–	21	19.0
Octane	21.8	–	–	26	19.7
Decane	23.9	–	–	35	19.8
Dodecane	25.4	–	–	42	19.3
Hexadecane	27.6	–	–	46	19.8
n-Alkybenzenes					
Hexylbenzene	30.0	–	–	52	19.6
Benzene	28.9	–	–	46	20.7
Halocarbons					
Methylene iodide	50.8	52	33.2	88	13.6
sym-Tetrabromomethane	49.7	–	–	79	17.6
Tetrachlorobiphenyl	45.3	–	–	81	15.1
α-bromonaphthalene	44.6	35	36.9	73	18.6
Hexachloropropylene	38.1	–	–	65	19.3
Perchlorocyclopentadiene	37.5	–	–	67	18.1
sym-Tetrachloroethane	36.3	10	35.7	56	22.1
Hexachlorobutadiene	36.0	–	–	60	20.2
Tetrachloroethylene	31.7	–	–	49	21.7
Carbon tetrachloride	26.8	–	–	36	21.9

Esters

Tricresyl phosphate	40.9	34	34.2	75	16.2
Benzylphenyl undecanoate	37.7	28	33.4	67	18.2
Bis(2-ethylhexyl)phthalate	31.2	5	31.3	63	16.5
Bis(2-ethylhexyl)sebacate	31.3	-	-	62	16.8

Miscellaneous

t-Butyl naphthalene	33.7	7	33.4	65	17.0
Formamide	58.2	77	21.8	92	13.5
Glycerol	63.4	79	22.5	100	10.8
Water	72.8	94	15.8	108	8.7
Mercury	485.0	-	-	150	2.2

Table 2.   COMPARISON AMONG SURFACE TENSION VALUES OBTAINED BY VARIOUS METHODS FOR
           SOLID POLYMERS, ORGANIC SOLIDS AND MONOLAYERS

Material	Zisman's Critical Surface Tension γ_c dyne/cm	Surface Tension, γ_S, dyne/cm at 20° C				
		By Eq. of State Plot, $\gamma_{c,max}$	From Melt Data	From Liquid Homologs	By Harmonic Mean Equation	By Geometric Mean Equation
Polymers						
Polyethylene	31	35.9	35.7	34.7	36.1	33.2
Polytetrafluoro- ethylene	18	22.6	-	23.9	22.5	19.1
Polytrifluoro- ethylene	22	29.5	-	-	27.3	23.9
Poly(vinylidene fluoride)	25	36.5	-	-	33.2	30.3
Poly(vinyl fluoride)	28	37.5	-	-	38.4	36.7
Poly(vinylidene chloride)	40	45.2	-	-	45.4	45.0
Poly(vinyl chloride)	39	43.8	-	-	41.9	41.5
Polystyrene	33	43.0	40.7	-	42.6	42.0
Poly(methyl methacrylate)	39	42.5	41.4	-	41.2	40.2
Polychlorotri- fluoroethylene	31	32.1	30.9	-	30.1	27.5

TFE/CTFE 60/40 (a)	24	26.2	–	–	25.2	20.9
TFE/CTFE 80/20 (a)	20	22.1	–	–	21.3	16.5
E/TFE 50/50 (b)	26.5	27.1	–	–	27.6	24.1
Nylon 66	46	43.8	46.5	–	44.7	47.0
Poly(ethylene terephthalate)	43	44.0	44.6	–	42.1	41.3
Organic Solids						
Paraffin Wax	23	32.0	35.0	–	31.0	25.4
Hexatriacontane	21	23.0	–	31.4	23.6	19.1
Monolayers (Adsorbed on Platinum)						
Octadecylamine	22	30.5	–	–	28.1	25.8
Perfluorolauric acid	5.6	14.5	–	–	–	–
Perfluorodecanoic acid	11	26.2	–	–	19.4	14.1
$-CF_2H$ Surface (mono-hydroperfluoro-undecanoic acid)	15	26.5	–	–	–	–
$=CCl_2$ Surface (Perchloro-2,4-pentadienoic acid)	43	49.5	–	–	48.4	46.2

(a) TFE/CTFE = copolymer of tetrafluoroethylene and chlorotrifluoroethylene.
(b) E/TFE = copolymer of ethylene and tetrafluoroethylene.

Paraffin Wax: The γ_S at 20°C is 32.0 dyne/cm by the equation of state method, 35.0 dyne/cm from melt data (31) and 31.0 dyne/cm by the harmonic-mean method. In contrast, the geometric-mean method (12) gives 23.7 dyne/cm; the Fowkes' equation gives 25.5 dyne/cm; Zisman's critical surface tension (4) is 15-22 dyne/cm.

The surface tension of molten paraffin wax at its melting point of 65°C has been directly measured to be 32 dyn/cm by Padday (31). As the molten paraffin wax solidifies and cools to 20°C, its surface tension should increase. Thus, it is noteworthy that the γ_S values for solid paraffin wax at 20°C obtained by the equation of state method, the melt data and the harmonic-mean method are all in the range of 31 to 35 dyn/cm, consistent with expectation. On the contrary, the Fowkes' value of 25.5 dyn/cm and the Zisman's critical surface tension of 15 to 23 dyn/cm seem much too low.

III.   CONCLUSION

The present analysis clearly shows that the proposed equation of state method gives accurate results. The surface tensions obtained by the equation of state method, the harmonic-mean method, the melt data and the liquid homologs agree remarkably well with one another. In contrast, those obtained by the geometric-mean method and Zisman's critical surface tensions are often too low.

REFERENCES

1.   H.W. Fox and W.A. Zisman, J. Colloid Sci. 5, 514 (1950).
2.   H.W. Fox and W.A. Zisman, J. Colloid Sci. 7, 109 (1952).
3.   H.W. Fox and W.A. Zisman, J. Colloid Sci. 7, 428 (1952).
4.   W.A. Zisman, Advan. Chem. Ser. 43, 1 (1964).
5.   S. Wu, J. Polym. Sci. C34, 19 (1971).
6.   S. Wu, J. Adhesion 5, 39 (1973); also in "Recent Advances in Adhesion", L.H. Lee, ed., pp. 45-61, Gordon and Breach, New York, 1973.
7.   S. Wu, J. Macromol. Sci. C10, 1 (1974).
8.   S. Wu, in "Polymer Blends", Vol. 1, D.R. Paul and S. Newman, eds., pp. 243-293, Academic Press, New York, 1978.
9.   S. Wu and K.J. Brzozowski, J. Colloid Interface Sci. 37, 686 (1971).
10. S. Wu, J. Phys. Chem. 74, 632 (1970).
11. D.K. Owens and R.C. Wendt, J. Appl. Polym. Sci. 13, 1741 (1969).
12. D.H. Kaelble, "Physical Chemistry of Adhesion", Wiley-Interscience, New York, 1971.
13. D. Patterson and A.K. Rastogi, J. Phys. Chem. 74, 1067 (1970).
14. K.S. Siow and D. Patterson, Macromol. 4, 26 (1971).

15. R.J. Roe, Proc. Nat. Acad. Sci. U.S. 56, 819 (1966).
16. E. Helfand and A.M. Sapse, J. Chem. Phys. 62, 1327 (1975).
17. H.W. Kammer, Z. Phys. Chem. Leipzig 258, 1149 (1977).
18. C.W. Stewart and C.A. von Frankenburg, J. Polym. Sci. A-2, 6, 1686 (1968).
19. H.W. Kammer, Z. Phys. Chem. Leipzig 255, 607 (1974).
20. R.J. Good and E. Elbing, Ind. Eng. Chem. 62, 54 (1970).
21. J.R. Dann, J. Colloid Interface Sci. 32, 302 (1970).
22. F. Schulman and W.A. Zisman, J. Colloid Sci. 7, 465 (1952).
23. R.E. Johnson, Jr. and R.H. Dettre, J. Colloid Interface Sci. 21, 610 (1966).
24. E.G. Shafrin and W.A. Zisman, J. Phys. Chem. 71, 1309 (1967).
25. F.M. Fowkes, in "Chemistry and Physics of Interfaces", pp. 1-12, American Chemical Society, Washington, D.C., 1965.
26. P.C. Heimenz, "Principles of Colloid and Surface Chemistry", Marcel Dekker, New York, 1977.
27. R.J. Good, in "Treatise on Adhesion and Adhesives", Vol. 1, R.L. Patrick, ed., pp. 9-68, Marcel Dekker, New York, 1967.
28. D.G. LeGrand and G.L. Gaines, Jr., J. Colloid Interface Sci. 31, 162 (1969).
29. S. Wu, J. Colloid Interface Sci. 31, 153 (1969).
30. R.H. Dettre and R.E. Johnson, Jr., J. Phys. Chem. 71, 1529 (1967); J. Colloid Interface Sci. 31, 568 (1969).
31. J.F. Padday, Proc. 2nd International Cong. Surface Activity 3, 136 (1957).

A Role of Molecular Forces in Adhesive Interactions of Polymers

V.A. Belyi, V.A. Smurugov and A.I. Sviridyonok

Institute of Mechanics of Metal-Polymer Systems of

Byelorussian Academy of Sciences, Gomel, U.S.S.R.

ABSTRACT

The present paper discusses a possibility of developing new ways of analytical and experimental study of molecular interactions within polymer/polymer and metal/polymer contacts starting from the general theory of molecular van der Waals forces. We consider that polymers belong to a class of materials possessing a distinct selective absorption of radiation. Formulas are suggested for calculation of forces of interaction between solid surfaces having discrete absorption spectra. It is shown that the intensity of molecular interaction can be estimated by the number, intensity, and mutual overlapping of absorption bands in the spectra of materials in contact. A method of estimation of the contribution of certain spectral ranges is suggested and an analysis is given of the role of infrared region. Experimental and calculated data are presented indicating the future of the research method chosen.

INTRODUCTION

Up till now there is no common opinion about the role of molecular forces in adhesive interaction of polymeric bodies (1-12). The need in a developed adhesion theory stimulates the study of this problem.

It is of interest to analyze the electromagnetic theory of molecular interactions betwen solid surfaces based on notions about fluctuating electromagnetic fields always present in any body and beyond it (13-16). From this theory it follows that the character and intensity of molecular attraction are determined by the di-

electric properties of materials. Herein, depending on the correlation between the clearance ℓ, between the surfaces, and major
wavelengths λ_o in absorption spectra of materials, the calculating
formulas include either dielectric permeabilities of imaginary
frequencies $\varepsilon_1(i\xi)$, $\varepsilon_2(i\xi)$ or their values at $\xi = 0$, i.e., static
dielectric permeabilities of materials ε_{10}, ε_{20}. The dielectric
permeability $\varepsilon(i\xi)$ is related to the imaginary part of the complex
dielectric permeability of the real component of the frequency
through Kramers-Kronig relation (17):

$$\varepsilon(i\xi) = 1 + \frac{2}{\pi} \int_0^\infty \left[\frac{\varepsilon''(\omega)\omega}{\omega^2 + \xi^2} \right] d\omega \tag{1}$$

where ω is frequency (Hz).

Difficulties met on application of the theory to solution of
practical problems are usually connected with the necessity of
accurate knowledge of the function of clearance between surfaces
and with uncertainty of $\varepsilon''(\omega)$ for most materials, i.e., their major
macroscopic characteristic, that determines the intensity of the
interaction. It is important that the dielectric permeabilities
included into the expression of force are integral values of
function $\varepsilon''(\omega)$ related to the nature of materials, their chemical
composition, molecular structure, etc.

As applied to polymeric materials, one might say that their
absorption spectra look like bands of one or another width where
regions of transparency of the material exist along with absorption
regions of electromagnetic fields. Function $\varepsilon''(\omega)$ for them can be
represented as in Fig. 1, if negligibly small absorption be neglected in the regions of transparency and if its value is taken as
constant in the absorption band. The fact that $\varepsilon''(\omega)$ is zero on
certain areas allows expression (1) to pass from integration over an
infinite region to a sum of integrals with finite limits that bound
the absorption regions:

$$\varepsilon(i\xi) = 1 + \frac{2}{\pi} \left[\int_{\omega_1}^{\omega_1 + \Delta\omega_1} \left(\frac{\varepsilon''(\omega_1)\omega}{\omega^2 + \xi^2} \right) d\omega + \int_{\omega_2}^{\omega_2 + \Delta\omega_2} \left(\frac{\varepsilon''(\omega_2)\omega}{\omega^2 + \xi^2} \right) d\omega + .. + \int_{\omega_n}^{\omega_n + \Delta\omega_n} \left(\frac{\varepsilon''(\omega_n)\omega}{\omega^2 + \xi^2} \right) d\omega \right]$$

The constancy of function $\varepsilon''(\omega)$ within each interval of integration
gives an opportunity to put it beyond the integral sign; then
integration over the part $(\omega_1 \div \omega_1 + \Delta\omega_1)$, for instance, gives

$$\frac{1}{2}\left[\ln\frac{(\omega_1+\Delta\omega_1)^2 + \xi^2}{\omega_1^2 + \xi^2}\right]$$

which with relatively narrow absorption bands ($\Delta\omega$ is small) approximates

$$\frac{\omega_1 \Delta\omega_1}{\omega_1^2 + \xi^2}$$

Herein, expression (1) has the form:

$$\varepsilon(i\xi) = 1 + \frac{2}{\pi}\sum_n \varepsilon''(\omega_n)\left[\frac{\Delta\omega_n\ \omega_n}{\omega_n^2 + \xi^2}\right] \qquad (2)$$

where ω_n, frequency of the beginning of nth-absorption band (Hz); $\Delta\omega_n$, width of the absorption band; $\varepsilon''(\omega_n)$, value of $\varepsilon''(\omega)$ within nth-band; n, the number of absorption bands in the spectrum of a material.

Thus, there is a possibility to estimate the contribution of an individual absorption band, or individual spectral region to dielectric permeability of a material and, accordingly, to the interaction force. Herein, most vivid is the role of peculiar properties

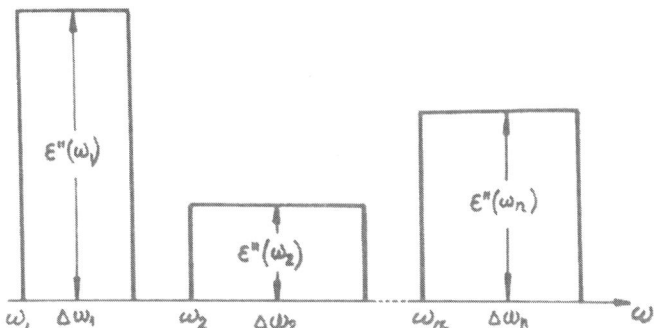

Fig. 1.   A schematic of function $\varepsilon''(\omega)$ for polymeric materials.

of materials associated with chemical composition and molecular
structure that determine the intensity and position of absorption
bands on the frequency axis. As applied to polymeric materials, of
greatest interest is the study of absorption spectra in the infrared
region (2 to 25μm) where exist major absorption bands associated
with excitation of molecular vibrations allowing an estimation of
the contribution of this frequency range.

EXPERIMENTAL RESULTS

     To check these suppositions, the following polymeric materials
were selected: polytetrafluoroethylene (PTFE), polyethylene (PE),
polypropylene (PP), polymethyl methacrylate (PMMA), polycaproamide
(PCA), polyethyleneterephthalate (PETP), for which molecular and
supermolecular structures, properties, and the shape of absorption
spectra, accordingly, differ considerably. Absorption spectra of
films of different thicknesses prepared from the above polymers
were obtained with a spectrophotometer UR-20.   The values of
absorption indices χ were calculated according to known techniques
for each absorption band, and widths $\Delta\omega$ of those bands determined.
Values of ε" were calculated from expression ε" = 2nχ (18), where n,
refractive index of a material in the optical region; its anomaly
within the absorption band has been neglected.

     Based on the data obtained when processing absorption spectra
were estimated values of $\varepsilon(i\xi)$ at ξ = 0. These values are given in
Table 1.   Table 1 shows that values of $\varepsilon(i\xi)$ obtained in such a way
for certain polymers slightly approach the values of ε_o, which
indicates a small contribution of IR-regions and a possible large
contribution of ($\Delta\varepsilon$) of other spectral regions in this case. At the
same time, the variations in polymers properties (density, melting
temperature) that are determined by intermolecular interactions are
similar to those when going from one polymer to another in values
$\varepsilon(i\xi)$ estimated for IR-regions.

     Recording of dielectric permeabilities $\varepsilon(i\xi)$ for contacting
materials according to expression (2) where index n belongs to body
1, index κ to body 2, and considering the statement from Levin and
Rytov (19) about primary contribution of spectral regions to adhesive force when absorption bands overlap allows obtaining calculating formulas for various types of contacts.

     When two polymeric bodies are in contact, the expression for
bonding force F will have the form:

$$F \approx \frac{\hbar}{8\pi^2 \ell^3} \int\limits_0^\infty \left\{ \frac{\sum\limits_i \varepsilon_1''(\omega_i)\varepsilon_2''(\omega_i) \dfrac{\Delta\omega_i^2 \omega_i^2}{(\omega_i^2 + \xi^2)^2}}{\pi^2 + \pi\left[\sum\limits_n \varepsilon_1''(\omega_n)\dfrac{\Delta\omega_n \omega_n}{\omega_n^2 + \xi^2} + \sum\limits_\kappa \varepsilon_2''(\omega_\kappa)\dfrac{\Delta\omega_\kappa \omega_\kappa}{\omega_\kappa^2 + \xi^2}\right] + \sum\limits_i \varepsilon_1''(\omega_i)\varepsilon_2''(\omega_i)\dfrac{\Delta\omega_i^2 \Delta\omega_i^2}{(\omega_i^2 + \xi^2)^2}} \right\} d\xi$$

(3)

where ω_i, frequency of beginning of i-regions of overlapping of absorption bands (Hz); $\Delta\omega_i$, width of this region; $\varepsilon_1''(\omega_i)$ and $\varepsilon_2''(\omega_i)$, are values of $\varepsilon''(\omega)$ for the first and second bodies, and ℓ is the distance between the surfaces, respectively, within this frequency interval. When materials are the same, then:

$$F \approx \frac{\hbar}{8\pi^2 \ell^3} \int\limits_0^\infty \left(\frac{\sum\limits_n \varepsilon_1''(\omega_n) \dfrac{\Delta\omega_n \omega_n}{\omega_n^2 + \xi^2}}{\pi + \sum\limits_n \varepsilon_1''(\omega_n) \dfrac{\Delta\omega_n \omega_n}{\omega_n^2 + \xi^2}} \right)^2 d\xi \qquad (4)$$

Table 1. Estimated Contribution of IR-Regions to Dielectric
Permeabilities of Materials

Materials	$\varepsilon(i\xi)$	ε_o	$\Delta\varepsilon$	Density, g/cm^3	m.p. $^\circ$C
PE	1.015	2.27	1.255	0.92	115
PP	1.021	2.26	1.239	0.92	170
PMMA	1.048	2.84	1.792	1.17	-
PCA	1.157	2.34	1.183	1.14	220
PETP	1.354	3.70	2.346	1.40	260
PTFE	1.722	2.10	0.378	2.20	327

Formulas (3) and (4) allow an estimation of the effect of various factors on the intensity of molecular interactions and underline its strong dependence on the number, intensity, and mutual overlapping of absorption bands in the spectra of the contacting materials.

For various polymeric materials, the values of sums were estimated at $\xi = 0$:

$$\sum_i \epsilon_1''(\omega_i) \epsilon_2''(\omega_i) \frac{\Delta\omega_i^2 \omega_i^2}{(\omega_i^2 + \xi^2)^2}$$

included into Eq. (3). For polymeric materials in contact with PMMA, these values were: for PP = 0.0245; PE = 0.0450; PCA = 0.138; PETP = 1.06. These values and the sums $\sum_n$, $\sum_k$ calculated for each material using its spectra were put into Eq. (3) to obtain those values of subintegral expression (A) which character of variations, when going from one pair of materials to another, would determine the variations in the force of molecular bonding, Table 2. Similar calculations were carried out for combinations of various polymeric materials.

Experimentally the force of molecular bonds of contacting surfaces was estimated through the friction force of thin polymeric films rubbed against PMMA and PTFE rollers having roughness R_a = 0.25 to 0.2μm with the contact angle π rad. The tension of the film was $25N/m^2$; sliding speed 0.075 m/s; the temperature in the friction zone was only 5° higher than that of the environment.

Table 2. Variations in the Force of Molecular Interactions When Polymeric Materials Are in Contact

Materials of friction pairs	A	Friction force, $10^{-2}N$	Materials of friction on pairs	A	Friction force, $10^{-2}N$
PMMA - PP	0.00245	4.31	PTFE - PE	0.0012	4.28
PMMA - PE	0.0045	4.45	PTFE - PP	0.0038	4.13
PMMA - PCA	0.0138	5.00	PTFE - PCA	0.21	4.62
PMMA - PETP	0.082	5.49	PTFE - PETP	0.39	6.98

Table 2 shows that the variations in molecular interactions which were studied using absorption spectra according to the technique described, coincide qualitatively (under the chosen experimental conditions preventing strong chemical interactions between surfaces, their melting, seizure, etc.) with the mode of variations in the friction force. A lack of proportionality between the calculated values of A and the friction measured, can probably be explained by that during calculations not the whole spectral region was considered, but only the infrared one. Table 1 gives a more clear picture about the extent of its contribution.

The results of the calculation of bonding forces given in Table 3 for polymeric surfaces separated by a clearance 0.1μm are comparable, in the order of value, with those available in the literature and with experimental data (3,15).

From the view of the theory considered the molecular bonding between contacting surfaces might be improved by enriching the spectral composition of fluctuating fields by creating, for instance, new types of bonds in materials, new functional groups, etc. Thus, when PE films were gamma-irradiated there was a band 1690-1750 cm^{-1} in the absorption spectra which intensity increased with the dose. The study of friction of irradiated films rubbing against steel under conditions excluding large losses for mechani-

Table 3. Values of Forces of Molecular Interactions
Between Surfaces from Different Estimations
(Clearance ℓ = 0.1μm)

Materials	F, N/m^2	Remarks
PTFE - PETP	$\sim 84 \cdot 10^{-5}$	From formula (3)
PMMA - PE	$\sim 13 \cdot 10^{-5}$	From formula (3)
Quartz plates	$\sim 2 \cdot 10^{-5}$	From Lifshits' formula (15)
Quartz plates	$\sim (2-9) \cdot 10^{-5}$	From Deryagin and Abricosova (15)
Metallic plates	~ 1.3	Values of absorption indices determining $\varepsilon''(\omega)$ are approximately 10^5 times higher for metals than for polymers

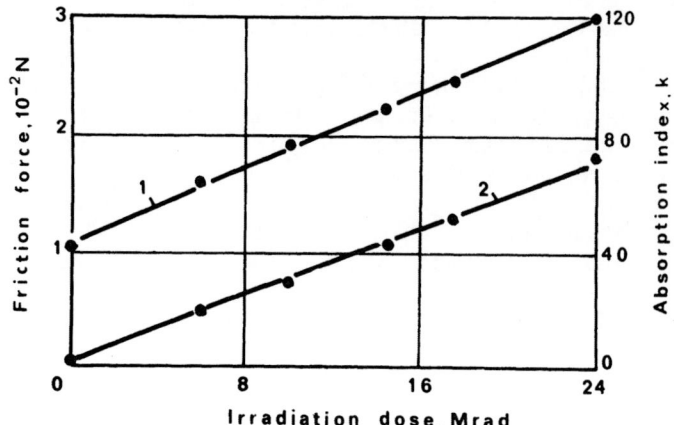

Fig. 2.   Variations in friction force and absorption
index within 1690 - 1750 cm^{-1} after rubbing
of gamma-irradiated PE films against steel.

cal deformation and tribochemical reactions at P = 10 N/m^2,
V = 0.075 m/s, showed (11) that the presence of an additional
component in the spectrum of fluctuating fields caused an increase
in the friction force; herein, it grows with increasing intensity of
thé absorption bands (Figure 2).   These observations are in good
agreement with those available in the literature when the normal
adhesion (9) and friction force (20) increased with the oxidation of
the polymer.

CONCLUSIONS

        A satisfactory qualitative conformation between calculated and
experimental data obtained by the present authors provides future
applications of the trend chosen for the study of the role of
molecular forces in the adhesion interactions of polymers.

REFERENCES

1.   B.V. Deryagin, What is Friction?, AN SSSR, Moscow (1963).
2.   B.V. Deryagin, N.A. Krotova and V.P. Smilga, Adhesion of
        Solids, Nauka, Moscow (1973).
3.   A.S. Akhmatov, Molecular Physics of Boundary Friction. Fiz-
        matgiz, Moscow (1963).
4.   V.A. Belyi, A.I. Sviridyonok, M.I. Petrokovets and V.G. Sav-
        kin, Friction and Wear of Polymer-Based Materials, Nauka
        i Tekhnika, Minsk (1976).

5. F.P. Bowden and D. Tabor, Friction and Lubrication of Solids (Russian translation), Moscow (1968).

6. A.A. Berlin and V.E. Basin, Fundamentals of Polymer Adhesion, Khimia, Moscow (1969).

7. S.B. Ainbinder and A.Ya. Loginova, "Adhesive Interaction Between Metal-Polymer and Polymer-Polymer Pairs", Mekh. Polim. 5 (1978).

8. E.V. Teodorovich, "Contribution of Macroscopic van der Waals Interactions to Friction Force", Izv. VUZov, Fizika, No. 11 (1976).

9. V.A. Belyi, N.I. Egorenkov and Yu. M. Pleskachevskii, Adhesion of Metals to Polymers, Nauka i Tekhnika, Minsk (1971).

10. L.H. Lee, "Effect of Surface Energetics on Polymers Friction and Wear", in Advances in Polymer Friction and Wear, Plenum Press, New York 5A (1974).

11. V.A. Belyi, A.I. Sviridyonok and V.A. Smurugov, "Molecular Changes in Surface Layers of Polymer Materials During Rubbing", in Kurzfassungen der Berichte, Eurotrieb-77, Düsseldorf 2/3 (1977).

12. V.A. Belyi, V.A. Smurugov and A.I. Sviridyonok, "Molecular Interaction of Polymers Within Region of Friction Contact", DAN SSSR 241, No. 3 (1978).

13. E.M. Lifshits, "Theory of Molecular Attraction Forces Between Solids", ZhETF 29, No. 1 (1965).

14. I.E. Dzyaloshinskii, E.M. Lifshits and L.P. Pitaevskii, "General Theory of van der Waals Forces", UFN, No. 3 (1961).

15. B.V. Deryagin, I.I. Abricosova and E.M. Lifshits, "Molecular Attraction of Condensed Bodies", UFN, No. 3 (1958).

16. Yu. S.Barash and V.L. Ginsburg, "Electromagnetic Fluctuations within a Substance and Molecular (van der Waals) Forces Between Bodies", UFN, No. 1 (1975).

17. L.D. Landau and E.M. Lifshits, Electrodynamics of Continuums, Nauka, Moscow (1959).

18. N.A. Borisevich, V.G. Vereshchagin and M.A. Validov, Infrared Filters, Nauka i Tekhnika, Minsk (1971).

19. M.A. Levin and S.M. Rytov, Theory of Equilibrium Heat Fluctuations in Electrodynamics, Nauka, Moscow (1967).

20. I.A. Gribova, A.V. Vinogradov, S.A. Pavlov, A.N. Chumaevskii, L.V. Dubrovina, A.P. Krasnov, V.A. Khomutov and V.V. Korshak, "A Study of Physico-Chemical Processes During Rubbing of Antifrictional Materials", Mekh. Polim., No. 4 (1974).

21. D. Tabor, "Surface Forces and Surface Interactions", J. Coll. Interface Sci. 58, No. 1 (1957).

Discussion

On the Paper by T. Hata and T. Kasemura

L.H. Lee (Xerox Corporation): I would like to take this opportunity to thank Professor Hata for this excellent presentation. Most of the data presented here have not been published in the western journals. This review serves to bring all of us up to date about various aspects of surface and interfacial tensions of polymer solids.

A. Silberberg (Weizmann Institute, Israel): Have you tried to separate the surface tension into entropic and ethalpic parts? The approach through the particular assigning to each group on equal appearance opportunity at the surface implies a truly random occupation of the surface. This is probably true for very short and very long chains but might break down with very anisotropic molecules and with copolymers of certain types. It would be good to have some information in each case to what extent one might be getting some ordering of the contact layer.

T. Hata (Gunma University, Japan): We have measured the temperature dependence of surface tension for all samples, obtaining surface entropy S^S and surface energy (enthalpy) E^S according to the relation

$$\gamma = E^S - S^S T.$$

However, I did not present here these data and did not analyze the change of γ by these factors, because of the limited time for the presentation. Generally speaking, when I said "selective adsorption" in the paper, it was always accompanied by the decrease of surface entropy as well as surface energy. The change of them was quite drastic especially for block copolymers similarly as shown in γ. It was noteworthy, however, that, for homologous series of the similar type $X\text{-}R_n\text{-}X'$ the molecular weight dependence of γ is well represented by the relation derived from the Sugden's equation based on the additivity of parachor over the whole range of molecular weight.

On the Paper by F.M. Fowkes

A. Jonath (Lockheed Palo Alto Research Laboratory): What is temperature dependence of acid-base interaction as compared to dipole-dipole interaction?

F.M. Fowkes (Lehigh University): The temperature-dependence of any interaction is measured by its enthalpy (ΔH). Since acid-base interactions appear to be far stronger than dipole interactions, they should have a greater temperature dependence.

A. Jonath: What is electric field dependence, if measured, of the acid-base interaction?

F.M. Fowkes: Acid-base interactions at interfaces always involve a local electric field on the order of 10^5 volts/cm with the base side positive and the acid side negative. An applied field of opposite sign and greater field strength might weaken the bond, but no such experiments have been tried.

F. Leo (De Soto Inc.): Will solvent-polymer combination with the most interaction (acid-base) have the highest viscosity?

F.M. Fowkes: Two interactions control polymer swelling and viscosity; dispersion free interactions and acid-base interactions. Maximum viscosity is observed with good match of dispersion forces (δ^d) plus strong acid-base interaction.

F. Leo: Keeping in mind, "OSHA" and Air Pollution Regulations, do you have any information on acid-base interactions of ketones (MEK, MIBK, cyclohexanone), esters (butyl acetate, cellosolve acetate, ethyl acetate), commonly used in preparing the polyester-polyisocyanate coatings?

F.M. Fowkes: The ketones and esters are all bases; the esters a bit stronger than the ketones. I suppose that these solvents could bond to isocyanate or urethane groups or to any free acid groups. These solvents should not be so basic as to interfere with bonding of the basic polymer to acidic surfaces such as silica, iron, etc.

F. Leo: Also what effect does structural changes in polyester or polyisocyanate have on acid-base interaction with surfaces (i.e. for use as adhesives)?

F.M. Fowkes: Structure can sometimes be important in shielding acidic or basic groups from each other. A well-shield group cannot interact with solvents or surfaces.

F. Leo: What is R-NCO before it becomes a urethane (while still -NCO), acidic or basic?

F.M. Fowkes: I have not measured the acidity or basicity of isocyanates, but would expect the oxygen to be an electron donor and the adjacent carbon to be an electron-acceptor, thus providing intermolecular association somewhat like a hydrogen-bond.

F. Leo: For a polyester with a finite acid number, at what points does this basic polyester become acidic with increasing acid number? Also if the acid is an aromatic acid (e.g. isophthalic acid) instead of an aliphatic acid (e.g. adipic acids), will this change over from base to acid occur sooner?

F.M. Fowkes: A polyester will be basic, and with increasing acid number, the basicity will tend to remain and the polyester will have both acidic and basic properties. The resulting internal acid-base interaction could compete effectively against other acids or bases if they are weaker. Thus, acid groups in a polyester could preferentially bond to the ester groups and minimize interaction with weakly acidic solvents (like CH_2CL_2) or with weakly acidic substrates such as silica. Stronger acids such as isophthalic should be more effective than weaker acids such as adipic.

D.W. Dwight (Virginia Polytechnic Institute): What is the basic significance of the Drago constants C and E?

F.M. Fowkes: The modern theory of Hard and Soft Lewis Acids and Bases (HSAB) shows that two properties are important for bases (electron-donor "strength" and polarizability of donor site) and two for acids (electron acceptor strength and polarizability of acceptor site). The "C" constants vary mostly with polarizability and the "E" constants vary mostly with inner shell electron binding energies.

P. Datta (RCA Laboratories): The copolymers of acrylic acid and ethylene are known to form domain structure of acrylic acid and ethylene. Were the surfaces of your copolymers analyzed with ESCA or other spectroscopic technique?

F.M. Fowkes: No.

P. Datta: How surface energy and contact angle measurements are affected by domain structures of the copolymer and acrylic acid?

F.M. Fowkes: Note in figures that W_A^{ab} increases with % acid, but less than proportionally, suggesting a smaller proportion is available when acid content is greater.

On the Paper by Souheng Wu

F.M. Fowkes (Lehigh University): I believe Dr. Wu's wettability data should be considered just a small part of the field of physical chemistry and therefore not be contradictory of well-established general principles, one of which is the geometric mean averaging of intermolecular energies, established by Berthelot, van Lane, Scatchard, Hildebrand, Leonard-Jones, and London. Dr. Wu has admitted in earlier papers that his averaging is not realistic in the light of London equations; what purpose is served by continuing this approach?

Secondly, Dr. Wu (and others) use the technique of defining γ_c or $\gamma_{c(max)}$ to be the surface tension of the solid and then without any proof assume they have established a truth. I agree that in some cases the value of $\gamma_{\ell s}$ might be zero, but until it is proved, such an assumption is very speculative. I suggest studying liquids on mercury where γ_{12} can be measured and where at least water gives a finite contact angle.

Thirdly, Dr. Wu suggests that the surface tension of polyethylene can be estimated by extrapolation of values for the liquid vs. temperature; this is unrealistic for the molar volumes are different and intermolecular distances are shorter in the solid.

Fourthly, he and many others may not have heard my two papers (not yet in print) on the specific interactions between long chain hydrocarbons which reduce γ_ℓ^d to values of several dynes and below γ_ℓ. It is therefore not surprising that γ_{PE}^d is less than γ_{PE}.

I heartily agree that γ_c values are different with different liquids on the same solid. That is what I show in my Figs. 1 and 2, but J.R. Dann made this quite clear in 1970 (see my reference No. 20).

Souheng Wu (du Pont Company): Firstly, the harmonic-mean relation for dispersion force interaction has a sound theoretical basis and is in fact derived from the London theory, as I have discussed in detail in J. of Adhesion, 5, 39 (1973) and in "Recent Advances in Adhesion", L.H. Lee, ed., pp. 45-61, Gordon and Breach, 1973. Both the harmonic-mean and the geometric-mean relations are special cases of the London theory; each has its range of applicability and limitation. I have shown previously that the harmonic-mean relation applies more accurately for polymers than the geometric-mean relation by testing them with the directly measured surface and interfacial tensions of polymer liquids and melts, as I have discussed in detail before and summarized in J. of Adhesion, $\underline{5}$, 39 (1973), in "Recent Advances in Adhesion", L.H. Lee, ed., pp. 45-61, Gordon and Breach, 1973, in J. Macromol. Sci., C$\underline{10}$, 1 (1974), and in "Polymer Blends", Vol. 1, D.R. Paul and S. Newman, eds., pp. 243-293, Academic Press, 1978. Therefore, the harmonic-mean relation is theoretically sound, and completely consistent with the London theory and experimentally realistic. This is exactly the reason why the harmonic-mean relation is preferred and used. If these are not good reasons, what are good reasons? There has never been an admission to the contrary. Please also note that earlier this morning, Dr. Hata confirmed that the harmonic-mean relation is better than the geometric-mean relation in his lecture.

Secondly, γ_c or $\gamma_{c,max}$ has $\underline{\text{never}}$ been $\underline{\text{defined}}$ as the surface tension of a solid. On the contrary, I have shown that γ_c or $\gamma_{c,max}$ cannot equal the surface tension $\underline{\text{in general}}$. I have shown that $\gamma_{c,\phi,max}$ equals the surface tension both theoretically and experimentally in the present paper. γ_{LS} cannot be zero $\underline{\text{in general}}$. This has been known for many years and by many people. The fact that γ_{LS} cannot generally be zero is exactly the crux of the problem, and is exactly what the present paper is about. I have measured the surface and interfacial tensions of many pairs of incompatible polymer liquids and melts. These data would provide a rigorous test of my method.

Thirdly, the extrapolation of surface tension from the liquid state to the amorphous solid state is based on the well-established corresponding state concept of Guggenheim, Katayama, Prigogine and others. The differences in molar volumes and intermolecular distances are accounted for in such an approach. Furthermore, remember that only amorphous solid surfaces are dealt with here. Amorphous solids may be regarded to have "liquid-like" structures. Extrapolation from the liquid state is still one of the best methods, if not the best method, for estimating the surface tension of a solid having an amorphous surface.

Fourthly, the present approach is to address to the problem of variable γ_c values. I am pleased to see that Dr. Fowkes "heartily" agrees that γ_c values are variable. Dr. Dann made this quite clear in his earlier paper as I quoted in the text. I believe that the present equation of state approach successfully handles this problem.

A. Silberberg (Weizmann Institute, Israel): I would be interested in knowing how the surface of your polymeric specimens were prepared. One may expect differences when these surfaces are formed in air, in contact with metal or with some other forming surface. Were these surfaces formed by machining, i.e. by mechanical manipulation. Their chemistry might be expected to be different in each of these cases.

Souheng Wu: All the contact angle data used are taken from the works by Dr. Zisman and his coworkers, as noted in the text. The surfaces used were reported to be representative of the materials in amorphous state free from contamination and scratches.

PART TWO

CHARACTERIZATION OF ADHESIVE INTERFACES

Introductory Remarks

Lieng-Huang Lee

Wilson Center for Technology

Xerox Corporation

Webster, New York 14580

The most important advance during the last several years is the characterization of surface by various new techniques (1), e.g., ESCA, AES, SIM, etc. With new techniques, we do not have to speculate on what is at the interface; instead, we can actually see what is there before and after adhesive bonding. In fact, because of the new characterization techniques, we may have to modify some adhesion theories formulated in the past.

In this session, we shall have five papers discussing characterization of adhesive interfaces. My paper will deal with photoacoustic spectroscopy, a new technique developed by A. Rosencwaig (2). This technique can be used to identify polymers, dyes and other compounds containing chromophores at the interface.

The second paper, by J.S. Solomon and D. Hanlin, is about the applications of AES (Auger electron spectroscopy) and ESCA (electron spectroscopy for chemical analysis) to the characterization of an adhesive-adherend bond joint.   Since Mr. Solomon* has not arrived, I shall read one of my papers previously presented to the Society of Photographic Scientists and Engineers. This paper is about electroadhesion, and its title is "A Surface Interaction Model for Triboelectrification of Toner-Carrier Pair" (3). I am glad that I did bring extra sets of slides after I heard about the UAL strike.

---

* Dr. Solomon finally arrived on Wednesday and his paper was delivered on that morning to the Fifth Session.

The third paper discusses techniques for adhesive vs. adhesive failure . One of the techniques used was β-ray back-scattering. The count rate could be correlated with the fractional area that failed adhesively and cohesively. This paper, by Dr. T. Smith and P. Smith, was to be presented by Dr. Smith. Though Dr. Smith works at Thousand Oaks, California, he, too, could not fly out from Los Angeles. (When we realized that several scheduled speakers could not be present, we started to draft participants for "Quickies". One of our audience, Dr. R. Van der Linden from Belgium, presented a very interesting "Quickie" on corona discharge on low density polyethylene. As a result, his paper is now included in this Proceedgings.)

The fourth paper is related to the locus-of-failure analysis in adhesive bonding research and will be presented by Dr. David Dwight. Dr. Dwight will discuss the use of SEM, EDAX (energy dispersive analysis of X-rays) and ESCA for the failure analysis. The last paper of this session is about the measurements of dielectric relaxation gradients in an adhesive bond and will be presented by Dr. John L. Crowley. Following Dr. Crowley's presentation, Dr. Jonath will present an interesting "Quickie" on "New Approaches to the Understanding of Adhesive Interface Phenomena and Bond Strength."

At any rate, we shall make this session as full as if we did not have the strike.

REFERENCES

1.   L.H. Lee, Ed., Characterization of Metal and Polymer Surfaces, Academic Press, New York (1977).
2.   A. Rosencwaig, Phys. Today 28, No. 1, 23 (1975).
3.   L.H. Lee, Photo. Sci. and Eng. 22, 228, July/August, 1978.

Photoacoustic Spectroscopy for the Study of Adhesion and Adsorption of Dyes and Polymers

Lieng-Huang Lee

Webster Research Center, Xerox Corporation

Webster, New York   14580

ABSTRACT

Photoacoustic spectroscopy (PAS) has recently been applied for the determination of the surface and bulk structure of solids. PAS enables one to obtain spectra, similar to absorption spectra, on any type of solid materials or composites. One of our studies involves the examination of adhesion and adsorption of dyes and polymers. This paper briefly describes basic principles, the instrument and some of the new techniques developed in our laboratory for the study of solid structure. The advantages and limitations of this analytical method will be discussed. Future applications of this technique for coating, dyeing, printing, and related technologies will be mentioned.

INTRODUCTION

During the last several years, photoacoustic spectroscopy (1-5) (PAS) has been employed to study the composition and structure of solids. Since many solids tend to scatter light and are either insoluble in a solvent or different in structure in solution, this technique offers advantages over many existing spectroscopic methods in elucidating the structure of solid materials. By varying the modulating frequency, one can study the bulk as well as the surface. Thus, PAS may be applied to study adhesion and adsorption of dyes and polymers on surfaces. Some of our preliminary studies will be discussed in a later section of this paper.

PAS as developed is not as versatile as originally claimed (1). There are limitations such as the saturation problem (6,7). Several methods have been reported in the literature (7) to alleviate this problem; however, not all of them are successful. We shall describe some of the improved methods used by us.

The applications of PAS can be greatly enhanced if the spectrum can be extended to the infrared. Recently, Kirkbright (8) and Adams (9) reported the extension of the spectrum to the near IR (2.5 microns and 3.4 microns, respectively). With these extensions, the applications (8) have been broadened for the characterization of pigments, blood, fats, thin-film coatings, polymers and lubricants. New spectrometers with near IR capabilities are now offered by EG&G Princeton Applied Research (10), Gilford Instruments (11) and EDT Research (England) (12). It is likely that the use of the laser (13,14) can bring forth a wider infrared range for future spectrometers.

Principles of Photoacoustic Spectroscopy

In the spectrometer, the sample is placed inside a specially designed closed cell containing air (or other gases) and a sensitive microphone. The solid is then illuminated with chopped or modulated monochromatic light. Rosencwaig (15) and Parker (16) theorize that the primary source of the acoustic signal arises from the periodic heat flow from the solid to the surrounding gas. Nonradiative de-excitation processes can convert a part or all of the light absorbed by the solid into heat. The periodic flow of this heat into the surrounding gas produces pressure fluctuations in the cell and thus generates sound detected by the microphone. The signal from the microphone is then recorded as a function of the wavelength of the incident light.

Although the magnitude of the PAS signal is a function of the heat evolved or the light absorbed by the samples, there are other factors (15) involved, e.g., thermal properties of the sample and of the cell and the chopping frequency. The interrelationship of these parameters may appear to be somewhat complicated; however, in general, the PAS spectrum corresponds closely to the optical absorption spectrum of the sample, provided that the sample is essentially in the same state in solid as in liquid or solution. Our work further confirms this relationship.

Three important parameters are specified by the theory (15): the sample thickness ℓ, the optical absorption length μ_b and the thermal diffusion length μ_s; $\mu_b = 1/\beta$, where β is the optical absorption coefficient, and $\mu_s = (\frac{2\alpha_s}{\omega})^{\frac{1}{2}}$, where α_s is the thermal diffusivity of the sample, and ω is the modulating frequency in radians per second. The thermal diffusion length determines how far a heat wave of frequency ω will travel before excessive damping occurs. The thermal diffusion length increases with the decrease in modulating frequency. The photoacoustic spectrometer responds to the total amount of light absorbed in the sample within a depth roughly corresponding to the thermal diffusion length. A sample may be considered to be photoacoustically opaque if substantially all of the light is absorbed within this depth.

The thermal diffusion length of air at 50 Hz is approximately 0.34 mm. The ranges of other photoacoustic parameters (16) are listed in Table 1.

Table 1. Photoacoustic Parameters

Reflectivity R	4-90%
Absorption coefficient β	$0-10^6$ cm^{-1}
Thermal diffusivity α_s	$10^{-3} - 10^3$ cm^2/sec
Thermal diffusion length, μ_s	$10^{-3} - 1.0$ cm
Thermal damping constant a_s	$1-10^3$ cm^{-1}
Chopping frequency f	$10-10^4$ Hz

EXPERIMENTAL

Instrument

The instrument is the Model 6001 Photoacoustic Spectrometer manufactured by EG&G Princeton Applied Research. The optical system (18) of this instrument is shown in Fig. 1. The light source is a 1kW xenon arc lamp which is electronically modulated. The monochromator incorporates a turret of three concave holographic gratings covering the spectral range of 200 nm to 26 μm. Relative aperture is F/4.2. Grating selection is automatic and under microprocessor control. Spectral bandwidth of the monochromator is controlled by the slit selection.

The sample cell is of fused silica. Samples (solid or liquid) are easily loaded into a rectangular receptacle (4x8x2 mm) which is inserted into a cuvette.

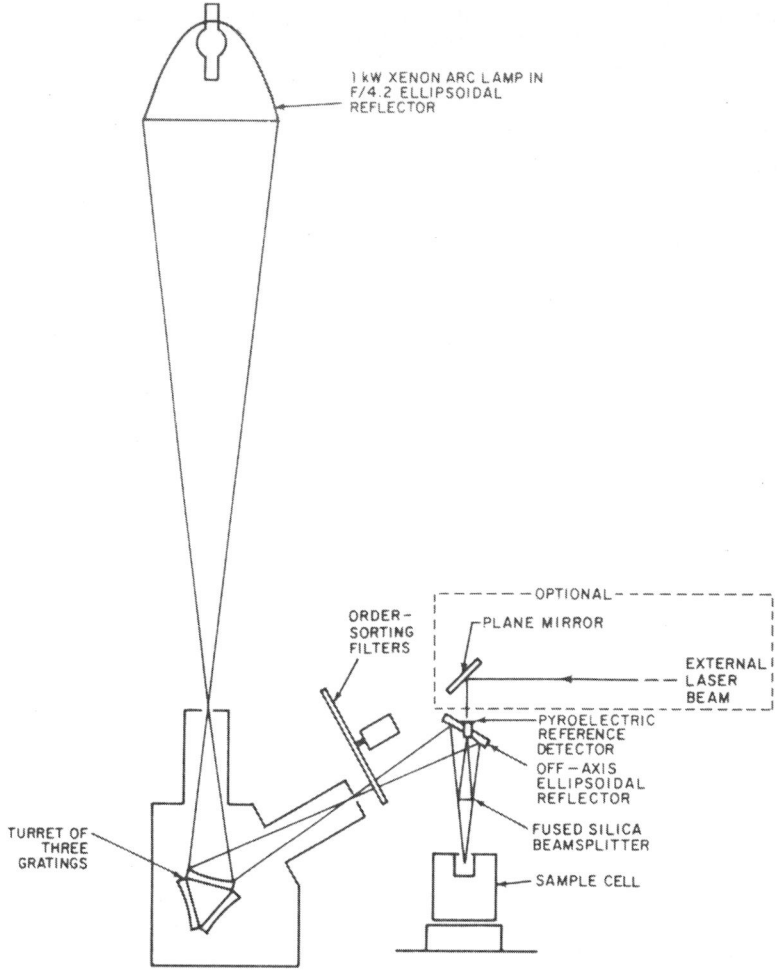

Fig. 1. Optical system of the model 6001 photo-
          acoustic spectrometer by EG&G Princeton
          Applied Research.

In the electronic system, there are two two-phase lock-in amplifiers which separately process the microphone and pyroelectric
signals. The entire system is under microprocessor control, with
keyboard entry of experimental parameters. A 14-bit analog-digital
converter encodes the outputs of the two lock-in amplifiers and the
output of a follower potentiometer coupled to the sinebar mechanism
of the monochromator. Two 12-bit digital-analog converters serve
to output data to the X-Y recorder.

RESULTS AND DISCUSSION

PA Spectra of Polymers

   Several polymers, e.g., polyester (Fig. 2), polycarbonate (Fig. 3), polystyrene (Fig. 4), polymethyl methacrylate (Fig. 5), have absorption mainly in the UV region (Fig. 6). This absorption range facilitates the study of adsorption of coloring dyes on the polymer surface. In this paper, we shall illustrate the applications with Mylar ® polyester, gelatin, and cellulose fibers (Fig. 7).

PA Spectra of Dyes

   Since dyes have strong absorption in the visible region, the PA spectra of solid dyes nearly all have saturation problems. Saturation (7,15) is closely related to the optical absorption coefficient, β. For a sample with a high optical absorption coefficient, saturation of the PA signal can occur. For a sample with a low optical coefficient, e.g., $\beta < 10^3$ cm^{-1}, the PA spectrum in the wavelength region will yield the true optical absorption spectrum. In the wavelength region where $\beta > 10^3$ cm^{-1}, saturation will occur and the true optical absorption peak height will not be observed in the PA spectrum.

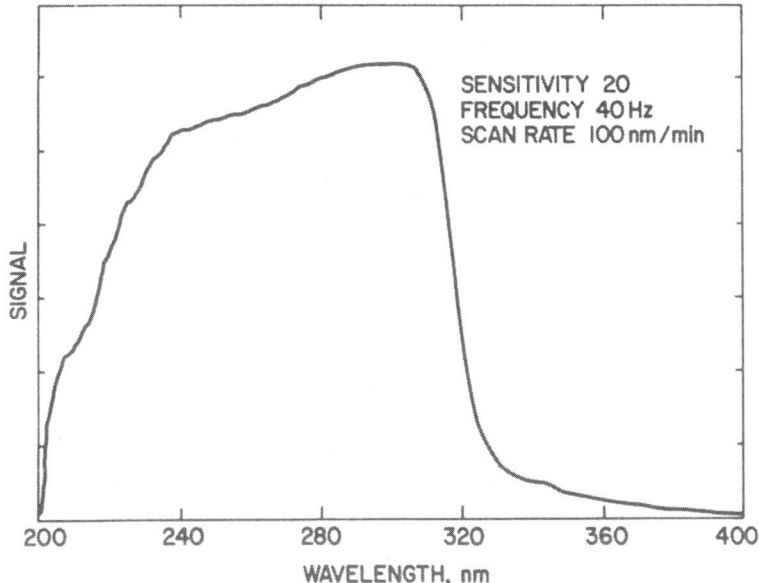

Fig. 2. Photoacoustic spectrum of mylar film.

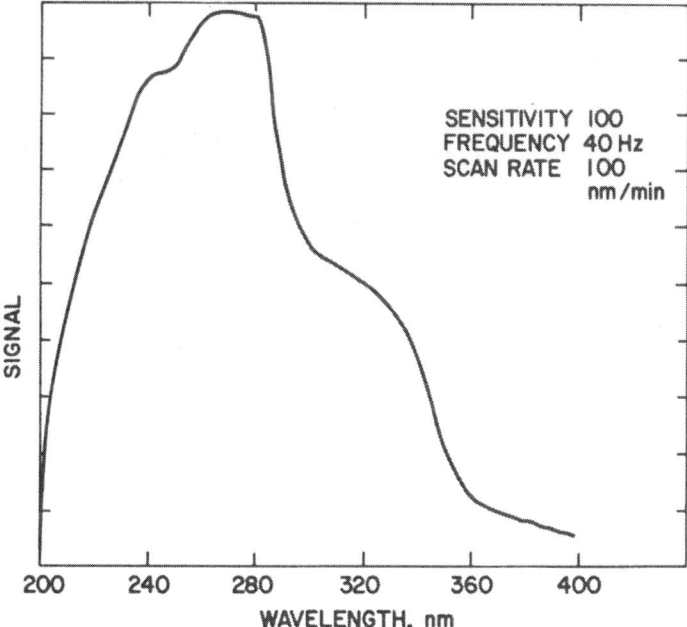

Fig. 3.   Photoacoustic spectrum of polycarbonate.

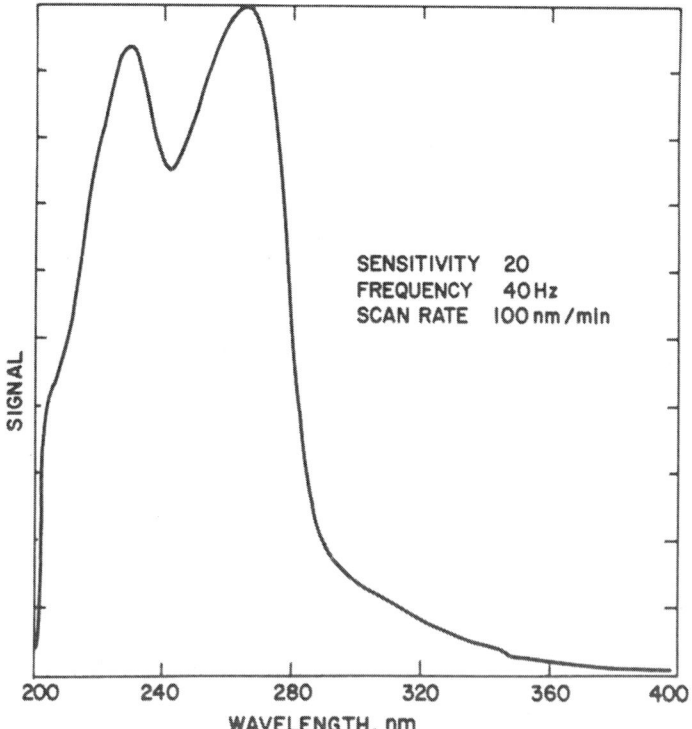

Fig. 4.   Photoacoustic spectrum of polystyrene.

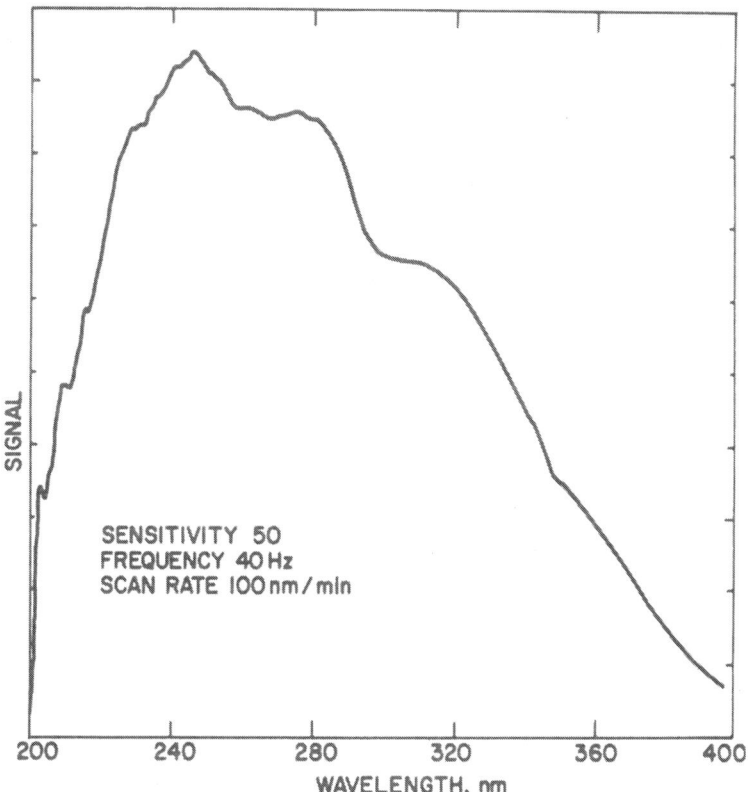

Fig. 5. Photoacoustic spectrum of polymethyl
        methacrylate.

        To alleviate the saturation problem, one can apply several
sample preparation techniques.

1. If the solid is soluble in a solvent, a liquid or solution
    (6,13,19) can be used instead of the solid.

2. The solution, e.g. methyl violet 2B, can be applied as a
    drop on a cellulosic paper (Fig. 8), or on a TLC plate.

3. The solution can be solidified into a gel, e.g., potas-
    sium dichromate, $K_2Cr_2O_7$, in gelatin (Fig. 9).

4. The solid can be diluted with a nonabsorbing powder,
    e.g., MgO (8). The mixture, e.g., α-metal free phthalo-
    cyanine and MgO, is co-ground (Fig. 10).

        Since PA spectra may contain results due to both absorption and
reflection loss, one should be careful in interpreting the spectra.

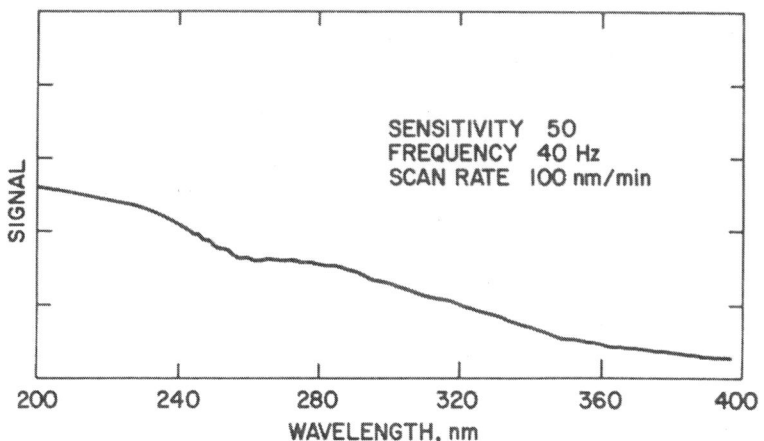

Fig. 6.   Photoacoustic spectrum of cellulose.

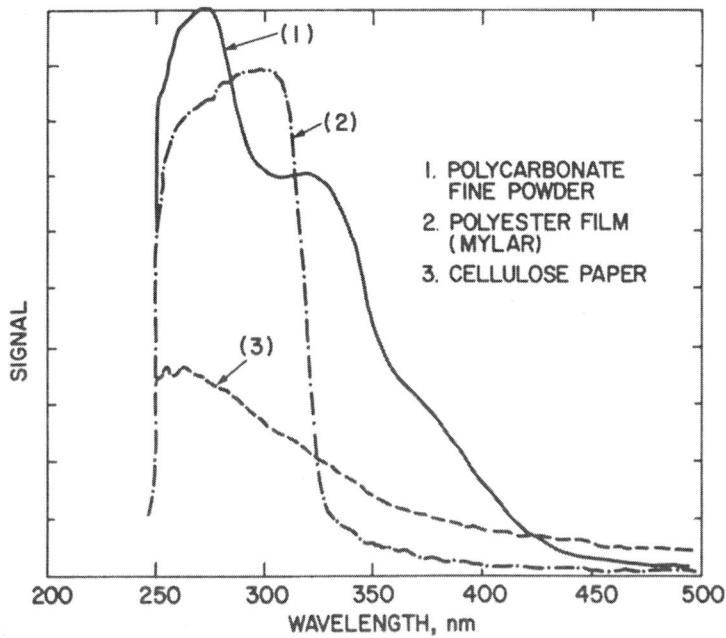

Fig. 7.   Photoacoustic spectra of polymers

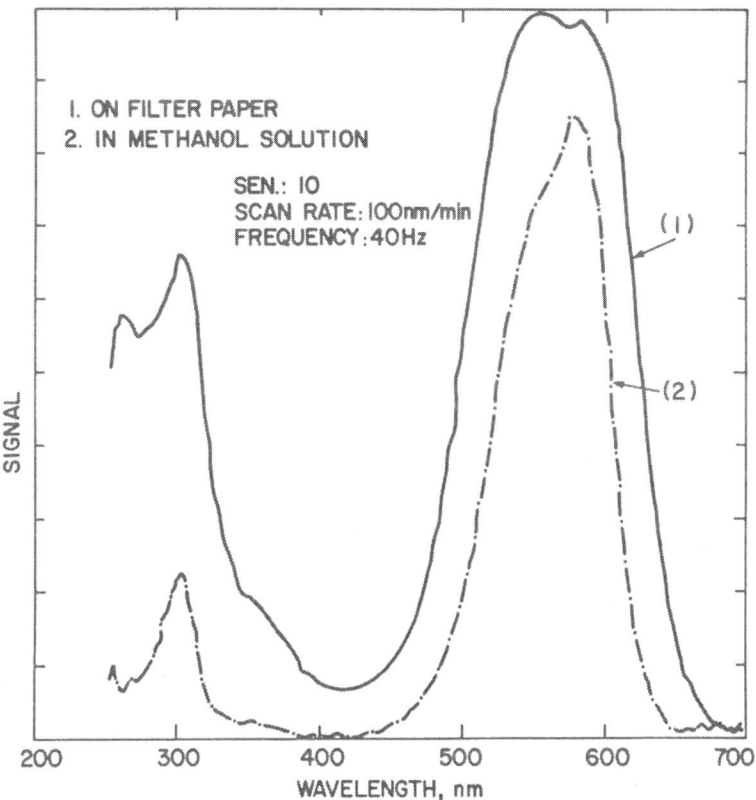

Fig. 8. Photoacoustic spectra of methyl violet 2B
on filter paper and in methanol solution.

Correlation Between UV-Visible and PAS Absorption Peaks of Dyes

The absorption spectra of dye solutions in the UV-visible
range and the PA absorption signals have been carefully compared.
Although there are some exceptions, the general agreement between
the two types of spectra is very good.   The results are shown in
Table 2.   There are some minor shifts of absorption peaks, but the
characterization with PA spectra appears to be well substantiated.
In Table 2, the PA absorption peaks of those dyes absorbed on paper
agree, in general, with those obtained directly from dye solutions.
This result suggests that PAS can be properly used for the study of
adsorption of dyes or polymers.

Adsorption of Dyes on Polyester

The adsorption of dye on a polyester surface was used to
illustrate the general application of PAS for the adsorption study.

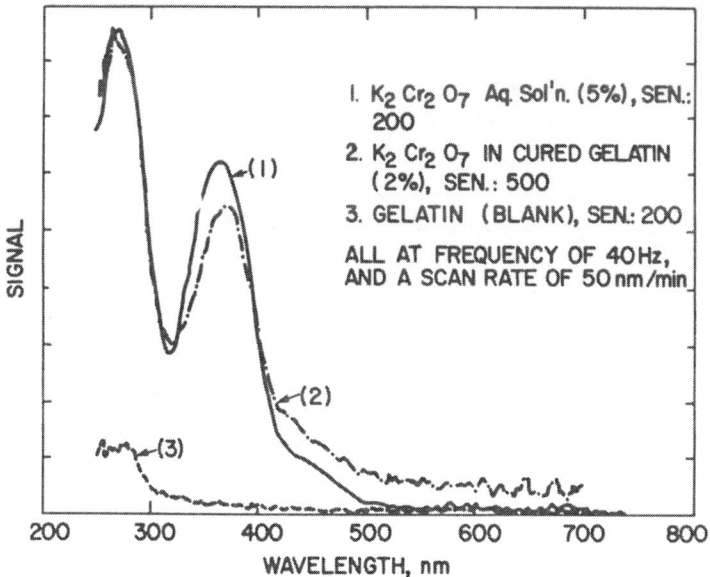

Fig. 9. Photoacoustic spectra of potassium dichromate
         in solution and in gelatin.

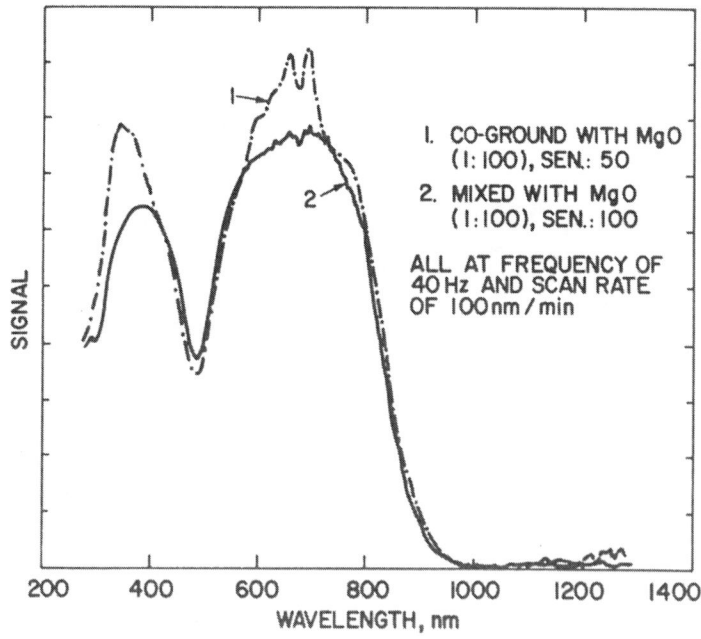

Fig. 10. Photoacoustic spectra of alpha metal-free
          phthalocyanine in magnesium oxide.

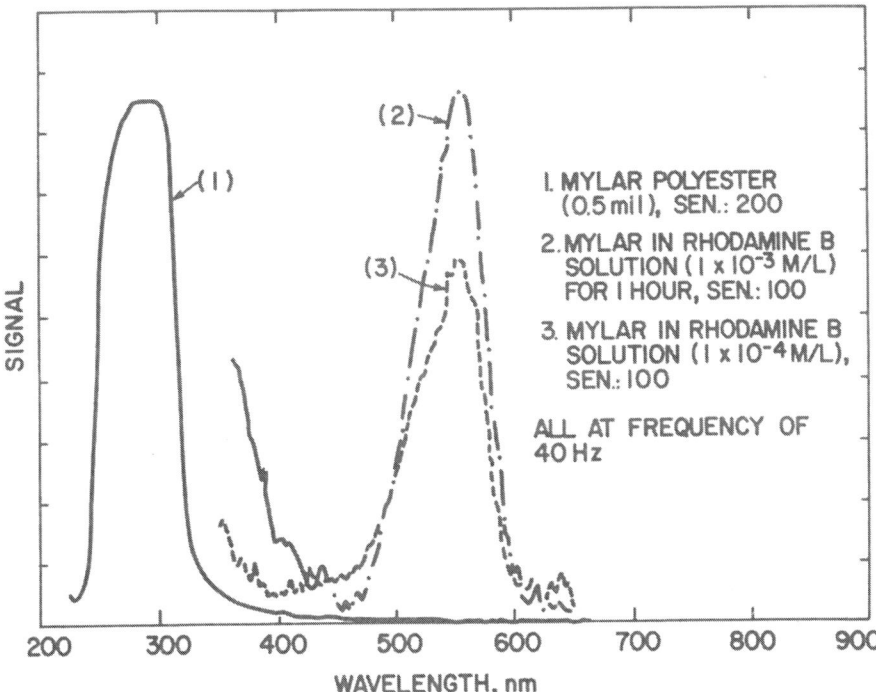

Fig. 11.   Photoacoustic spectra of adsorbed rhodamine
           B on mylar.

A thin Mylar polyester film (12.5 x 1x5 cm) was placed in each
of the following rhodamine B solutions: (a) $1x10^{-3}$ mole/ℓ , (b)
$1x10^{-4}$ mole/ℓ, (c) $1x10^{-5}$ mole/ℓ. The adsorption took place at room
temperature for one hour. The adsorbed films were dried and then
used for the PA analysis. The films from all three solutions showed
the absorption peak of the adsorbed dye, rhodamine B at 555 nm (Fig.
11). The advantage of PAS in this case is the examination of the
adsorbed species without any separation of the substrate.

PAS For Multilayer Films

Multilayer films can be formed in nature, e.g., the green
leaves, or by lamination or adhesion. The examination of the
composition of various layers without any destructive treatment is
one of the advantages of the technique. Examination of the protec-
tive layer of leaves with PAS reveals the waxy material synthesized
by the plant to protect the interior photosynthetic structure.

Table 2. Comparison of UV-Visible Absorption Peaks with PAS Signals of Dyes

Dyes	UV-Visible Absorption Peaks, (in solution) (nm)	PAS Signal (in solution) (nm)	PAS Signal (on filter paper) (nm)
Alkali blue, 4B	586	592	595
Aniline blue, SS	582	590	599
Anthraquinone violet R	548-550	536	554
Azure A	637	617	603
Biebrich scarlet (H_2O sol.)	506	509	512
Crystal violet	593	589	555-595
Cyanine acid blue	572	580	596
4',5',-Dibromofluorescein	520	524	526
Eosin B	515	522	530
Indulin (SS)	595	598	600
Methyl green	632	630-645	645
Methyl red	423-519 (dep. on pH)	515	483
Rhodamine B	(276,315)548,556	546	560

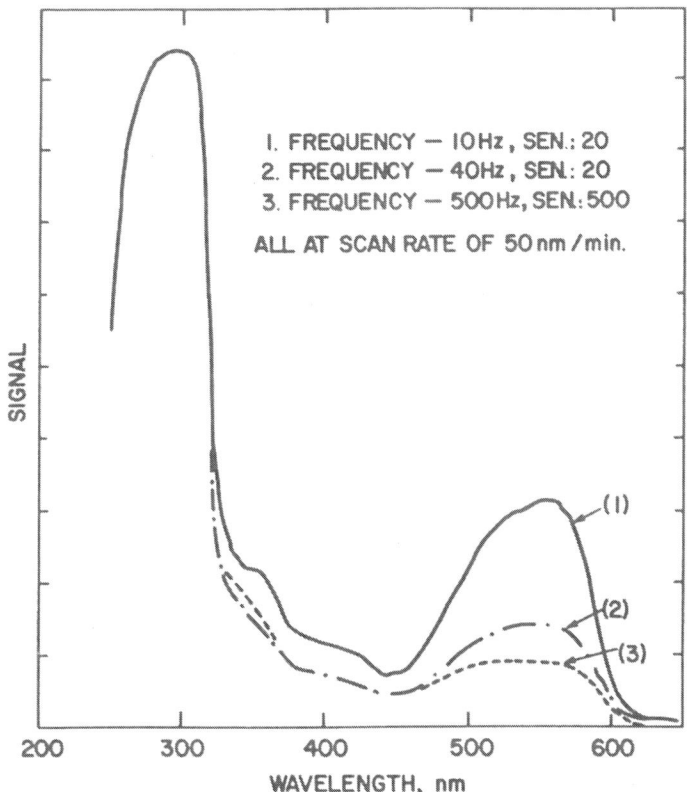

1. FREQUENCY — 10 Hz, SEN.: 20
2. FREQUENCY — 40 Hz, SEN.: 20
3. FREQUENCY — 500 Hz, SEN.: 500

ALL AT SCAN RATE OF 50 nm/min.

Fig. 12. Photoacoustic spectra of laminated paper with Rhodamine B at various chopping frequencies.

The examination of the solid surface is achieved by the increase in the modulating frequency or the decrease in the thermal diffusion length, as discussed in one of the earlier sections. The same technique is used to reveal the surface layer of a laminated film. A polyester film (12.5μ) was laminated on a filter paper coated with a drop of rhodamine B dye. By increasing the modulating frequency, the absorption peak of the underlayer dye decreased, as shown in Fig. 12. In other words, the thermal diffusion length is shortened and less dye in the bulk is detected by PAS. In practice, this technique in changing the modulating frequency is not as straightforward as it sounds. The results should be interpreted with caution.

Future Applications of PAS to Adhesion and Adsorption

In this paper, we briefly demonstrate the applications of PAS to adsorption and adhesion studies, and the limitations of the technique, e.g., the saturation problem. With a further broadening of PAS into the IR range (13,14) the applications can be expanded for the identification of organic functional groups. We can visualize future applications of PAS in the following areas:

1. Adsorption and adhesion of biopolymers on various substrates.

2. Adhesion of coloring materials on polymers or other substrates.

3. Identification of inks, dyes and pigments in a polymer matrix.

4. Photoadhesion phenomena.

5. Photoadsorption and photodesorption phenomena.

Technologically, PAS will find uses in the characterization of coatings, inks, and dyes for textiles.

CONCLUSIONS

Recent developments in photoacoustic spectroscopy for solid materials indicate new applications for the study of adhesion and adsorption of polymers and biopolymers. Though there are still problems associated with this technique, progress has been made, especially in regard to the saturation problems. The comparison between the optical absorption spectra and PA signals confirms the general agreement. With the current state of the art, PAS can be qualitatively used to characterize dyes or polymers as isolated or as adsorbed species. However, because of the nature of the PA effect, the interpretation of spectra should be exercised with caution.

ACKNOWLEDGEMENTS

The author would like to thank Mr. H. Reichard of EG&G Princeton Applied Research for discussions about the techniques in using the Instrument, Dr. A. Rosencwaig of the Lawrence Livermore Laboratory and Dr. R. Staley of the Massachusetts Institute of Technology for the comments on the manuscript, and P. Alden for technical assistance.

REFERENCES

1. A. Rosencwaig, "Photoacoustic Spectroscopy of Solids", Phys. Today 28, No. 1, 23 (1975).
2. A. Rosencwaig, "Photoacoustic Spectroscopy - A New Tool for Investigation of Solids", Ana. Chem. 47, No. 6, 592A, May 1976.
3. R. Harshbarger and M.B. Robin, "The Opto-acoustic Effect. Revival of an Old Technique for Molecular Spectroscopy", Accounts of Chem. Res. 6, 329 (1973).
4. A. Rosencwaig, "Solid State Photoacoustic Spectroscopy", p. 193 in Optoacoustic Spectroscopy and Detection, Ed. by Y.H. Pao, Academic Press (1977).
5. F.W. Karasek, "Photoacoustic Spectroscopy", Res./Dev., p. 33, September 1977.
6. J.F. McClelland and R.N. Kniseley, "Signal Saturation Effects in Photoacoustic Spectroscopy with Applicability to Solid and Liquid Samples", Appl. Phys. Lett. 28, No. 8, 467 (1976).
7. R.B. Somoano, "Photoacoustic Spectroscopy of Condensed Matter", Angew. Chem. Int. Ed. Eng. 17, 238 (1978).
8. G.F. Kirkbright, "Optoacoustic Spectrometry and Its Application to Examination of Solid and Liquid Samples", paper presented to the Photoacoustic Spectroscopy Symposium, Boston, October 1978.
9. M.J. Adams, "Optoacoustic Spectrometry of Solid Samples in the Near Infrared", paper presented to the Photoacoustic Spectroscopy Symposium, Boston, October 1978.
10. H.S. Reichard, "A Microprocessor-Based Photoacoustic Spectrometer", paper presented to the Photoacoustic Spectroscopy Symposium, Boston, October 1978.
11. R.E. Blank and T.D. Wakefield II, "A Newly Developed Photoacoustic Spectrometer with Selected Applications", paper presented to the Photoacoustic Spectroscopy Symposium, Boston, October 1978.
12. D.J. Johnson and G.F. Kirkbright, "A New UV-Visible and Near-IR Optoacoustic Spectrometer for Solids and Liquids", paper presented to the Photoacoustic Spectroscopy Symposium, October 1978.
13. P.E. Nordal and S.O. Kanstad, "Photoacoustic Spectroscopy on Ammonium Sulfate and Glucose Powders and Their Aqueous Solutions Using a CO_2 Laser", Opt. Commun. 22, 185 (1977).
14. M.J.D. Low and G.A. Parodi, "Infrared Photoacoustic Spectroscopy of Surface Species", paper presented to the Spring Meeting of the American Chemical Society, April 1979.
15. A. Rosencwaig and A. Gersho, "Theory of the Photoacoustic Effect with Solids", J. Appl. Phys. 47, 64 (1976).
16. J.G. Parker, "Optical Absorption in Glass: Investigation Using an Acoustic Technique", Appl. Opt. 12, 2974 (1973).

17.   J.F. McClelland and R.N. Kniseley, "Photoacoustic Spectroscopy
      with Condensed Samples", Appl. Opt. 15, 1658 (1976).
18.   Technical Bulletin, "The Model 6001 Photoacoustic Spectro-
      meter", EG&G Princeton Applied Research (1978).
19.   J.W.P. Lin and L.P. Dudek, "Photoacoustic Spectroscopy. I.
      Signal Saturation Effect and Techniques for Obtaining
      Good Solid Spectra", paper submitted to J. Anal. Chem.

Adhesive-Adherend Bond Joint Characterization by Auger Electron Spectroscopy and Photoelectron Spectroscopy

J.S. Solomon and D. Hanlin

University of Dayton Research Institute

300 College Park Avenue, Dayton, Ohio   45469

and

N.T. McDevitt

Air Force Materials Laboratory (MBM)

Wright-Patterson Air Force Base, Ohio   45433

ABSTRACT

Chemical and physical information about intact adhesive bond joints is usually inferred from data obtained from each isolated component (adhesive and adherend) prior to bonding or after bond failure. A technique is presented in which the chemical state as well as the elemental distribution can be obtained from intact bond joints using conventional surface characterization techniques such as Auger electron and photoelectron spectroscopies. The technique involves the use of thin film adherend adhesively bonded structures. The thin film adherends ($5x10^{-7}$m) are prepared by vacuum deposition, which, after various treatments and subsequent bonding, can be ion beam etched away until the adherend-adhesive interface is reached for characterization. A number of bonding parameters have been investigated using this technique, including effects of adherend surface treatments (i.e. anodization, protection primers, corrosion inhibitors) and cure conditions (i.e. time, temperature and pressure).

INTRODUCTION

Adhesive bonding offers considerable weight and economic
advantages over conventional fabrication processes. Aircraft and
automotive applications of adhesive bonding have spawned considerable research into various related subjects involving a number of
scientific disciplines. Just a few years ago, adhesive bonding was
considered more of an art than a science and the majority of bonding
evaluations were based strictly on empirical results. With the
growing interest in adhesive bonding, the direction of related
research has been shifted somewhat away from the empirical to a more
systematic scientific approach.

Of empirical interest in any potential application of adhesively bonded structures are strength and durability. Consequently, those parameters which affect strength and durability are
at the heart of adhesive bonding research. Included in this
research are the material and physical properties of adhesives, the
bulk and surface characteristics of adherends, and the nature of the
forces involved in bonding. One of the recognized facts about an
adhesive bond is that these parameters are all interrelated and can
undergo changes once a bond joint is formed. Therefore, much
information can be gained if the same quantities (chemical and
physical) could be investigated within the bond joint as well as
from the separate components.

The chemical and physical characterization of the adherend
surface and adhesive can be performed using a number of surface and
energetic characterization techniques. However, these techniques
are generally applied to the characterization of the isolated bond
joint components before bonding or after failure (1-4). The reason
being, the size and location of the bond joint do not fulfill
conventional surface technique requirements.

Normally, when one thinks of an adhesive bond, the bond line is
thought of as a two dimensional "interface" between the adhesive and
adherend. In reality, the bond joint is more than this. It is a
three-dimensional region which has been called an "interphase" (5).
This region incorporates all the material from some point in the
bulk adhesive toward and through the actual boundary between adhesive and adherend to the point where the local properties approach
those of the bulk adherend. This generally amounts to a cross
section hundreds of nanometers in thickness, located perhaps,
several millimeters or more below the outer surface of one of the
adherends.

One method of assessing the chemical and physical properties
of an adhesive bond is to induce failure within the interphase and
characterize the failed surfaces (2,3). However, as reported by
Dwight (1), real fracture surfaces do not always exhibit a clear

failure mode. When failure is entirely cohesive within the adhesive, surface analysis can be extremely difficult because of
(1) charging effects and (2) uncertainty as to the location of the
interphase with respect to the locus of failure. When mixed mode
failures are encountered, difficulties also arise since both surfaces may appear the same, when in fact, some areas on the surfaces
may only be covered with a very thin carbon containing layer (2).
In the latter case, an ion beam etch would quickly remove the thin
carbon layer and reveal the local mode of failure. This approach
still leaves an uncertainty as to the size and location of the
interphase.

If the interphase volume was at or near the surface, composition profiling techniques could be employed to examine the bond
joint in an undisturbed state. This would remove some of the
difficulties and uncertainties described above. Auger in-depth
profile analysis of thin films is a well established technique with
sufficient depth resolution to provide composition and elemental
distribution information from thin films (6) and has been used
successfully in the past to provide chemical information from both
adherend and adhesive surfaces (7-9). The technique, however, is
not very practical for examining very thick specimens, such as
intact adhesively bonded structures, because of the relatively slow
etching rate of the ion beam and depth resolution dependence on
thickness (10,11). Consequently, unless one or both adherends of
an adhesively bonded structure are extremely thin ($\sim$1μm) surface
probing techniques cannot be realistically applied. They can be
applied, however, if vacuum deposited films could be used for the
adherend. Since a relatively thick material is required for
bonding, the substrate on which the metal film is deposited would
serve as a "secondary" adherend. After bonding, the "primary"
adherend would be separated from its supporting substrate and
analyzed by Auger electron spectroscopy (AES) and x-ray photoelectron spectroscopy (XPS) for a number of related parameters. For
example, Auger in-depth profiles through a bond interphase could
provide information concerning adhesive penetration as a function
of surface preparation, elemental distributions within an interphase, and effects of corrosion inhibiting primers on adhesive
penetration. The state of chemical bonding or changes in chemical
bonding within an interphase could be characterized by XPS.

SPECIMEN PREPARATION

Since there is considerable interest in adhesively bonded
structures of aluminum and aluminum alloys in the aircraft industry, aluminum was chosen as the "primary" adherend for the initial
study. The problem of separating the deposited aluminum from its
supporting substrate after bonding was easily solved.

The presence of fluorine on titanium surfaces has been suggested as the primary cause of adhesive failure (failure at the
adhesive-adherend juncture as compared to cohesive failure which is
failure totally within the adhesive) (12,13). Our previous work
involving surface preparations for adhesive bonding of titanium
confirmed this finding. In addition to the presence of fluorine,
removal of surface oxides almost guaranteed an adhesive failure
mode. Therefore, to enable separation of the bonded thin film
aluminum adherend from its supporting substrate, titanium or 6A1-
4V-Ti were used as supporting substrates. These "secondary" adherends were pretreated with an oxide removing etch containing fluorine (HNO_3/HF) prior to vacuum deposition (14,15). Aluminum films
of approximately 0.5 to 1μm were then vacuum deposited onto the
etched substrates.

At present, the surface preparation of aluminum for adhesive
bonding receiving the highest interest in the aircraft industry is
anodization (16). Therefore, the deposited aluminum films, together with the supporting substrates, were anodized at constant
potentials, which could produce anodic oxide films less than 0.4μm
(4000Å) in thickness. Figure 1 is a scanning electron microscope
(SEM) micrograph at 50,000x magnification of an anodic aluminum
oxide film grown on aluminum at 40 volts for 20 minutes at room
temperature in 1.0M H_3PO_4. The columnar structure (17-19) of the
porous anodic oxide is very evident. Figure 2 contains an SEM
micrograph of the cross section of the anodic oxide film grown on a
vacuum deposited aluminum film which was anodized for 5 minutes at
the conditions stated above. The pore diameters of the anodic oxide
film shown in Figs. 1 and 2 are approximately 400Å. Comparing Figs.
1 and 2, no distinguishable differences are evident between the
anodic film grown on bulk aluminum and one grown on the thin film.
This is in agreement with Dell'Oca (20).

SEM micrographs of anodic oxide films grown on aluminum
alloys, such as 2024 aluminum, were also compared to micrographs of
anodic oxides grown on bulk and thin film aluminum. The conclusion
was that the anodic oxides grown on the alloys were structurally
indistinguishable from those grown on bulk or thin film aluminum.
Therefore, the anodized thin film aluminum model adhesively bonded
structures could be used to simulate the bonding of thick anodized
aluminum and aluminum alloy panels.

The anodized aluminum films attached to "secondary" substrates
were bonded to another anodized surface in an autoclave chamber (21). The adhesives used were 3M's AF143 (22), and American
Cyanamid's FM73 and FM400 thermo-setting epoxy adhesive films (23).
A typical adhesively bonded structure is illustrated in Fig. 3.
Figure 4 illustrates the orientation of the bonded structure for ion
beam etching and analysis after the secondary adherend is delaminated from the aluminum film.

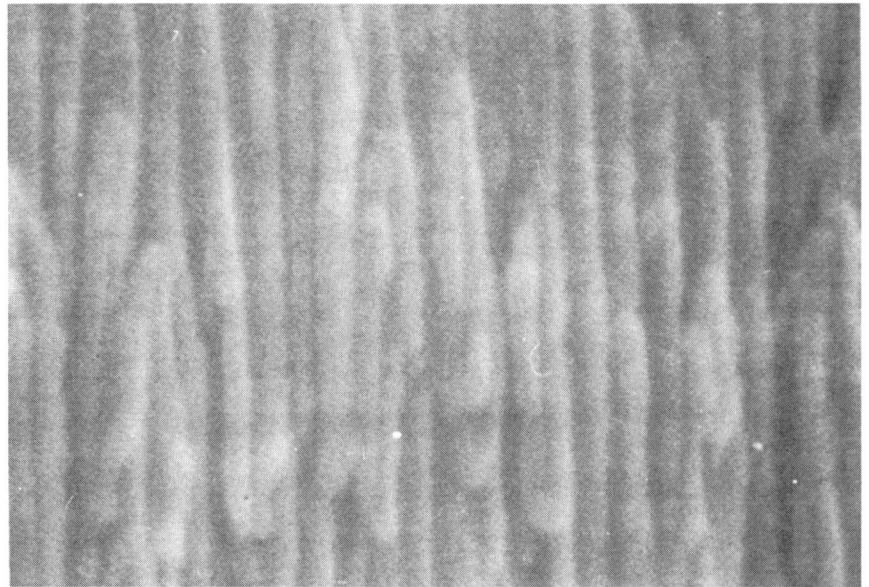

Fig. 1.   Scanning electron micrographs at 50,000x
          magnification of a cross section from an
          anodic oxide film formed on aluminum with
          1.0M H_3PO_4 at 40 volts for 20 minutes.

EXPERIMENTAL

        The Auger instrumentation consisted of a Physical Electronics
Industries (PHI) Inc. model 540A Thin Film Analyzer equipped with a
single pass cylindrical mirror analyzer (CMA) with a resolution
$\Delta E/E \sim 0.6\%$.   The coaxial electron gun was operated with a 4keV
potential at 0.1 to 0.5μA beam current. A peak-to-peak modulation
of 5eV was applied to the analyzer for phase sensitive detection and
the resultant N'(E) data was digitally recorded and computer pro-
cessed as previously reported (24).   An electron flood gun biased
at -18V with respect to the specimen was used to minimize possible
charging effects. The ion beam was generated with a PHI model 94-
191 Sputter Ion Gun which was operated with a beam potential of 2keV
and an ion current density of approximately 1μA/mm^2. When charging
effects were evident at the onset of analysis, the electron flood
gun and ion gun (at reduced current) were turned on prior to data
collection.   Since the region of interest was well below the
aluminum metal surface, any ion etching during this time was not
detrimental to the analysis.

        The x-ray photoelectron spectrometer was an AEI model ES-100
with a magnesium x-ray source. The description of this instrument

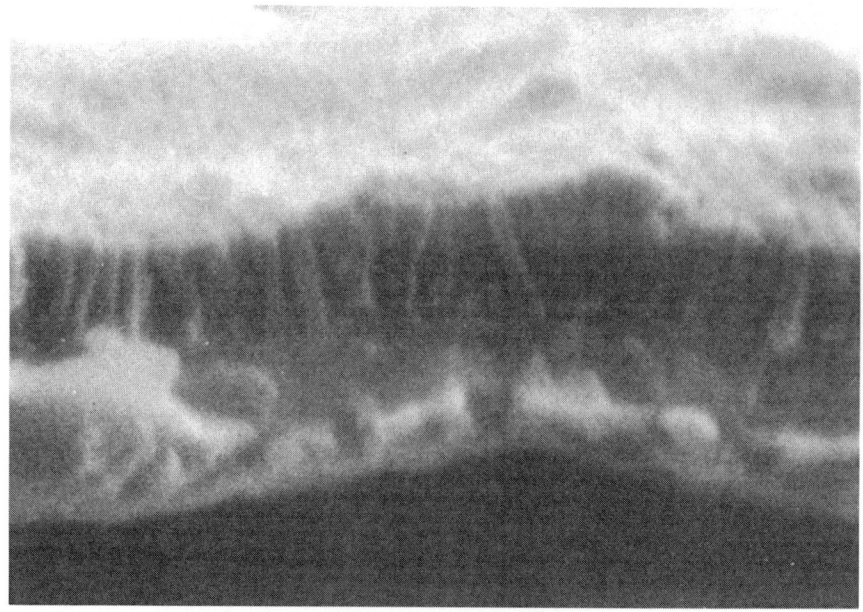

Fig. 2.   Scanning electron micrographs at 50,000x
          magnification of a cross section from an
          anodic oxide film formed on a 900 nanometer
          thick vacuum deposited aluminum film anodized
          in 1.0M H_3PO_4 at 40 volts for 5 minutes.

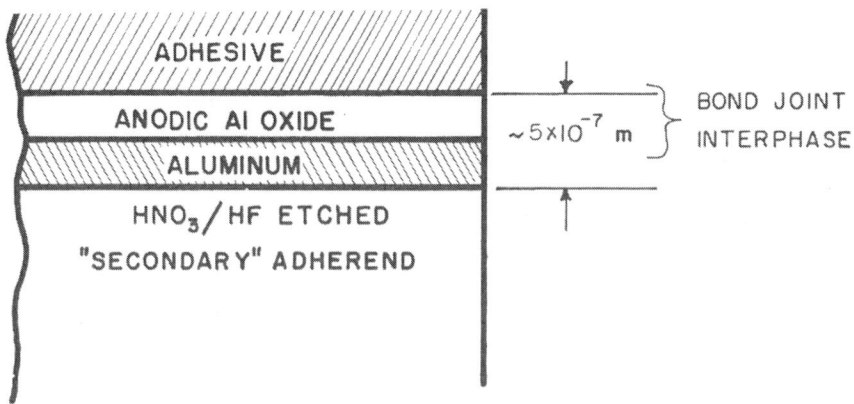

Fig. 3.   Illustration of the bond joint cross section
          of an adhesively bonded anodized thin film
          aluminum adherend attached to a supporting
          "secondary" adherend.

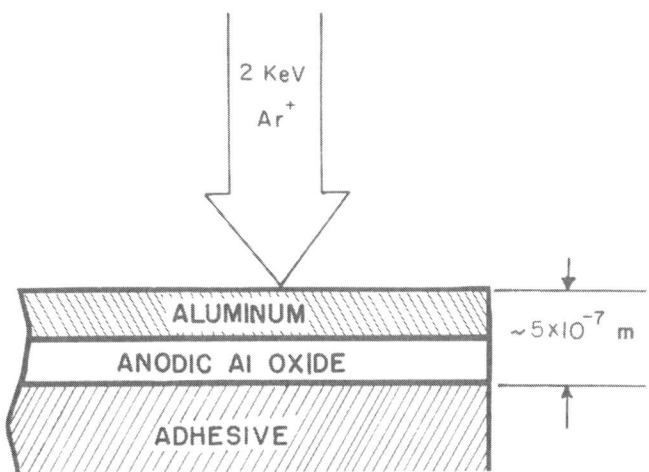

Fig. 4. Illustration of the bond joint cross section shown in Fig. 3 with "secondary" adherend removed for ion beam etching and subsequent AES and XPS analyses.

was previously reported (25).

Vacuum deposition of aluminum was performed in a Varian Associates F-12E ultra high vacuum system from a hot tungsten conical
filament at a background pressure of approximately 1.3×10^{-6} Pa
(1×10^{-8} torr). A Kronos Inc. 301 Film Thickness monitor was used
to measure the thickness of deposited aluminum.

RESULTS AND DISCUSSIONS

Elemental Distribution

Two thin film adhesively bonded structures were prepared using
aluminum films anodized in phosphoric acid and phthalic acid,
resulting in a porous anodic film in the case of phosphoric acid and
a dense nonporous oxide film with the phthalic acid. After bonding
with AF143, the supporting substrates were separated from the
aluminum thin films. AES spectra from the delaminated surfaces in
Fig. 5 show a clean and complete separation. Auger in-depth
elemental profiles through the respective interphases are shown in
Fig. 6. The method to construct Auger in-depth profiles was
previously reported (24).

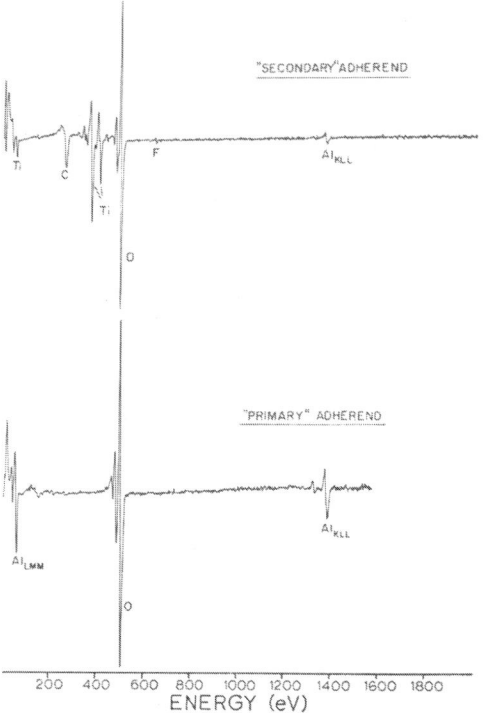

Fig. 5. Auger electron spectra of delaminated thin
         film aluminum "primary" adherend and HNO$_3$-HF
         etched 6A1-4V-Ti alloy "secondary" adherend
         surfaces.

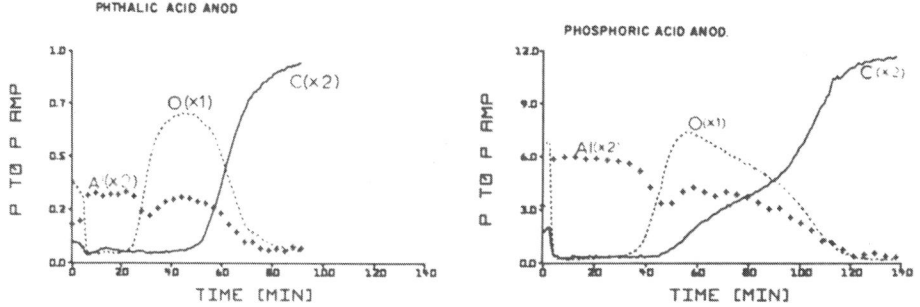

Fig. 6. Auger elemental profiles of Al, O, and C, as
         a function of ion sputter time, through the
         interphase regions of bonded phthalic and
         phosphoric acid anodized thin film aluminum
         adherend structures.

The abscissa in Fig. 6 is a measurement of sputtering time which can be converted to a depth measurement if sputtering rates are known. As reported by Kirschner and Etzkorn (26), converting sputtering time to depth is not a straightforward task when multi-layers of different elements and chemical states are profiled. The problem is the sometimes large variance in sputter yields for pure materials, mixtures, and oxides. Wehner has reported that there is nearly a factor of two difference between the sputter yield of pure aluminum and aluminum oxide (27). In the case of a porous anodic oxide film, there is a thickness dependency on sputter rate because the pore diameter decreases with thickness (28) and a more dense barrier oxide layer is present between the pore base and the metal (19). Sputter rate determination is further complicated within an interphase because of the mixture of the adhesive with the oxide and the oxide with the metal. Thus, the anodic oxide layer of the phosphoric acid anodized aluminum film, as determined by the oxygen profile in Fig. 6, appears to be 1.5 times thicker than the phthalic acid anodic oxide, when in fact, the phthalic acid anodic film is $0.1\mu m$ (1000Å) thick while the phosphoric acid anodic film is $0.45\mu m$ (4500Å) or 4.5 times thicker.

In addition to sputter rate considerations, attempts to relate thickness to sputtering time can result in very large errors if depth resolution is ignored. Depth resolution has been defined as the quality of an in-depth profile compared to an ideal rectangular signal as determined by the broadening of the trailing edges of the profile (26,29). Figure 7 contains the smoothed and normalized in-depth profiles of (A) aluminum, oxygen, and carbon from an adhesively bonded phosphoric acid anodized aluminum film and (B) aluminum, oxygen, and titanium from an identical unbonded anodized film on a 6A1-4V-Ti substrate. Based on predetermined sputtering rates for anodically grown Al_2O_3, the oxide film thickness, as determined from Fig. 7B is approximately $0.14\mu m$ (1400Å).

The standard convention for determining depth resolution is to divide the sputtering time, Δt, between the points at which 16% and 84% of a particular $dN(E)/dE$ signal are reached, by the total sputtering time up to the point at which 50% of the signal is recorded. The depth resolution, $\Delta t/t$, based on the oxygen profiles in Fig. 7, is 31% (7A) and 23% (7B). The latter is in good agreement with Mathiew and Landolt's value of 22% for an Al/Al_2O_3 interface in anodic oxide films greater than 4×10^{-8}m (400Å) thick (11). The difference in depth resolution between the oxygen profiles in Fig. 7 is most likely due to greater roughness effects produced by sputtering (29) and, to a lesser degree, the change in sputtering rate with the presence of carbon. The problem of identifying the exact point of the interface between layers as represented by the profiles, when profiles are subject in poor depth resolution (26), was easy to solve in the case of aluminum-aluminum oxide because the Al_{KLL}

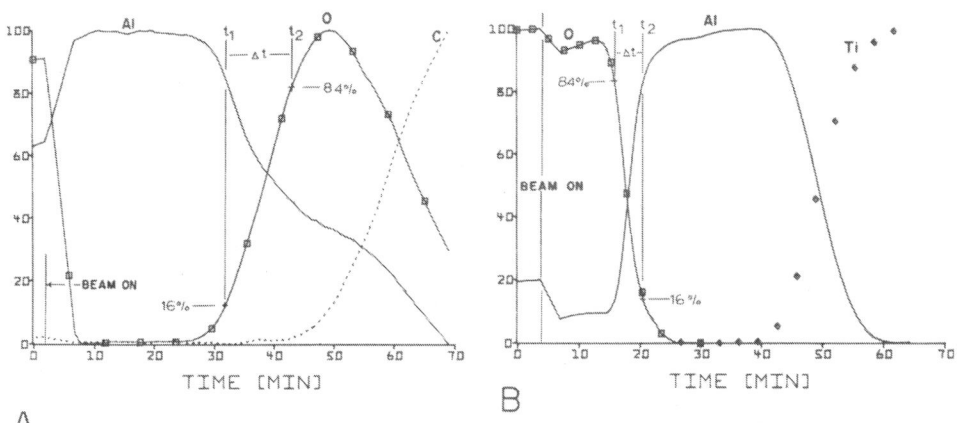

A

B

Fig. 7.   Normalized Auger elemental profiles as a
function of ion sputter time of (A) Al, O,
and C through the interphase of a bonded
phosphoric acid anodized thin film aluminum
adherend and (B) Al, O, and Ti through an
unbonded piece of the aluminum adherend
profiled in A to the supporting titanium
alloy substrate.

profile goes through a minimum at the metal-oxide interface (30).
This artifact serves as a consistent marker for the purpose of
thickness calculations.   This point also corresponds to the midpoint of the positive going slope of the oxygen profile between the
16% and 84% levels.   Therefore, when the Al_{KLL} artifact is not
evident the oxygen profile can be used.   The interface between the
metal oxide and the adhesive was determined empirically to be the
midpoint of the negative going slope of the oxygen profile, between
the 84% and 16% levels.

Returning to the profiles in Fig. 6, attention is directed to
the carbon profiles, which represent the adhesive in the interphase.   In the case of the dense, nonporous anodic film, there was
very little adhesive flow into the oxide.   But it appears that
adhesive is present throughout the entire porous oxide film in
decreasing amounts towards the interface of the oxide and metal.
This, of course, should be expected if the anodic oxide density
increases towards the oxide-metal interface (19).   One element
present in the anodic film formed with phosphoric acid, but not
profiled in Fig. 6, is phosphorous.   Thompson et al. have reported
the presence of the phosphate in microcrystallites along the walls
of the columnar Al_2O_3 with higher concentrations at the oxide

surface or pore openings (17). The AES spectrum from the surface of
a phosphoric acid anodized aluminum sheet in Fig. 8 does show the
presence of phosphorous, and the profile of phosphorous in Fig. 9
agrees with the model proposed by Thompson et al. Figure 10
contains the Auger in-depth profiles of Al, 0, C, and P from a 20
volt, 5 minute, 1.0m H_3PO_4 anodized aluminum film bonded with FM400.
The phosphorous profile in Fig. 10 no longer shows a higher concentration of that element at the oxide surface, but rather a Gaussian
like distribution throughout the oxide with a slightly higher
concentration towards the oxide-adhesive interface. Therefore, the
difference between the phosphorous profiles in Figs. 9 and 10 imply
that, during the cure cycle, the phosphate at the surface moved
ahead of the adhesive into the pore vacancies.

Comparison of the profiles in Fig. 10 with those from the
phosphoric acid anodized specimen in Fig. 6 shows less adhesive
penetration into the oxide in Fig. 10. Profiles from other specimens with thinner oxide layers, bonded with FM400, indicate a limit
of penetration, probably related to the presence of inorganic
filler material in the adhesive.

In-depth profiles from an interphase can be used to assess cure
conditions such as temperature, time, and pressure. Too rapid
curing can result in inadequate adhesive flow and wetting. The
carbon profiles in Fig. 11 reflect subtle differences in adhesive
penetration with a 10° difference in cure temperature. In this case
the aluminum was anodized at 10 volts in phosphoric acid and bonded
with FM73.

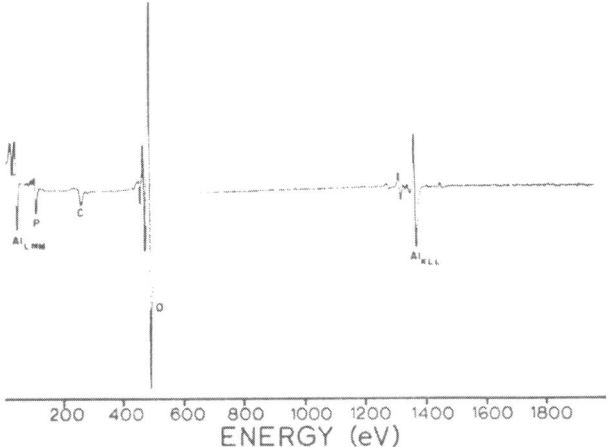

Fig. 8. Auger electron spectrum of the surface of
phosphoric acid anodized aluminum.

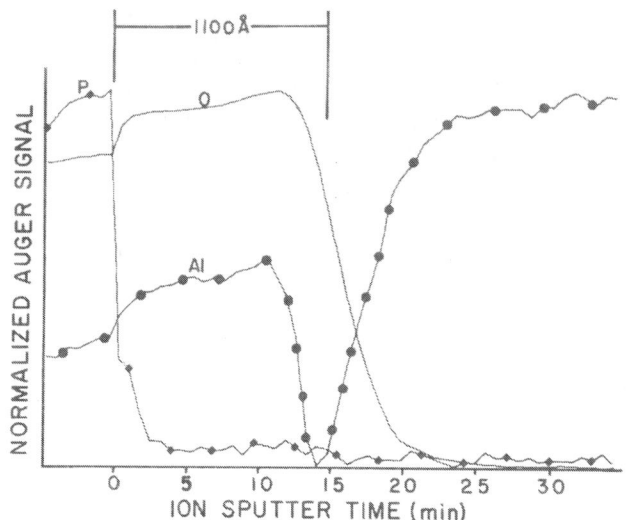

Fig. 9.   Normalized elemental profiles of Al, O,
          and P, as a function of ion sputter time,
          through the anodic oxide layer on phosphoric
          acid anodized aluminum.

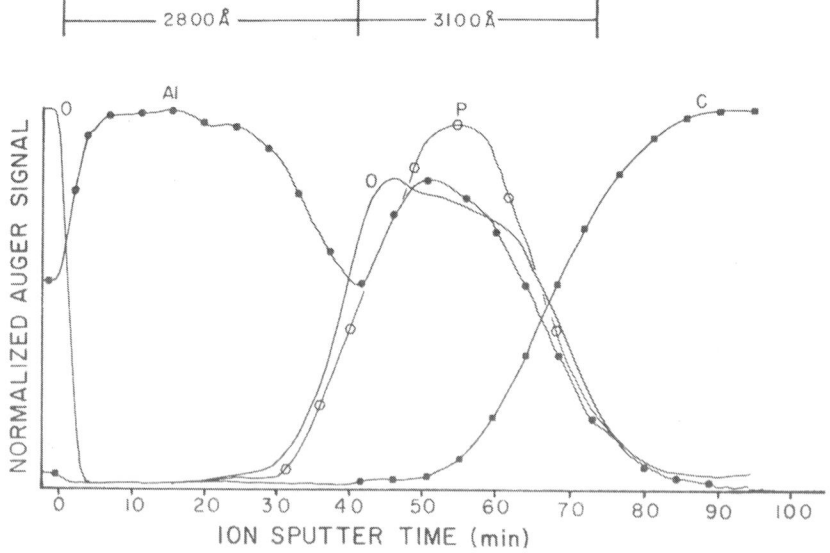

Fig. 10. Normalized Auger elemental profiles as a
          function of ion sputter time of Al, O, P,
          and C through the interphase of 20 volt,
          5 minute, 1.0M H3PO4 anodized thin film
          aluminum bonded with FM400.

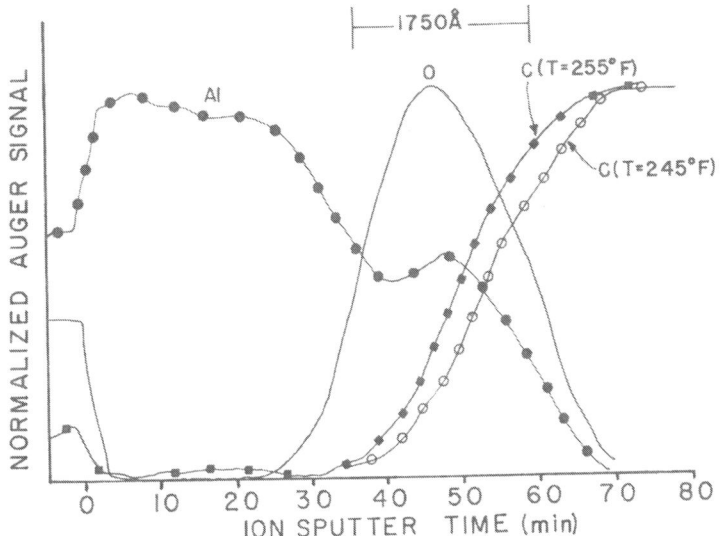

Fig. 11. Normalized Auger elemental profiles as a
          function of ion sputter time of Al, 0, and
          C through the interphase of 10 volt, 1.0M
          H3PO4 anodized thin film aluminum bonded
          with FM73 at cure temperatures of 245°F
          versus 255°F.

          A major problem associated with aluminum alloys, especially
when adhesively bonded, is corrosion.   Consequently, corrosion
inhibiting primers are often used to protect the alloy surfaces.
This, of course, presents problems to adhesive bonding since the
presence of a primer may not be conducive to good bonding. Marceau
(31) reported adhesive failure modes in phosphoric acid anodized
2024-T3 clad aluminum bonded structures when BR227 (23) was used as
a corrosion inhibiting coating.   He reported the presence of high
molecular weight molecules at the oxide-adhesive interface which
might have been too large to penetrate the pore voids in the anodic
oxide.   When anodized aluminum is coated with a silane primer,
normally used to promote adhesive on glass or glass fibers, failure
modes of bonded structures of these primed anodized aluminum panels
are usually adhesive at the metal oxide-adhesive interface.   The
elemental profiles in Fig. 12 from the interphase regions of a
silane primed and unprimed bonded aluminum thin film adherend
structures reveal the role of the primer, as represented by the
silicon profile, as preventing adequate penetration of the adhesive
into the oxide layer.   In fact, adhesive penetration in the silane
primed specimen appears to be limited to the primer coating.

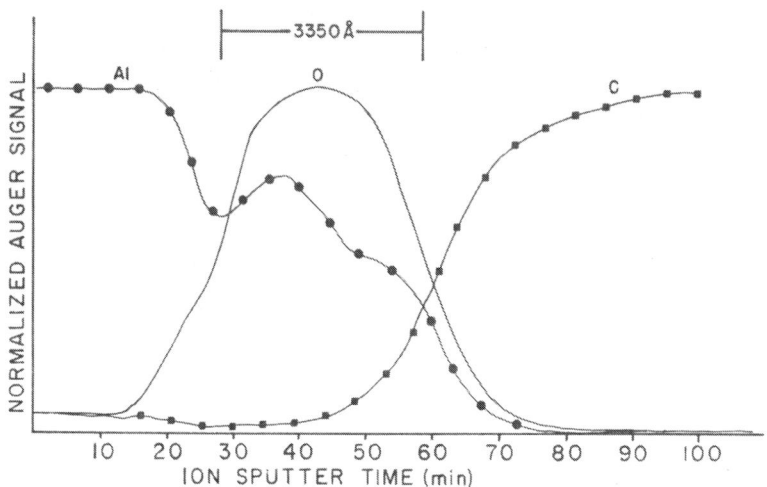

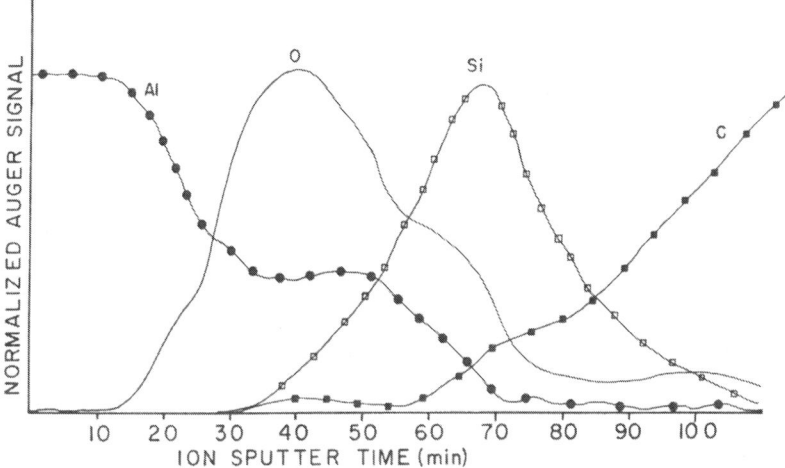

Fig. 12.   Normalized Auger elemental profiles as
a function of ion sputter time of Al, O,
and C through the interphase regions of
identical phosphoric acid anodized bonded
thin film aluminum structures in which
one was coated with a silane primer prior
to bonding.

A primer that has been used successfully for FM73 bonded structures of anodized aluminum alloys is BR127 (23). This is a modified epoxy phenolic primer containing strontium chromate which quickly settles upon application (32). Figure 13 contains Auger elemental profiles of the interphase of a thin film aluminum bonded specimen in which the aluminum was phosphoric acid anodized and coated with BR127 prior to bonding. Its profile shows carbon to be present throughout the oxide layer. To the right of the profiles in Fig. 13 are positive secondary ion mass spectra (SIMS) recorded simultaneously at the specified time intervals during ion beam etching. SIMS was used to monitor Sr because of its sensitivity for the SR^+ ions (33). At low Sr concentrations, such as that in BR127, the Auger signal strength was insufficient for profile analysis. The SIMS data shows Sr at the oxide-adhesive interface and not throughout the anodic oxide. This possibility infers that the carbon detected within the interphase is associated with the cured primer and not the adhesive. SEM micrographs confirm the presence of large particles with mean separation distances of approximately 30 μm, which together with its epoxy solvent, could act as a barrier for adhesive penetration into the anodic oxide.

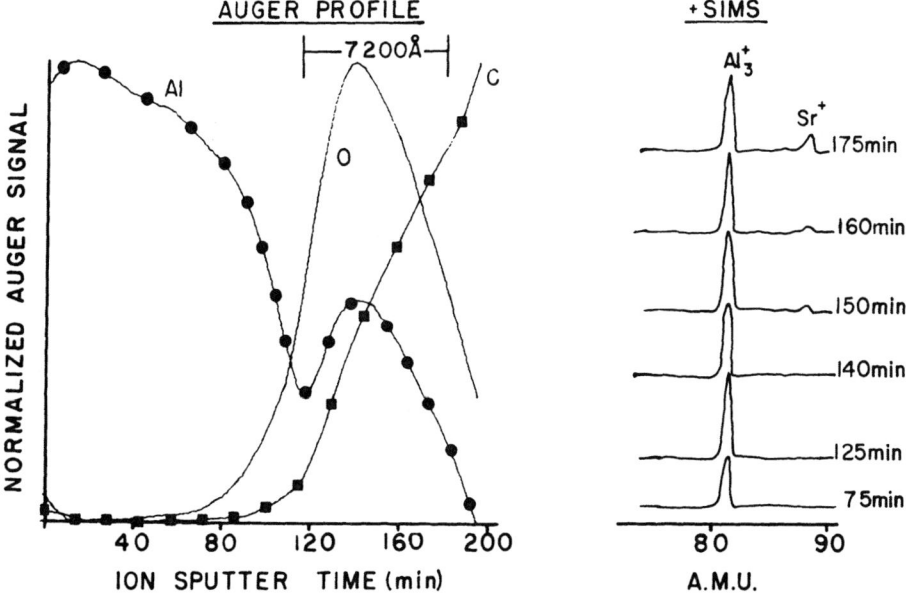

Fig. 13. Secondary mass spectra of Sr^+ (right) and normalized Auger elemental profiles, as a function of ion sputter time, of Al, O, and C (left) through the interphase of a BR127 primed and bonded 10 volt, 1.0M H3PO4 anodized thin film aluminum structure.

Chemical Bonding

       One of the major questions concerning the bond joint is what,
if any, chemical bonding occurs between adhesive and the oxide layer
of the adherend and to what extent this contributes to overall bond
strength and durability? In their work with the adhesive bonding of
aluminum with Epon 828, Lin and Bell suggested that chemical bonding
between the adhesive and aluminum during cure and the decrease in
adhesive bond strength may be related to a decrease in chemical
bonding (34).

       One of the best means available to study chemical bonding is
photoelectron spectroscopy. When the exciting mechanism is an x-
ray beam, the technique is referred to as x-ray photoelectron
spectroscopy (XPS). AES can also provide some chemical bonding
information, but that capability is somewhat limited. For example,
Fig. 14 contains the low energy Si_{LMM} AES spectra from (A) a surface
coated with the silane primer previously discussed, and from (B) a
region within the interphase region of the primed and bonded
specimen profiled in Fig. 12. The difference in shape between these
two spectra gives some indication of a change in the chemical
bonding of Si in the pure primer versus Si in the primer-adhesive
interphase. Likewise, the carbon AES spectra in Fig. 15 were
recorded at the oxide-adhesive interphase of the phosphoric acid
anodized specimen profiled in Fig. 6, and at a point within the bulk
adhesive. Of course, extreme caution must be exercised in applying
any interpretation to the differences between the spectra in Figs.
14 and 15 because of a number of unknowns, the foremost being
electron and ion beam effects (35). XPS, on the other hand, does
not introduce beam damage effects. Therefore, an identical bond
joint model was analyzed by (XPS) and the resultant carbon 1s
spectra are shown in Fig. 16. The solid line trace is the C_{1s} from
the bulk adhesive and the dashed line trace is the C_{1s} from within
the oxide layer. The main peaks with a binding energy of approximately 285 eV reflect carbon-carbon and carbon-hydrogen bonding
while unresolved features contributing to the asymmetry on the high
binding energy side of the C_{1s} peaks reflect carbon-oxygen bonding (36). The broadening of the C_{1s} from the interphase region
where aluminum oxide is present indicates an increase in carbon-
oxygen bonding. Thus, the differences in the C_{KLL} spectra in Fig.
15 probably reflect real differences in the chemical bonding state
of carbon, although possible ion beam effects are unknown.

SUMMARY AND CONCLUSIONS

       Surface characterization techniques, such as AES and XPS, have
been used to characterize the interphase region of an intact
adhesively bonded structure using thin film adherends which were

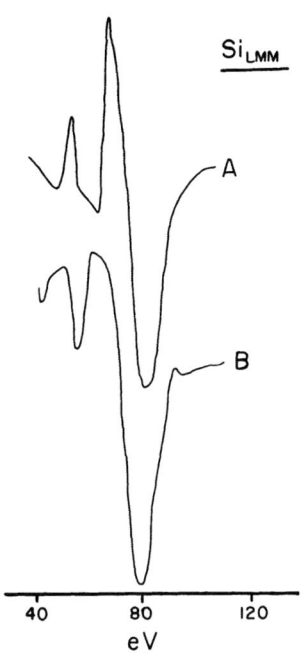

Fig. 14. Si$_{LMM}$ Auger electron
spectra from (A) a silane
primed surface and (B) interphase region of the silane
primed specimen profiled in
Fig. 12.

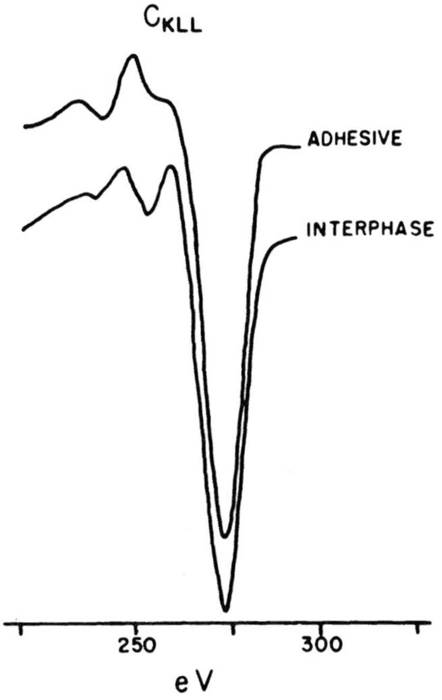

Fig. 15. C$_{KLL}$ Auger electron
spectra from the interphase
and bulk adhesive regions of
the phosphoric acid anodized
specimen profiled in Fig. 8.

thin enough to permit conventional depth profiling through the
interphase region. Attempts in the past to obtain chemical information from the interphase region of a ruptured or failed bonded
structure with thick adherends using AES and XPS have been limited
because of severe charging at the onset of the analysis when the
surfaces are either nonconducting or there is uncertainty as to
failure location with respect to the interphase. It has been
demonstrated that elemental distribution and chemical bonding
information obtained from an interphase can be used to evaluate the
effects of various bonding parameters and adherend surface conditions on the performance of an adhesively bonded structure. When
combined with mechanical test data, chemical and physical information from the interphase should lead to clearer understandings of
the nature of a particular failure mode in an adhesive bond.

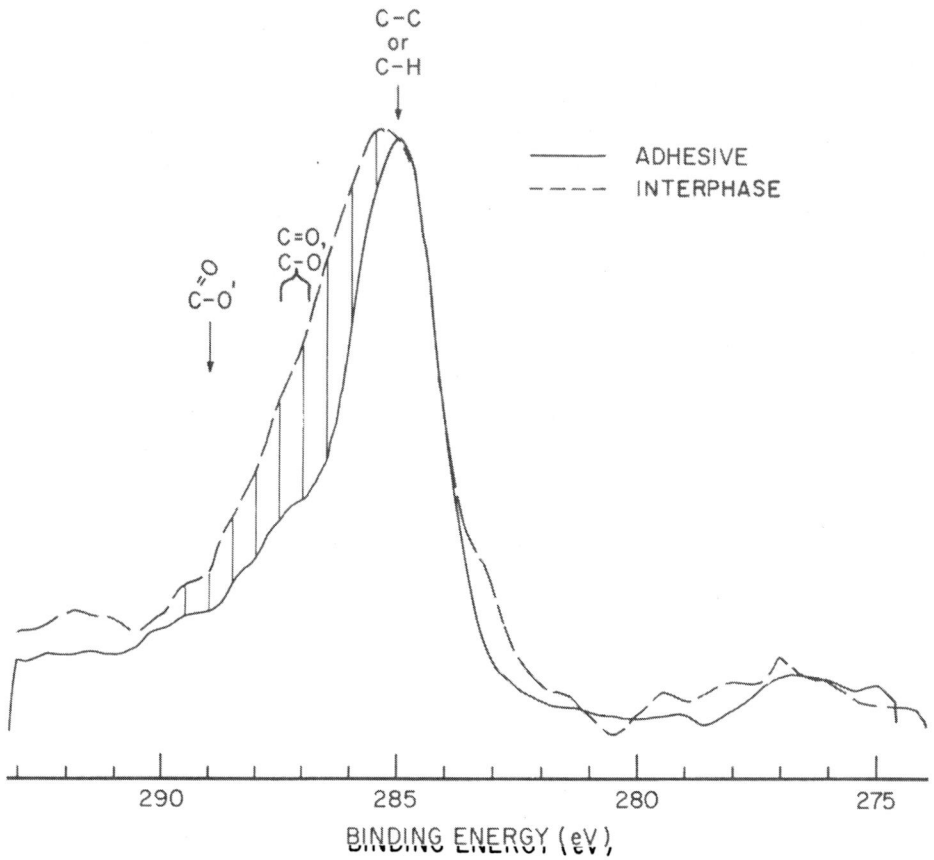

Fig. 16.   C$_{1s}$ photoelectron spectra from the interphase and bulk adhesive regions of the
phosphoric acid anodized specimen profiled
in Fig. 8.

ACKNOWLEDGEMENTS

      Research was sponsored in part by the Air Force Materials
Laboratory, Air Force Systems Command, United States Air Force,
Contract F33615-76-C-5185.

      The authors wish to thank G. Hammer and J. Peters for the XPS
data and W. Moddeman and J.T. Grant for helpful discussions and data
interpretation.

REFERENCES

1. D.W. Dwight, J. Colloid Interface Sci. 59, 447 (1977).
2. W.L. Baun, J. Adhesion 7, 261 (1976).
3. W.L. Baun, Experimental Methods to Determine Locus of Failure
    and Bond Failure Mechanism in Adhesive Joints and
    Coating-Substrate Combination, in "Characterization of
    Metal and Polymer Surfaces", Vol. 1, K.L. Mettal, ed.,
    Academic Press, New York (1977).
4. C.F. Garrett and E.F. Good, "Characterization of Bonding Sur-
    faces Using Surface Analytical Equipment", 4th Inter-
    national Symposium on Contamination Control, Washington,
    D.C., September 1978; proceedings to be published by
    Plenum Publishing Company, April 1979.
5. L.T. Drzal, "Summary of the Workshop Held on the Role of the
    Polymer Substrate Interphase in Structural Adhesion", Air
    Force Materials Laboratory Technical Report AFML-TR-77-
    129, 1977.
6. P.W. Palmberg, Journal of Vacuum Sci. and Technol. 9, 160
    (1972).
7. N.T. McDevitt, W.L. Baun, and J.S. Solomon, J. Electrochem.
    Soc. 123, 1058 (1976).
8. J.S. Solomon and W.L. Baun, "Surface Characterization of Con-
    tamination of Adhesive Bonding Materials", 4th Inter-
    national Symposium on Contamination Control, Washington,
    D.C., Spetember 1978; proceedings to be published by
    Plenum Publishing Company, April 1979.
9. W.L. Baun, N.T. McDevitt, and J.S. Solomon, Chemistry of Metal
    and Alloy Adherends by Secondary Ion Mass Spectroscopy,
    Ion Scattering Spectroscopy, and Auger Electron Spectro-
    scopy, in "Surface Analysis Techniques for Metallurgical
    Applications", ASTM STP 596, American Society for Testing
    and Materials, 86 (1976).
10. J.W. Coburn and C. Kay, Critical Review of Solid State Science
    4, 561 (1974).
11. H.J. Mathiew and D. Landolt, Depth Profile Analysis of Thin
    Oxide Films by Auger Electron Sepectroscopy, in "Proceed-
    ings of the 7th International Vacuum Congress and 3rd
    International Conference on Solid Surfaces" (Vienna
    1977), R. Debrozemsky, F. Rüdenauer, F. Viehböck, and A.
    Breth, ed., Dobrozemsky, Vienna (1977).
12. Y.H. Choo and O.F. Devereux, J. Electrochem. Soc. 123, 1868
    (1976).
13. R.E. Pawel, J.P. Penler, and C.A. Evans, J. Electrochem. Soc.
    119, 24 (1972).
14. W.L. Baun, "Surface Analysis of 2024 and 7075 Aluminum Alloys
    After Conditioning by Chemical Treatments", Air Force
    Materials Laboratory Technical Report AFML-TR-75-122,
    1975.

15. W.L. Baun and N.T. McDevitt, "Surface Characterization of Titanium and Titanium Alloys, Part II: Effect on Ti-6A1-4V Alloy of Laboratory Chemical Treatments", Air Force Materials Laboratory Technical Report AFML-TR-76-29, 1976.

16. H.S. Schwartz, SAMPE Journal 13, 2 (1977).

17. G.E. Thompson, R.C. Furneau, G.C. Wood, J.A. Richardson, and J.S. Goode, Nature 272, 433 (1978).

18. J.P. O'Sullivan and G.C. Wood, Proc. Roy. Soc. London, A317, 511 (1970).

19. J.U. Diggle, T.C. Downie, and C.W. Goulding, Chem. Rev. 69, 365 (1969).

20. C.J. Dell'Oca, Thin Solid Films 26, 371 (1975).

21. W.H. Gutmann, "Concise Guide to Structural Adhesives", Reinhold Publishing Company, New York, 37 (1961).

22. Minnesota Mining and Manufacturing ACIS Division, 3M Center, St. Paul, Minnesota 55101.

23. American Cyanamid Company, Bloomingdale Aerospace Products, Havre de Grace, Md. 21078.

24. J. Solomon and W.L. Baun, J. Vac. Sci. Technol. 12, 375 (1975).

25. T.N. Wittberg, J.R. Hoenigman, and W.E. Moddeman, J. Vac. Sci. Technol. 15, 348 (1978).

26. J. Kerchner and H.W. Etzkorn, Thin Film Analysis, from Sputter Profiles to Depth Profiles by Combined Auger/X-Ray Analysis, in "Proceedings of the 7th International Vacuum Congress and 3rd International Conference on Solid Surfaces" (Vienna 1977), R. Debrozemsky, F. Rüdenauer, F. Viehböck, and A. Breth, ed., Dobrozemsky, Vienna (1977).

27. G.K. Wehner, The Aspects of Sputtering in Surface Analysis Methods, in "Methods of Surface Analysis, A.U. Czanderna, ed., Elsevier Scientific Publishing Company, Amsterdam (1975).

28. G. Paolini, M. Masares, F. Sacchi, and M. Paganelli, J. Electrochem. Soc. 112, 32 (1965).

29. R. Honig and U. Harrington, Thin Solid Films 19, 43 (1973).

30. J. Solomon, Applied Spectroscopy 30, 46 (1976).

31. J.A. Marceau, SAMPE Quarterly 9, 1 (1978).

32. K.K. Rice, S.L. Lehmann, D.K. Klappratt, and C.L. Mahoney, "Exploratory Development of Corrosion Inhibiting Primers", Air Force Materials Laboratory Technical Report AFML-TR-77-121, July 1977.

33. A. Benninghoven, Surface Sci. 53, 593 (1975).

34. C.J. Lin and J.P. Bell, J. Appl. Polymer Sci. 16, 1721 (1972).

35. P. Braun, W. Faber, G. Betz, and F.P. Viehböck, Vacuum 27, 102 (1977).

36. K. Siegbahn, C. Nordling, G. Johansson, J. Hedeman, P.F. Heden, K. Hamrin, U. Gelius, T. Bergmark, L.O. Werme, R. Manne, and Y. Baer, "ESCA Applied to Free Molecules", North Holland, Amsterdam, 1969.

Techniques for Measuring Adhesive vs. Cohesive Failure

Tennyson Smith and Pierre Smith[*]

Rockwell International Science Center

Thousand Oaks, California   91360

ABSTRACT

Techniques have been developed for measuring the fractional area of an adhesive joint that has failed adhesively (near or at the substrate-adhesive interface) and cohesively (in the adhesive). These techniques involve β ray backscattering, light reflection and microscopy from an array of positions on the fracture surfaces of the failed joint.

The β ray count rate depends upon the elemental composition and is, therefore, different for the substrate and adhesive. The count rate correlates with the fractional area that failed adhesively and cohesively, for the localized spot that is monitored. A map of the count rate values correlates with microscopic photographs which also reveal adhesive vs. cohesive failure. The β ray and reflection techniques are conducive to automatic computerized mapping and averaging whereas the photographic technique is not.

INTRODUCTION

One of the common ways of characterizing an adhesive joint is to measure the force required to fracture a lap shear bonded couple. The lap shear bond strength is very useful for comparing adhesives and surface preparations. Perhaps of equal or more importance is the loci of fracture within the adhesive joint. Figure 1 outlines the various loci that are possible. In the case of metal adherends failure can occur in the adherend, in the metal oxide, in a weak

[*]Summer employee from Brigham Young University, Provo, Utah.

123

contamination boundary layer, in the primer, in the adhesive and at or near any of the boundaries between these phases. The fracture surface is a topographical map of the weakest regions within the joint. That is the fractograph is a map of those regions for which the local stress first surpassed the local failure stress. Discovery of the weakest regions can lead to corrective action in bond design and thus lead to stronger more durable joints.

The usual procedure is to report the lap shear strength and a qualitative statement that the failure was either cohesive or interfacial. This judgement is based on a visual observation that failure occurred at or near the adherend-adhesive interface (adhesive failure) if the surface has the appearance of the metal, or within the adhesive (cohesive failure) if the surface has the appearance of fractured adhesive. Although it is hoped that eventually quick and inexpensive techniques will be developed for characterizing the loci and mode of fracture in detail over the fracture surface, a simple means for evaluating the fraction of the surface that failed adhesively vs that which failed cohesively and mapping these regions would be a step forward.

This paper reports new techniques for measuring the fraction of interfacial failure, mapping these regions, and relates the fraction to lap shear bond strength for brittle fracture.

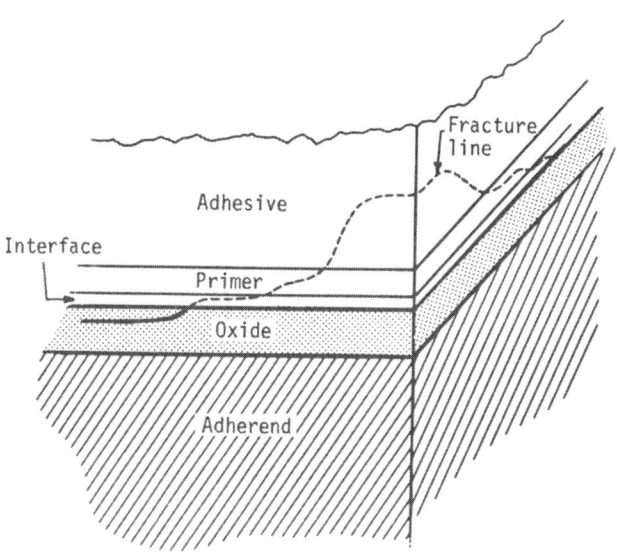

Fig. 1.   Schematic diagram of bond regions and
          fracture topography.

EXPERIMENTAL

Three techniques have been investigated for measuring quantitative values for the fraction or percent interfacial failure.
These three techniques are microscopic analysis of low magnification micrographs, β ray backscattering and light reflection.

The Microscopic Technique (1)

The mating surfaces of a fracture joint are placed under a low
power microscope and photographed to provide an enlargement of
about 15X by piecing together an array of small areas. An example
of one piece of an array is given in Fig. 2. The areas that appear
to be the metal substrate (adhesive failure) are easily discernible. The interfacial fractional area is determined by placing a
plastic grid over the photograph and counting small squares within
each interfacial region. To obtain the percent interfacial failure
the area of the squares counted is divided by the total area and
multiplied by 100. This technique is tedious and time consuming
(therefore expensive), but was done as a check on the other techniques which are more easily automated.

Fig. 2.   Photograph of microscopic view of lap shear
          bond fracture region. Reveals cohesive vs
          interfacial failure.

The β Backscattering Technique

A sample is irradiated with β rays (electrons) and the back-scattered current is measured with a Geiger-Mueller counter. The instrument used for this purpose has the trade name "Micro-Derm" (2). The backscattered current depends on the elemental composition of the sample. Thus, if the sample is a metal with very thin metal oxide and contamination layer, the backscattered current will be characteristic of the particular metal. For oxides and contamination of a few hundred Angstrom the effect of the oxygen and carbon will be negligible. If the sample is adhesive the current will be characteristic of the adhesive. For example, the metal mostly used in this study was Al 7075-T6 alloy which gave a β count of about 100 in 16 sec whether it was uncleaned, given the standard FPL etçh (sulfuric acid-dichoromate) or the phosphoric acid anodize (~4000Å hydroxide) treatment.

The adhesive was a modified epoxy (FM73 from American Cyanamid) which gave a count of about 35 in 16 sec. The percent interfacial failure was determined by measuring the β current at a large number of equally spaced positions over the fracture surface, summing the counts and dividing by the number of measurements (this number is referred to as c). The percent interfacial failure is determined by the following equation:

$$\% \text{ I.F.} = [(c-c_A)/(c_M-c_A)] \times 100 \tag{1}$$

where c_A is the β counts from the fracture surface adhesive and c_M is the β counts from the metal.

The Reflectance Technique

A sample is irradiated with light that has been focused to a point. The light is scanned over the fracture surface and the reflectance recorded for each position. The interfacial regions are reflective whereas the regions of cohesive fracture scatter the light strongly. We have used a HeNe laser beam for this purpose. An equation similar to Eq. 1 is used to calculate the % interfacial failure.

Computer Mapping (3)

The β backscattering and reflectance techniques are easily automated to produce computerized scanning maps of the fracture surface. This is done by a microcomputer operated X-Y positioning table that scans the sample beneath the sensor head and stores the count or reflectance value for each position. The computer is programmed to either calculate the % interfacial failure area or to produce a map of the fracture surface. The map is made by plotting a dot pattern with the density of dots proportional to the β count or reflectance signal value at a given position. The map is blank in the cohesive regions and dotted in the interfacial failure regions.

EXPERIMENTAL RESULTS

Calibration of β Backscattering with the Photographic Technique

Four samples of Al 7075-T6 alloy were phosphoric acid anodized (PAA) in 10% H_3PO_4 at 10V for 20 minutes and used as control samples to mate with samples that deviated from the proper process to deliberately produce weak bonds and increased interfacial failure. Table 1 gives the sample treatment, the deviation and the resultant % interface failure area as determined by the photographic and backscattering techniques. The last column in Table 1 gives the lap shear strength between the control sample and the damaged sample.

Samples 1B and 2B were both given the proper PAA treatment, sample 2A was exposed to the phosphoric acid for 20 minutes, but without applied voltage; sample 1F was PAA treated and then contaminated with stearic acid by placing a drop of pentane containing stearic acid and spinning the sample to obtain a uniform deposit. Samples 1Z, 2Z, 1D and 2D were given the standard FPL (sulfuric acid/dichromate) etch, then 1Z was contaminated with silicone vacuum grease and 1D was smudged with a clean cotton glove. Sample 1S and 2S were stainless steel samples that were anodized in sulfuric acid/dichromate solution.

Figure 3 is a plot of % interfacial area as measured by the Micro-Derm (β backscattering) vs. that determined by the photographic picture technique. A good correlation exists between the two techniques. The data would fall along the line in Fig. 3 if both techniques would have produced identical results. The fact that all of the β backscattering data yield greater % I.F. than the photographic technique is significant and expected. This is because it is difficult to estimate interfacial regions smaller than the grid size on the clear plastic that are undoubtedly present in the area that appears to be cohesive failure. The β backscattering and reflection techniques do not suffer from this limitation and thus pick up these areas as well as the larger ones that are visible.

Locus of Failure

Having devised a means of identifying the locus of cohesive vs. adhesive failure in a plane parallel to the adherend surface, by

Table 1. Correlation Between Interfacial Failure, as Measured by the Photographic and β Backscattering Technique, and Lapshear Bond Strength

Sample No.	Material	Treatment	Deviation	% Interfacial Failure Area Photographic	% Interfacial Failure Area Backscattering	Lapshear Bond Strength (Ksi)
1B	Al 7075-T6	PAA*	None	30	32.7	4.3
2B	"	PAA	None	10	13.4 / 46.1	
1A	"	PAA	None	15	17.4	3.5
2A	"	PAA	O.V.+	33	36.8 / 54.2	
2F	"	PAA	None	11	15.2	6.1
1F	"	PAA	stearic acid contamination	21	27.0 / 42.2	
2Z	"	FPL** etch	None	2	5.3	1.75
1Z	"	FPL etch	silicone grease contamination	80	90.4 / 95.7	
2D	"	FPL etch	None	6	14.0	4.7
1D	"	FPL etch	cotton glove smudge	49	54.0 / 68.0	
1S	SS	Anodize in H_2SO_4/Cr_2O_7	None	26	27.0	

* PAA = phosphoric acid anodize in 10% acid
** FPL etch = sulfuric acid/dichromate etch
+ Zero applied voltage

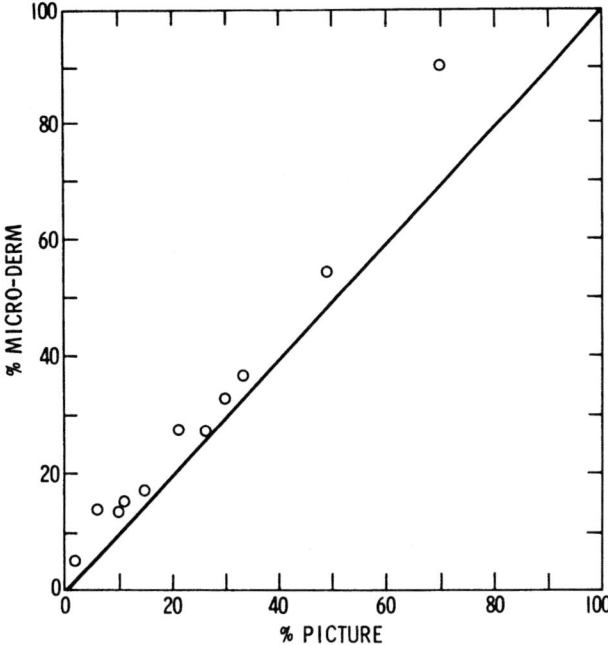

Fig. 3.   Measurement of Interfacial Failure
(picture vs. micro-derm).

micrographs or computer maps, it is of interest to examine the loci
of failure perpendicular to the surface of any particular map
position. From the low magnification photograph in Fig. 2 and the
SEM micrographs in Fig. 4, three loci are identified. These loci
are fracture in the polymer matrix, fracture at the matrix-fiber
interface and fracture at the metal-matrix interface.

The % interfacial failure was based on fracture at the metal-
matrix interface and although these areas appear, from photographic
and SEM pictures, to be metallic with roll groove pattern and grain
boundaries, these pictures cannot reveal the presence or absence of
very thin oxide, contamination or adhesive. Thus Auger electron
spectroscopy (Fig. 5) was performed at the spot indicated in Fig. 4.

The AES spectrograms (Fig. 5) were taken prior to Ar$^+$ sputtering and after Ar$^+$ sputtering for 2 minutes. Prior to Ar$^+$ sputtering
only a large carbon, a small oxygen and perhaps sulfur peaks are
observed. This indicates that the fracture surface is not metal or
metal oxide, but either adhesive or stearic acid contamination. The
short sputter time required to reveal the aluminum oxide is in favor
of stearic acid contamination as the locus of failure.

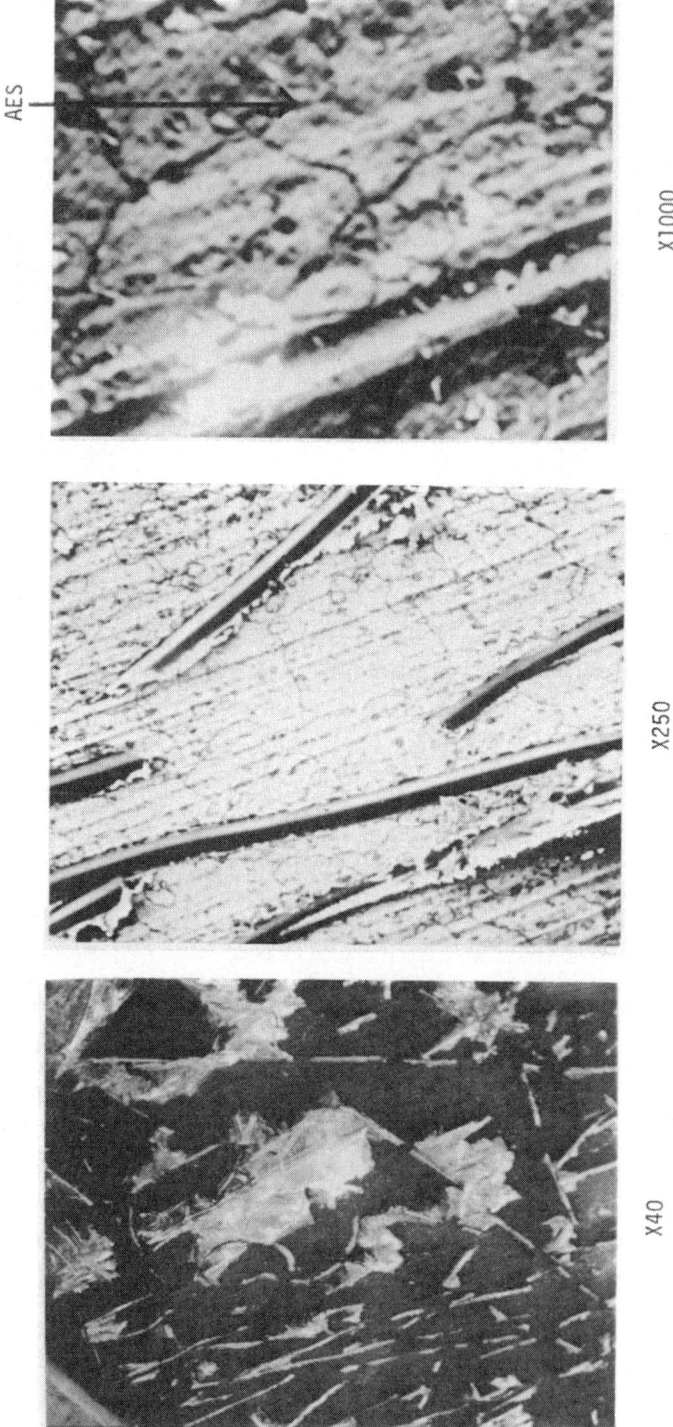

Fig. 4. SEM micrographs of interfacial failure area of 1F (Table 1).

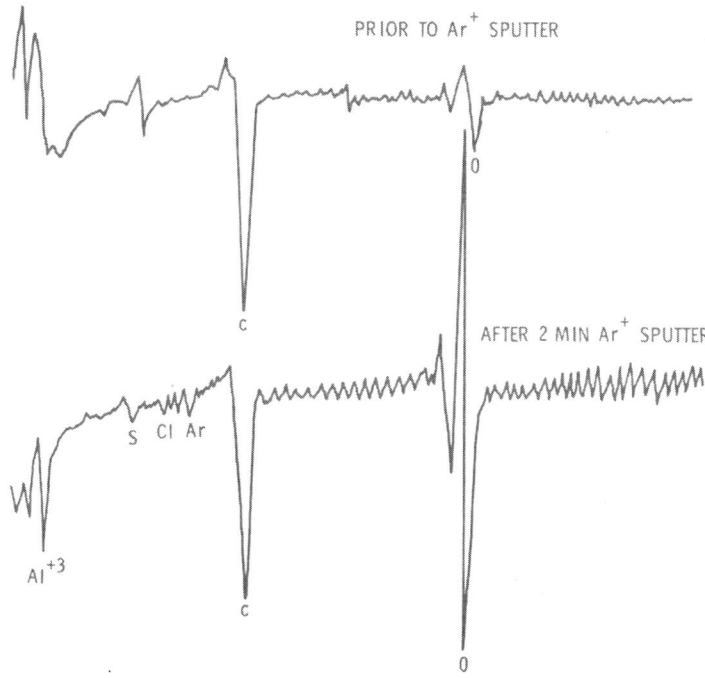

Fig. 5. Auger electron spectroscopy of interfacial
          failure regions indicated in Fig. 4.

Computer Mapping

    The equipment has not as yet been modified to map with respect
to β backscattering, but Fig. 6 is a reflection map of a fracture
surface. The dotted regions faithfully correspond to the visual
observed interfacial failure regions.

Relationship Between Bond Strength and Interfacial Failure

    One might expect that an increase in interfacial failure would
decrease the lap shear strength and this is observed in Table 1.
However, there is no obvious relationship between lap shear
strength and individual molecular bonds between adhesive and adherend because failure is initiated at a point (flaw) at which the
local stress first exceeds the local strength and then proceeds
sequentially from point to point as this criteria continues to be
met. Of course, once a crack is initiated this sequence of points
usually is at the crack tip, thus causing it to propagate along a
surface of least resistance.

1.21

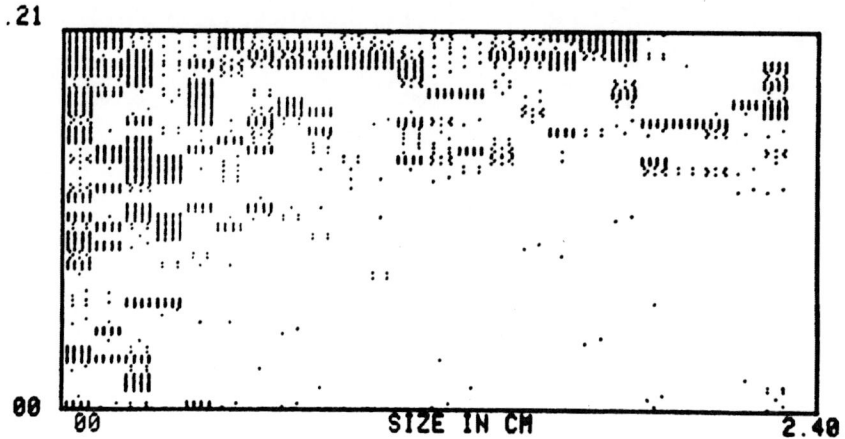

00

00   SIZE IN CM   2.40

Fig. 6.   Computer map of reflectivity of a fracture
          surface.   The dotted regions failed inter-
          facially.

        The question we would like to answer is: what relationship is
there between the composite materials, the configuration of the
joint and the stress to initiate failure?   However, we limit our
discussion to the question of why there is an approximately linear
relation between lap shear strength and the fraction of the bond
area that failed at the interface?

        In the schematic Fig. 7a, the thickness dimension of the
adhesive layer has been exaggerated.   To indicate a possible stress
distribution profile, curves have been drawn within the grid for
which the horizontal axis remains the position axis but for which
the vertical axis is used as a stress axis.   A section of the grid
near a region of probable fracture initiation is expanded in Figs.
7b and c.   In Fig. 7b a particular column has been selected to
identify possible materials (oxide, contamination, gas bubble,
primer and adhesive) that might occupy the net of elements between
adherends.

        The average shear stress σ (measured by the Instron tester)
will be the sum of the stresses σ_j associated with each column j
modified by an interaction parameter α due to interaction between
columns,

$$\sigma = \sum_j \alpha\,\sigma_j. \qquad\qquad\qquad\qquad (1)$$

For a given strain, σ_j will depend upon the compliance of each
column and will therefore depend upon the materials within the
column.   It should be noted, that although Eq. 1 relates the
measured stress and the stress distribution, it tells nothing about

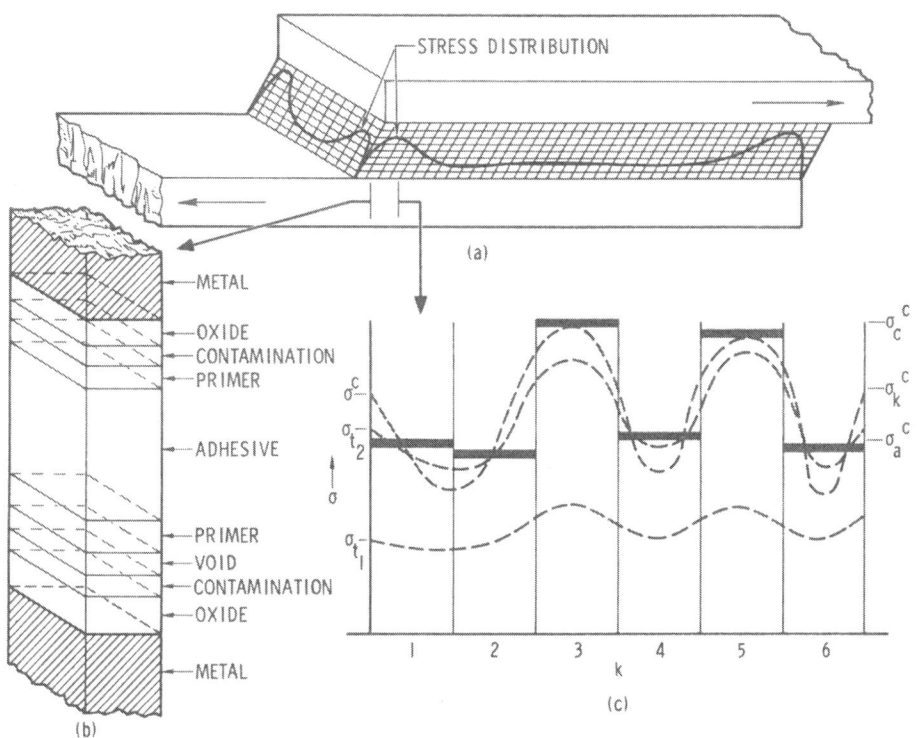

Fig. 7. Schematic representation of (a) lap shear
joint, (b) one element, and (c) stress profile
in the critical region.

the value of σ at which failure will occur (σ_F).

To estimate σ_F, we invoke the model in Fig. 7c. Consider a
collection of connected columns (k in number), each column having
critical stress values σ_i, associated with the ith material within
the column. For simplicity we consider only two regions, that near
the adherend interface and that in the adhesive. In Fig. 7c the
horizontal bars indicate the critical stress values near the interface have a spacial distribution related to the surface contamination in column k = 1, 2, 4 and 6. If the amount of contamination in
each column is about the same, the values of σ_j will be close, with
average value designated σ_a^c for adhesive failure. The bars at
higher levels are for failure in the adhesive with average cohesive
value σ_c^c.

Figure 7c is for the collection of columns in which the imposed
topographical stress distribution surface first contacts the topo-

graphical yield stress surface. Curves σ_{t_1}, σ_{t_2} and $\overset{c}{\sigma}$. indicate the hypothetical stress distribution long before, just prior to and at failure.   The compliance values of each column influence the distribution curve σ_t. As σ increases such that the distribution curve σ_{t_1} reaches the critical stress values for interfacial fail- ure, some adhesive failure occurs which changes the compliance and thus modifies the stress distribution curve to show some relaxation in the failed region.   As the failed regions grow the stress distribution curve is pinned in these regions, but continues to increase in the other regions. When the stress distribution curve contacts the upper cohesive levels, they fail and a flaw of critical size is created. The crack propagates causing failure in the whole joint.

It follows from this model that the average critical stress in the set of k columns will depend on the fraction of the area, of the k columns, that fail at the interface (θ_a) and the fraction that fails cohesively (θ_c), i.e.

$$\sigma_k^c = \sum_{j=1}^{k} \alpha \; \sigma_j^c \approx \alpha k \; [\theta_a \sigma_a^c + \theta_c \sigma_c^c] \qquad (2)$$

Since the average stress, σ_k, in the critical flaw region, should increase in proportion to the average stress for the total joint, we make the approximation

$$\sigma \approx \beta \; \sigma_k \qquad (3)$$

where β is a proportionality constant. It follows that

$$\sigma_F \approx \beta \sigma_k^c \equiv \alpha \; \beta k \; [\theta_a \sigma_a^c + \theta_c \sigma_c^c] \qquad (4)$$

and

$$\sigma_F \approx \theta_a (\sigma_a - \sigma_c) + \sigma_c \qquad (5)$$

where $\sigma_a \equiv \alpha \; \beta k \sigma_a^c \qquad (6)$

$$\sigma_c \equiv \alpha \; \beta k \sigma_c^c \qquad (7)$$

The results in Table 2 can be used to test Eq. 5. A set of PAA, Al7075-T6 samples were contaminated with silicone grease to varying degrees. These samples were then bonded to uncontaminated PAA samples, fractured and both fracture surfaces mapped with β backscattering for % interfacial failure (% I.F.). The % I.F. for

Table 2. Correlation Between Lapshear Strength and Percent Interfacial Failure
(PAA-Al 7075-T6-FM73)

% Interfacial Failure β Backscattering

Sample	Control	Contaminated with Silicone grease	Total	Lap Shear Strength (Ksi)
1	15	27	42	6.1
2	21	22	43	5.7
3	22	45	67	4.9
4	14	69	83	3.2
5	10	87	97	2.6

the control samples (curve 1), the contaminated samples (curve 2) and the sum of these (curve 3) are plotted vs. the bond strength in Fig. 8.

The data from the contaminated surface and the data for the total % I.F. fall near a visual straight fit in accordance with Eq. 5.   As failure at the weak interface increases, failure at the control interface naturally decreases (see lower curve, Fig. 8).

A plot of σ_F vs. % I.F. for which σ_c is constant (common adhesive) but σ_a varies (different interfacial failure mechanisms), should result in lines converging to a common value of σ_F at zero % I.F. but intercept at different values for 100% I.F. This is shown in Fig. 9 from the data in Table 1.

The intercepts at % I.F. = 100 and 0 give the extrapolated values of the fracture strength of the interface and the cohesive strength of the adhesive respectively.   From Fig. 8 the cohesive strength of the FM73 is about 8.6 Ksi for the 5 mil bond line thickness.   The fracture strength of the silicone grease contami- nated surface is about 2.3 Ksi. Since the cohesive strength should remain constant for a given adhesive and bond line thickness,

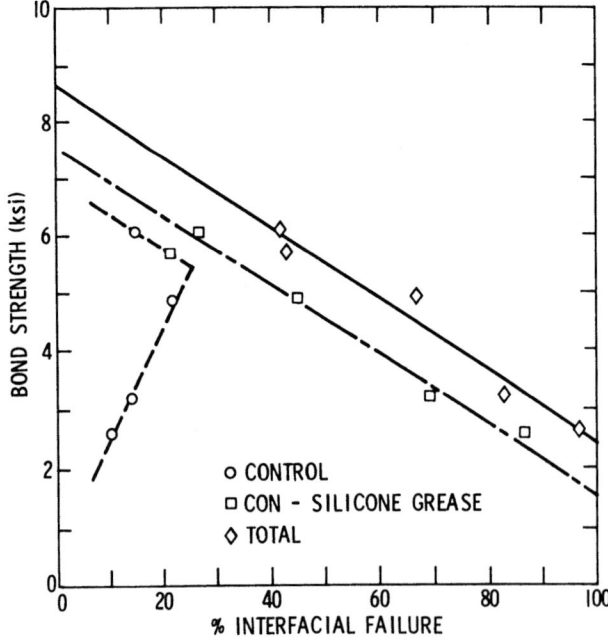

Fig. 8.   Plot of bond strength vs. % interfacial
         failure.

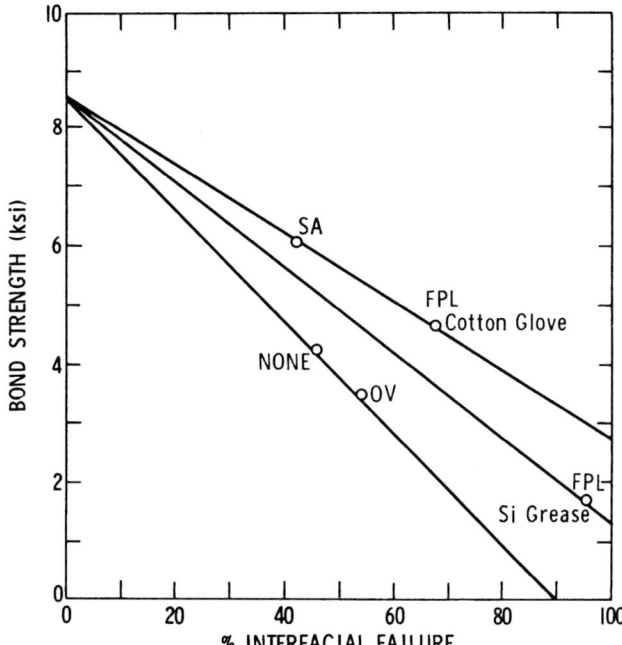

Fig. 9.   Plot of bond strength vs. % interfacial
          failure.

fracture strengths for various types of surface damage can be
obtained from the intercept at % I.F. = 100 as in Fig. 9. For the
lower curve Eq. 5 is not valid, as % I.F. approaches 100 because σ_F
cannot be negative.

Another example of the utility of Eq. 5 is given in Fig. 10,
from a previous report (1).   Aluminum samples were aged in humid
atmosphere with varying degrees of condensation. The thickness of
the oxide was measured by ellipsometry and ranged between about 200Å
and 600Å as the condensation increased. These samples were bonded
with a different adhesive from American Cyanamid (HT 424). This
adhesive is also a modified epoxy but contains a glass scrim rather
than a nylon scrim (in FM73). The larger diameter glass produces
bond line thickness (BLT) values of about 10 mil as compared to ~5
mil for FM73. Except for two points the data fall along a straight
line as expressed by Eq. 5. For this adhesive and BLT, σ_c ~ 3.8
Ksi and σ_a ~ 2.5 Ksi. The value of σ_c has been shown to be very
sensitive to BLT and depends on the presence of the scrim. It was
shown (4) that the shear strength of the matrix polymer in HT424 is
about 8.9 Ksi in good agreement with the value extrapolated from
Fig. 8 (i.e. 8.6 Ksi).

Curve 2 in Fig. 10 is a plot of oxide film thickness vs. %
I.F., indicating a direct correlation between oxide thickness, %

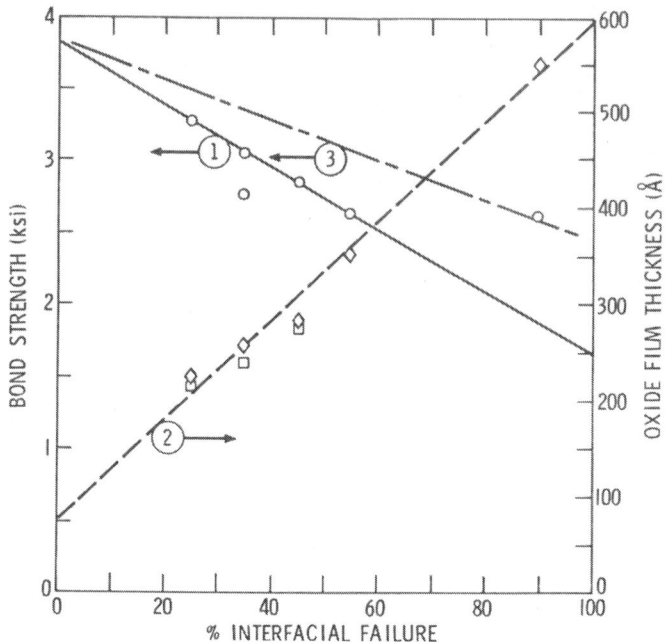

Fig. 10.   Bond strength and oxide thickness vs.
           % I.F.

I.F. and thus bond strength. It was demonstrated in this case that
interfacial failure was in the oxide film.   The probability of
failure increases in direct proportion to the number of flaws in the
oxide and this increases with film thickness.   Although the film
thickness at about 90% I.F. indicates that the bond strength should
fall on the theoretical curve 1 in Fig. 10, the index of refraction
of this film proved to be about 2.5 vs. about 2.0 for the other
films.   The larger index corresponds to a more dense oxide (fewer
voids or flaws) and the observed larger value of σ_a is therefore
predicted, i.e., curve 3 is appropriate for this oxide.

SUMMARY AND CONCLUSIONS

     New techniques are described for measuring the percent inter-
facial (adhesive failure vs. cohesive) failure.   The photographic
technique is tedious and time consuming, but is used to verify the
utility of the β backscattering and light reflection techniques
which are easily automated. These techniques can be used to produce
maps of interfacial failure regions, in addition to % interfacial
failure.

Plots of lap shear bond strength vs. percent interfacial failure confirm a theoretical analysis relating these parameters. An important consequence of this research is that one can extrapolate the data to obtain the cohesive strength of the adhesive and the fracture strength at the damaged surfaces. This procedure will allow the separation of adhesive intrinsic properties from surface preparation properties and thus lead to a better understanding of the strength of adhesive joints.

REFERENCES

1. T. Smith, "Mechanisms of Adhesive Failures Between Polymers and Metallic Substrates", 7th National SAMPE Technical Conference, Vol. 7, pp. 349 (1975).
2. United Process Assemblies, Inc., 60 Oak Drive, Syosset, New York 11791.
3. T. Smith, "Nondestructive Inspection of Phosphoric Acid Anodized Aluminum Panels for Contamination", AFML-TR-77-42.
4. T. Smith, "Mechanisms of Adhesive Failure Between Polymers and Metallic Substrates", AFML-TR-74-73, Part II.

Failure Characterization of a Structural Adhesive

David W. Dwight[*], Erol Sancaktar[+], and Halbert F. Brinson

College of Engineering

Virginia Polytechnic Institute and State University

Blacksburg, Virginia   24061

ABSTRACT

Strength vs. rate-of-strain data show accurate, viscoelastic time-dependence for a bulk epoxy adhesive (Metlbond 1113) in tension and shear, but not in the adhesively lap-bonded state. A detailed analysis of the lap shear joints, combining results from neutron radiography, scanning electron microscopy and roughness measurements, indicated the dominance of internal voids and stress concentrations not present in the bulk state. A qualitative fracture mechanics approach was used to reconcile the results. A general methodology to elucidate practical adhesive bonding problems is suggested that combines elastic, plastic and surface characterization.

INTRODUCTION

Bulk Properties of Adhesives

Adhesive technology is growing rapidly; composite and light-weight materials technology for aerospace and automotive applica-

---

[*] To whom correspondence should be addressed.
[+] Current address: Mechanical and Industrial Engineering Department, Clarkson College, Potsdam, New York 13676.

tions are leading the development. Of critical importance to the
designer are strength and durability under anticipated service
conditions.   Experience with organic, high polymer adhesives is
limited: for predicting the behavior in bulk or bonded state, static
tensile, dynamic fatigue, and fracture mechanics tests, among
others, are variously recommended.

We have been developing a unified approach that combines
microscopic information from modern surface analytical techniques
(1,2) with traditional viscoelastic studies of bulk tensile and
shear specimens. Using a structural epoxy adhesive (Metlbond 1113)
as a practical model, the fractographic results clarify scatter in
the lap shear strength data, identifying critical, uncontrolled
variables due to processing effects upon structure and properties.

Bulk stress-strain, strain rate, creep, relaxation, yield
and/or failure properties (including delayed failures) of a structural adhesive (Narmco Whittaker's Metlbond 1113 and 1113-2
adhesives) were measured first (3). These characterization studies
were performed on bulk materials using dog bone tensile coupons.
The data were correlated with a modified Bingham model and methods
described by Crochet (4), Naghdi and Murch (5), and Ludwik (6), to
produce linear viscoelastic models of the bulk adhesives.

It would be useful to determine a relationship between tensile
property data and the design data needed to predict the behavior of
lap-bonded joints. Thus, the present study was initiated to attempt
to find whether bulk tensile properties could be correlated in some
reasonable way to the shear behavior as obtained from bulk shear
coupons and lap shear adhesive joints using Metlbond 1113.   Of
further interest was the viscoelastic response of lap shear joints
of Metlbond 1113 tested with yield data plotted versus elastic
strain rate. Extensive fractographic analysis, combined with qualitative fracture mechanics arguments were employed to help explain
the results.

Adherend Surface Preparation

Since joint strength is a function of the adherend surface
preparation, chemical treatments of adherends were considered. Lin
and Bell (7) reported that the shear strength of thin-walled tubular
butt joints reached 9000 psi (62 mpa) after the joints were
machined, polished, vapor degreased, and treated with dilute sulfuric acid-potassium dichromate solution. Without chemical treatment the bond strengths decreased to 80% of that for chemically
treated joints. For those joints with aluminum surface degreased
only, the strength was 65% of the maximum bond strength. Good (8)
reports that surface treatment of materials (e.g. etching and
particle bombardment) will help to obtain a system for which the

interfacial region is not the weakest.   Dwight (1) reported
molecular-level details of adhesive bond strength to solid fluoropolymers and improvements when the surface was etched with sodium
complex solutions or exposed to an electrical discharge.   Joints of
high strength are obtained when the controlling factors in chemical
bonding can be optimized.

     In regard to surface roughness, Jennings (9) reports:
1) Roughening of the adherend surface can increase butt tensile or
shear strength of a joint with brittle adhesives.   2) Butt tensile
strength can be a function of the metallurgical state of the
adherend.   3) It is possible to have a joint stronger than the bulk
strength of the adhesive polymer.   4) A change in stress distribution at the adherend-adhesive interface will cause the joint
strength to vary with adherend treatment.   The effect of surface
roughness depends on the type of adhesive mechanism.   Mechanical
(interlocking) adhesion will be stronger when appropriate solidification is achieved between the pores of the adherend.

A system for which the interfacial region is not the weakest
requires good wetting of surfaces to eliminate interfacial flaws,
and strong interfacial forces should be created by covalent bonding
or chemisorption.   Also, the failure mode is strongly dependent on
the adherend surface treatment.   In one study, all joints without
chemical treatment failed adhesively at the resin-metal interface.
For those specimens with surfaces treated, both cohesive and adhesive types of failures were observed (7).

Locus of Failure and Mechanics Analysis

     Scanning Electron Microscopy (SEM) is presently the best
method of qualitative analysis of micromechanisms in polymer fracture.   Recent studies illustrated the effects of weak links such as
voids, corrosion, and interfacial separation (1,2).   Often fracture
patterns reflect stress distributions, as well as deformation and
fracture micromechanisms operating during crack propagation.   For
example, a fracture mechanics model for "rubber toughening" effects
in epoxy polymers has been developed from SEM photomicrographs
(10).

     To explicitly and quantitatively account for the separate
contributions of interfacial properties and bulk rheology, Andrews
and Kinloch (11) proposed that the adhesive failure energy,

$$\theta = \theta_0 \cdot f(R) \qquad\qquad\qquad (1)$$

where θ_0 depends only on the physical and chemical nature of the
fracture surface, and f is a function of the "reduced" rate of
failure propagation obtained from rate and temperature data using

the WLF equation. θ_0 is the work of bond fracture: the sum of the interfacial, adhesive and adherend failure energies. Experimental data required to test this theory are quantitative surface analysis to determine the fractions of interfacial and bulk failure, the intrinsic failure energies, and the overall joint failure energy measured as a function of temperature and rate of crack propagation.

Andrews and Kinloch prepared joints with different surface-energy substrates bonded to a single SBR rubber, cross-linked in situ with one initial crack located at the interface at the edge of the test specimen. Their theoretical development was corroborated by the results in these idealized specimens. However, in recent work they have employed qualitative surface analysis to elucidate the case of mild steel bonded with an epoxy adhesive (12). The nature and topography of fracture surfaces were examined visually and by scanning electron microscopy as well as electron probe microanalysis, using 5% titanium dioxide tracer in the adhesive. The initial locus-of-failure was cohesive through the adhesive, but after water immersion, a complex locus-of-failure was found: the path of fracture occurred between the oxide surface layer and the epoxy adhesive, alternating into the adhesive layer.

In an approach to relate failure loads to shear strain energy, Hart-Smith (3) developed an expression for adhesive joints with idealized elastic-plastic adhesives and elastic adherends, showing that bonded joint strengths could be characterized by the adhesive strain energy (shear strength per unit area of bond). For double lap joints with uniform shear distributions:

$$P = [8E \, d\tau(\tfrac{1}{2}\gamma_e + \gamma_p)2t]^{\frac{1}{2}} \qquad\qquad (2)$$

where

E = Adherend Young's Modulus

$\tau(\tfrac{1}{2}\gamma_e + \gamma_p)$ = Area under the stress-strain curve per unit volume of adhesive

$d\tau(\tfrac{1}{2}\gamma_e + \gamma_p)$ = shear strain energy per unit area of bond

d = Bondline thickness

t = Adherend thickness

γ_e, γ_p = Elastic and plastic adhesive shear strain

τ = Adhesive shear stress

P = Adherend load per unit width.

Generally, similar expressions arise from fracture mechanics analysis, using Griffith-Irwin theory; for a solid of elastic modulus E containing a crack of length L, the stress σ_f at which fracture will occur is given by

$$\sigma_f = K(EG/L)^{\frac{1}{2}} \qquad (3)$$

where $K \simeq 1$ is a constant and $G(= 2\gamma_s + \psi)$ is the total (irreversible) work dissipated per unit crack extension as new surface free energy (γ_s) and/or deformation and other fracture process (ψ).

The following sections describe several experiments with a practical epoxy/aluminum system and evaluate semi-quantitatively the pertinent interfacial and bulk mechanisms.

EXPERIMENTAL

Materials and Preparations

Surface preparation of the aluminum adherend was done according to the following specifications (14):

1)   Solvent wipe with methyl ethyl ketone.

2)   Vapor degrease in trichloroethylene by suspending in the vapor until condensation on the surface no longer occurs.

3)   Immerse for 20 minutes at $150^{\circ}F$ ($66^{\circ}C$) in a solution of 68% distilled or de-ionized water, 23% sulfuric acid (specific gravity 1.84), 9% sodium dichromate.

4)   Rinse thoroughly.

5)   Air dry so as to prevent retention of water. Thorough draining is required.

6)   Apply the primer at ambient temperature, agitating continuously during brushing, and air dry to a thickness of 0.00007 in. (1.78×10^{-3} mm) to 0.000025 in. (6.35×10^{-3} mm).

Single lap shear specimens were prepared with an overlap length of ~0.30 in. (7.6mm), adhesive thickness of ~0.006 in. (0.15mm). Aluminum adherends were 0.125 in. (3.17mm) thick for the ASTM (D 1002-72) recommended single lap geometry. Symmetrical single lap specimens were also prepared, for they were recommended by several authors (5,6) as a superior geometry which has lower stress concentrations and more uniform shear stresses across the glue line. These specimens differed from the single lap specimens in that the adherends were machined to half thickness across the overlap area to produce uniform specimen thickness and vertical load application line. Aluminum adherends for all specimens were 1.00 in. (2.54cm) wide and were bonded at 260°F (127°C) for 30 minutes under 50 psi (345 kpa) pressure. Metlbond is a structural epoxy adhesive commercially available from Narmco Materials Inc. (600 Victoria Street, Costa Mesa, California 92627) as a 100% solid film with (Metlbond 1113) or without (Metlbond 1113-2) a synthetic carrier cloth. (Machining, preparation and bonding of the adhesive lap joints were performed by the NASA-Langley Research Center.)

Neutron Radiography

Neutron radiographic examination of the symmetric lap specimens was done at the VPI & SU Research Reactor. Using a direct exposure method, x-ray film was placed in physical contact with a converter screen in a light-tight container placed behind the specimen. The converter screen was painted with a thin layer containing gadolinium, which has a high mass absorption coefficient for neutrons. (Other work (17) incorporated 5% Gd_2O_3 directly into Narmco 6800 adhesive, enabling detection of voids as fine as 0.02 in. (0.05cm).) Neutrons passing through the sample are absorbed by the converter screen, which emits radiation and exposes the film. Neutron flux is governed by thickness, material density, or absorption cross section, which can be determined from the image density on the exposed film. Differences make it possible to detect voids, flaws or other inhomogeneities in the glueline. The neutron flux emitted from the source was maintained between $10^5 n/cm^2$ and $10^6 n/cm^2$.

Mechanical Behavior Testing

A symmetric rail shear test (Fig. 1) was used for bulk shear studies. Specimens for this test were prepared with a ratio of 10:1 for the distance between tabs* (d) to length (1) in accordance with the analysis of Whitney, et al. (18). These specimens were sensitive to tab application: Metlbond 1113 specimens had tabs attached with Epoxy 907 adhesive and yielded at lower loads due to premature tab/specimen failure, relative to the Metlbond 1113-2 specimens that were glued with Eastman 910 adhesive. Loads were applied in opposite directions on the center vs. the outside tabs and most of the latter specimens produced 45° shear failures (Fig. 1).

---

*Tabs are ~0.75 in. x 5.2 in. (1.9 x 13.2cm) 0.074 in. (1.6mm) thick aluminum pieces glued onto symmetric rail shear test specimens to insure deformations in excess of elastic limits.

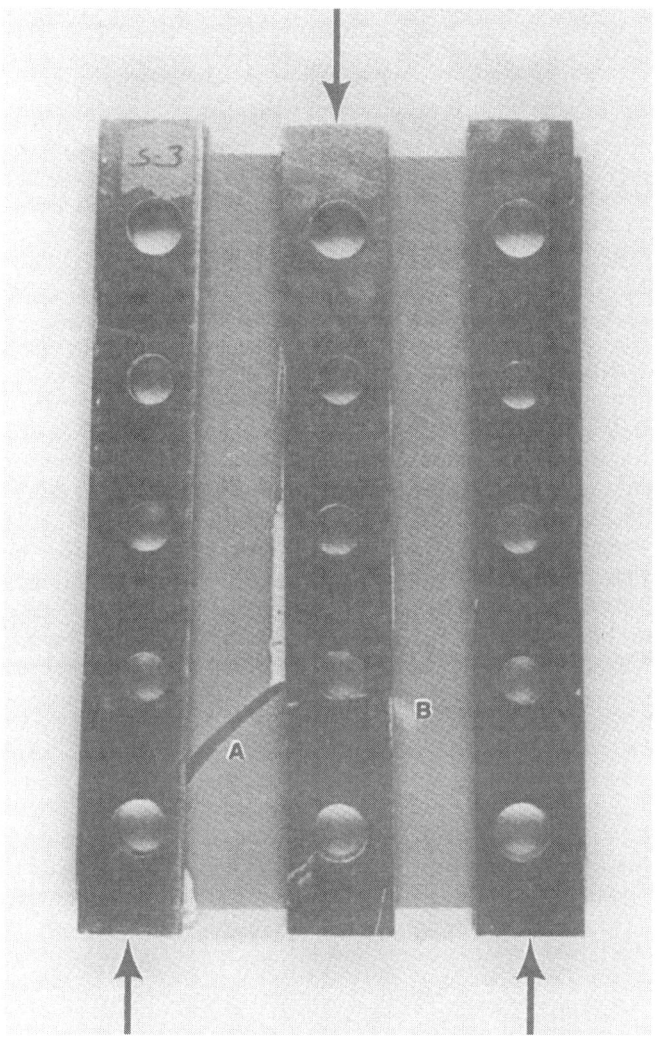

Fig. 1. Bulk shear test specimen. Load was applied
         to the metal tabs as indicated by the arrows.
         Typical 45° shear failure is shown at point
         (A) and stress whitening at (B).

     An Instron universal testing machine was used with constant
strain rates ranging from 0.002 in./min. (5.08 x 10^{-2}mm/min.) to 2.0
in./min. (5.08cm/min.). Adjustable grips were used for single lap
specimens to avoid excessive bending of the specimen). Tests were
performed at 72°F (22°C) and 50% relative humidity.

Surface Analysis

Scanning electron photomicrographs were obtained with an Advanced Metals Research Corporation Model 900 scanning electron microscope. Operating at 20kV, high magnification views (500X) gave the details of surface features, while survey scans at 50X provided a check on the distribution of representative features that characterize the surface. Specimens were cut to approximately 0.25 x 0.25 in. (6.4 x 6.4mm) with a band saw, and fastened to the SEM stubs with conductive, adhesive-coated copper tape. To enhance conductivity, a thin (~200Å) film of Au/Pd alloy was vacuum evaporated onto the samples. Photomicrographs were taken with the sample inclined to accentuate the appearance of small asperities.

In an attempt to quantify the lap joint deformation and fracture processes, the roughness of 0.24 in. (6.25mm) of representative fracture surfaces was measured with a Talysurf 4 with 20μm sensitivity, scanning at 0.72 in./min. (18.3mm/min.) stylus travel speed and vertical magnification of 243X. The analog output was conditioned through a Tektronix data processor that scanned the output at 20 msec. intervals, producing 1.024 discrete data points. The best fit to these points produced curves for comparison with calibrated standards, thus enabling a quantitative measure of surface roughness.

Talysurf profiles on representative sections of 21 fractured single lap specimens, combined with photographs of the fracture surfaces, support the qualitative findings from the SEM photomicrographs. Further interpretation of the Talysurf profiles was attempted by measuring the actual length of the fracture surface. To do this, horizontal sections that are either inherent voids or interfacial failure sites were subtracted first from the total profile length. Inclination angles of all remaining segments of the profile were determined and components were adjusted to the same scale. The adjusted profile lengths thus obtained are assumed to represent the fracture surfaces, comparatively. The amount of deformation and fracture surface created is assumed proportional to the fracture energy. Hence, the relative profile gives an idea about the contribution of deformation and fracture to the ultimate load.

RESULTS AND DISCUSSION

Strength in Bulk

Tensile and shear measurements on the bulk material (3) showed rate dependence, and revealed that elastic stresses and strains increased with an increase in strain rate. An empirical equation for shear was derived from bulk tensile results using $\tau = \sigma/2$ as maximum shear stress and Hooke's Law. Figure 2 shows an excellent fit using tensile constants and experimental data for yield stresses and strains of the bulk shear tests, in a Ludwik type equation,

$$Y_s = Y_s' + \tau' \log \left(\frac{\overset{o}{\gamma}}{\overset{o}{\gamma'}} \right) \tag{4}$$

where the yield stress, $\overset{o}{\gamma}$, is the elastic shear strain rate and $\overset{o}{\gamma'}$, Y_s, and τ' are experimentally determined material constants. Many more details of the mechanical property testing can be found elsewhere (3,19).

Uniformity of the shear data depended upon the load transfer mechanisms. Metlbond 1113 specimens without tabs had yield stresses comparable to the bulk tensile predictions using Tresca failure criteria. Metlbond 1113 and 1113-2 specimens with tabs, however, had higher values than calculated from tensile yield stresses. The yield strains were higher than values calculated from tensile yield strains (using $\gamma = (1 + \nu)\epsilon$). These results show that it is possible to obtain shear stresses and strains higher than

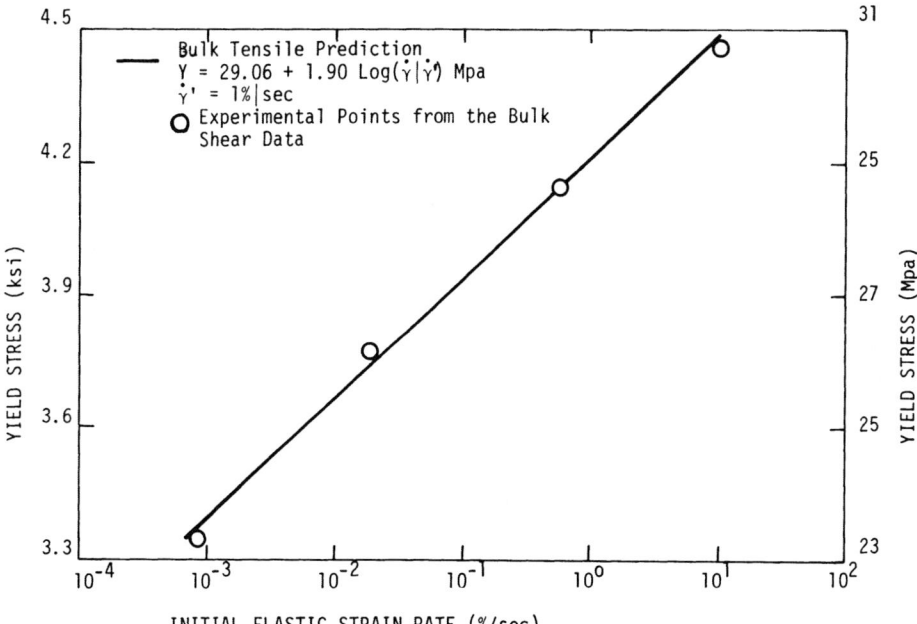

Fig. 2.   Rate dependence of strength, showing corre-
         lation between shear data and a Ludwik-type
         equation with material constants derived
         from tensile tests.

predicted by tensile data. Perhaps the same applies to the bonded state.

Comparison of the yield stresses in the bulk tensile and shear forms reveals an interesting phenomena. In the bulk tensile form, Metlbond 1113 specimens had higher yield stresses compared to the Metlbond 1113-2 specimens. This was expected, since Metlbond 1113 had a carrier cloth increasing its stiffness and strength. In the bulk shear form, however, Metlbond 1113-2 produced higher yield stresses than Metlbond 1113. This appeared to be due to the differences in tab application and load transfer mentioned earlier. Also, it is possible that shear forces accentuate fiber/matrix interfacial failure and do not have the benefit of fibers held in tension.

Hart-Smith suggested (13) that the failure load for a bonded joint is a function of the area under the stress strain curve (toughness) of the adhesive. Its magnitude will depend on whether the failure is brittle or ductile, and the latter case is an increasing function of the strain rate. Toughness will increase with the strain rate only up to the ductile/brittle transition, then toughness will decrease. With bulk materials we found that failure strains were smaller at higher strain rates.

The effect of carrier cloth on toughness is illustrated in Fig. 3 where the area under the tensile stress-strain curve is plotted as a function of the elastic strain rate. Maxima in toughness appear for both adhesive forms, but shifted from $\sim 10^{-3}\%$ sec^{-1} for epoxy alone to $\sim 10^{-1}\%$ sec^{-1} for the epoxy on carrier cloth. The high fiber stiffness apparently contributes resistance to rate effects. On the other hand, interfacial sites for crack initiation lower toughness by 20% on the average. The maxima indicate ductile-to-brittle transition phenomena, confirmed by fractography. In fact, the fracture profile-length results also show a maximum at approximately the same strain rates.

Joint Strength

In the adhesive bonded state there was scatter in the ultimate shear strength vs. strain rate for both single lap (Fig. 4) and symmetrical specimens. We feel the scatter results from different mechanisms acting simultaneously during deformation and failure of the joint. In other words, joint preparation (inherent voids and variable stress states), viscoelasticity (rate dependence), and fracture modes (brittle and ductile) may effect failure of the joint at random. The variation of stress in the lap shear geometry, as well as voids and fiber/matrix failures are important, but uncontrolled variables. Therefore, a complete study of bonded joints requires determination of rate effects in the bulk, combined with

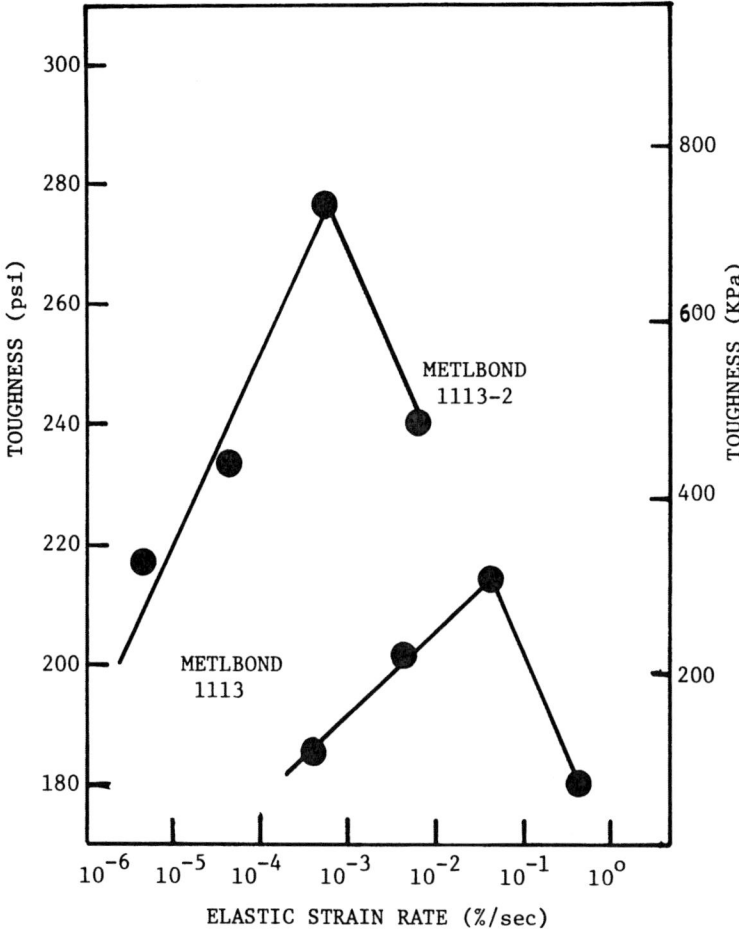

Fig. 3. Area under the stress-strain curve as a
function of test rate. The presence of
scrim cloth in Metlbond 1113 lowers the
toughness and shifts the ductile-to-brittle
transition to higher rates.

fracture behavior in the bonded state. All of these results are
necessary for a general model with parameters representing bulk
properties as well as the constraints and fracture mechanisms
unique to the bonded state of adhesives. The following section
describes our use of SEM to qualify these features. Preliminary
attempts were made to quantify the fractographic data using surface
profile measurements.

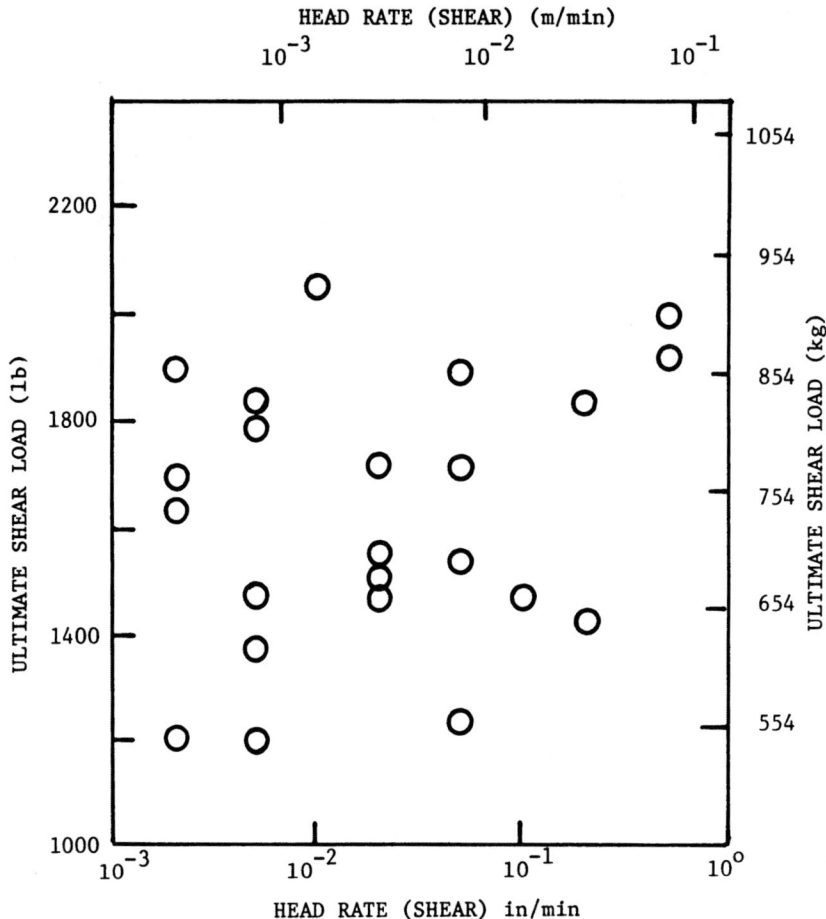

Fig. 4. Illustration of the unpredictable rate
         effects upon strength of Metlbond 1113
         in single lap shear adhesive bonds. (Cf.
         bulk behavior in Fig. 2).

Fractography

        Visual and photographic examination at low magnification of
the single lap joints revealed different failure mechanisms
including plastic deformation and matrix/primer interfacial failure.   Figures 5 and 6 juxtapose representative photographs of
fracture surfaces and Talysurf profiles that provide information on
the new surface area generated during deformation and fracture.
Profile lengths (after subtracting horizontal sections, assumed to
be inherent voids or interfacial failure sites) provide semiquanti-
tative estimates of the strength-affecting mechanisms, at least

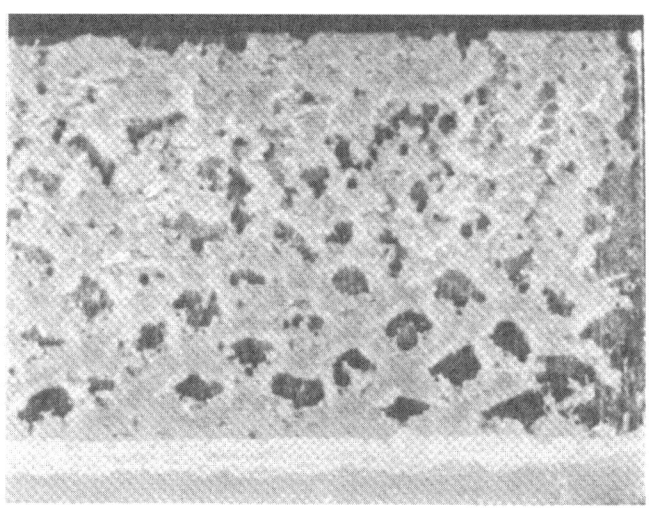

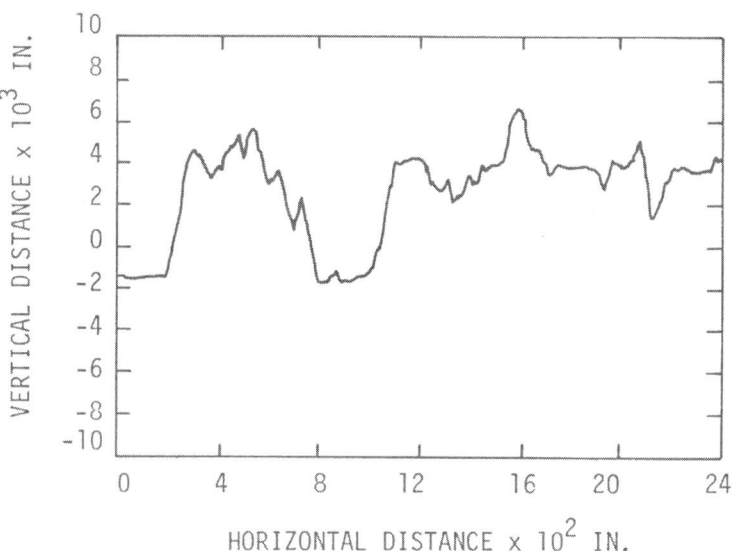

Fig. 5.   Optical pnotomicrograph and corresponding
          Talysurf profile of lap shear specimen #11.

differentiating bonding and fracture areas from voids.

    Figures 7 and 8 juxtapose two fracture surfaces with their
corresponding Neutron Radiographs. In Fig. 7 it is clear that a
large, initial void on the Radiograph is detectable on the fracture
surface. This particular void has an area of approximately 6

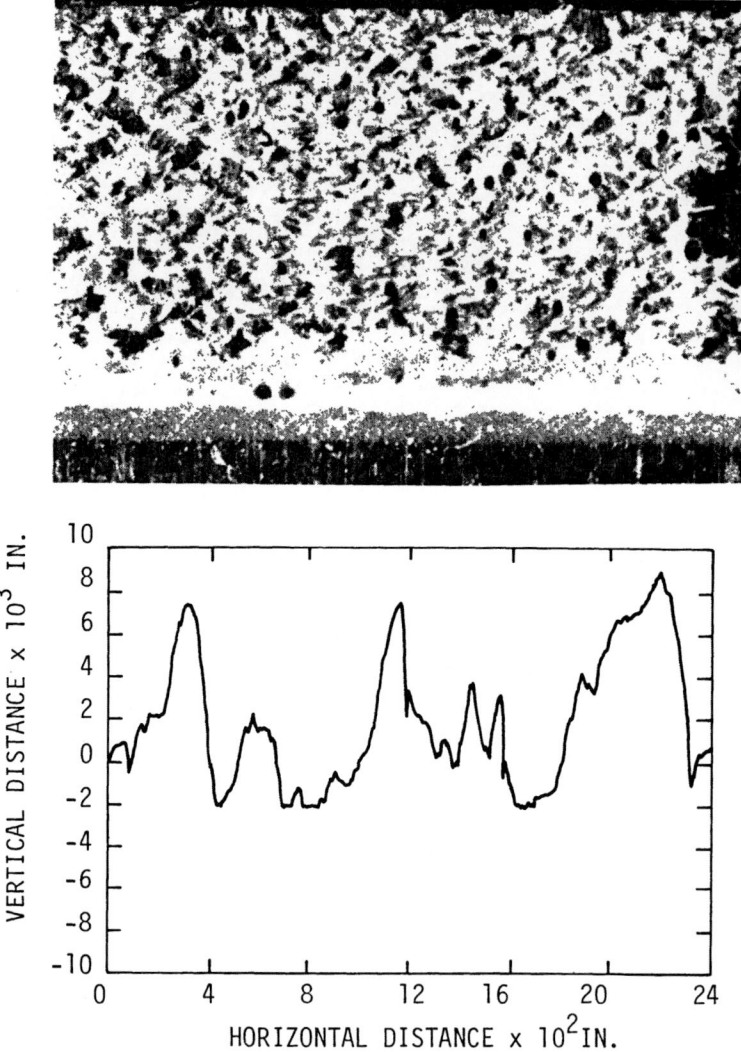

Fig. 6.   Optical photomicrograph and corresponding
          Talysurf profile of lap shear specimen #23.

percent of the overlap area. Other, smaller voids can be detected
on the NR photograph in a pattern similar to that on the fracture
surface. In contrast, Fig. 8 provides an example of the variability
of size, shape and distribution of initial voids, and there is a
correspondence to the defect structure apparent on the fracture
surface.   We believe that Neutron Radiography (or similar NDT
inspection) must be developed to the point that quantitative estimates of the relative defect structure can be made.   Clearly,

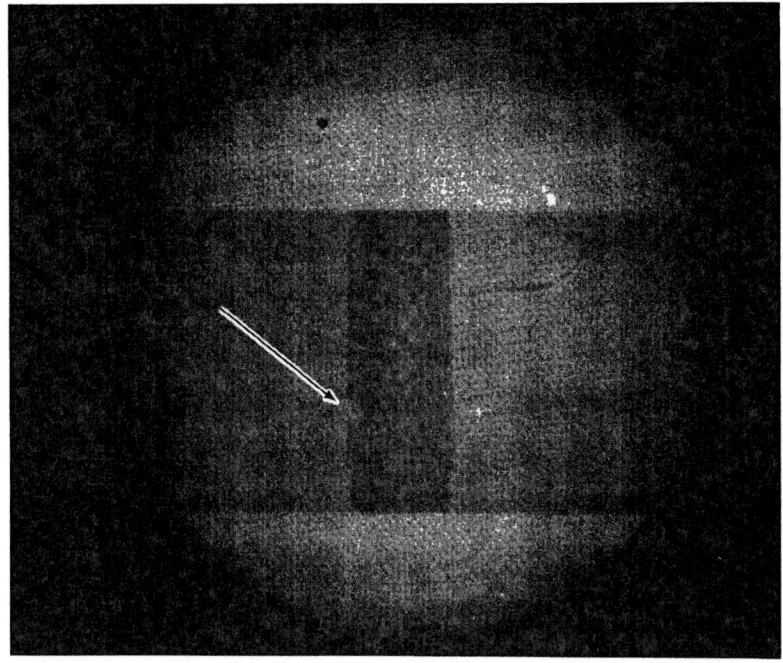

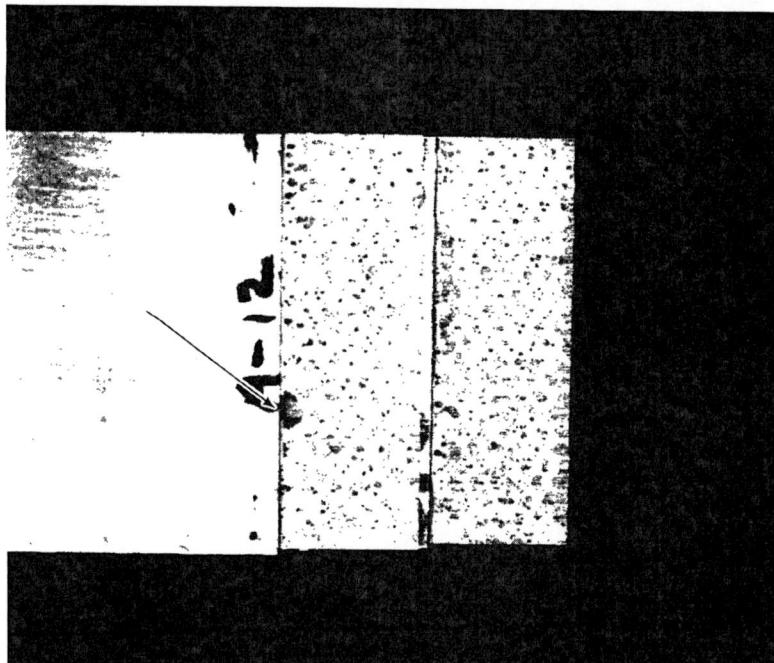

Fig. 7.   Neutron radiograph and corresponding
          optical photomicrograph showing the same
          void (arrows) before and after fracture.

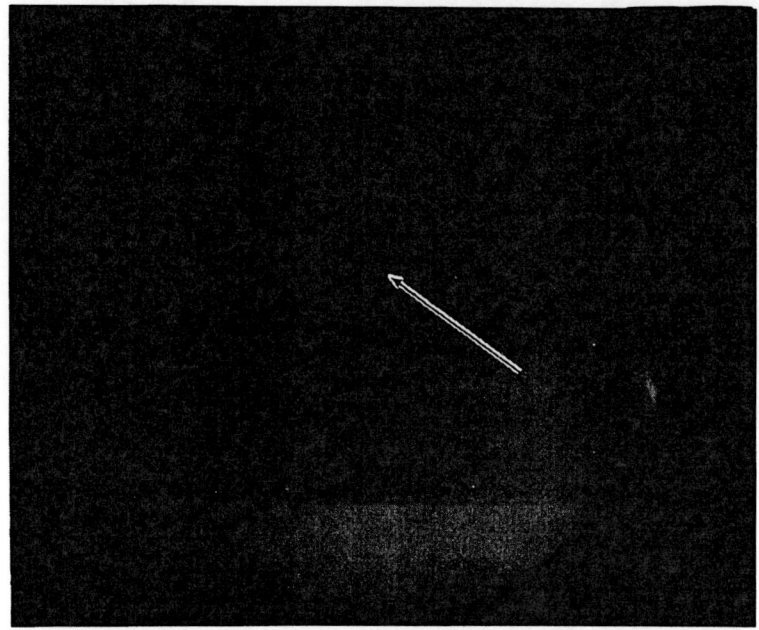

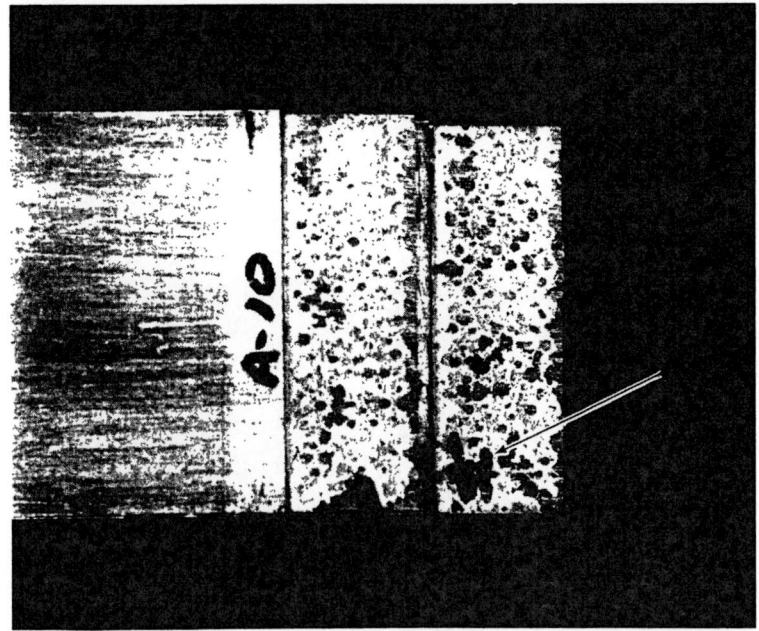

Fig. 8.   Neutron radiograph and corresponding opti-
          cal photomicrograph showing a different
          distribution of void volume than Fig. 7;
          The larger inherent voids lead to larger
          cavities on the fracture surface.

equations 1-3 require values for this term in order to calculate failure energies or loads for the overall joint.

Fractography of the tensile failure (Fig. 9) shows that the bulk adhesive fails with uniform horseshoe pattern of dimensions between 3.9×10^{-3} in., 0.1 mm and 2×10^{-2} in., 0.5 mm, typical of rubber-toughened epoxies (10). At high magnification, interfacial failure between the fibers of the carrier cloth and the matrix is clear, as is a ductile appearance due to spherical indentations ($\sim 1.0 \mu m$ diameter) thought to coincide with rubber inclusions.

Generally, our SEM study of nine fractured Metlbond 1113 single lap specimens showed plastic deformation in low strength joints (Fig. 10). On the other hand, Fig. 11 illustrates that high strength joints were characterized more by brittle cleavage cracks, matrix/primer interfacial failure, and visible fiber beds in the matrix with fibers extending out of the surface.

Examination of a high strength joint (2000 lb. - 907 kg) shows large, smooth interfacial failure sites connected with brittle cleavage crack belts. Figure 11 clearly shows that voids (A) and fiber/matrix interfacial failure (B) are dominant. We conclude that the relatively smooth surface shown on Fig. 11 did not absorb much energy in creating new surfaces by deformation and fracture. Therefore, in this case, the high strength of the joint probably was due to high strain rate. However, examination of another surface (Fig. 12) shows that the high strength obtained (1836 lb. - 933 kg) was due to high surface free energy of the many void edges and brittle cracks. This example shows the possibility of obtaining high strengths at low strain rates through increased fracture surface area. Talysurf profilometer scans of the three single lap specimens discussed above provided further insight into the qualitative findings of the SEM studies. The profile length (assumed to be proportional to the failure area) was relatively low for Fig. 11, but relatively high for Figs. 10 and 12.

Although we are unable to predict adhesive bond strengths, at least we can explain the scatter in Fig. 4 and the differences from bulk fracture on the basis of variable micromechanisms revealed by SEM.

It should be noted, however, that the rate effects and the fracture energies are not the only factors to affect the joint strength. Inherent voids and flaws and chemical impurities at the adhesive/adherend interface also had an effect on the joint strength. In fact EDAX (Energy Dispersive Analysis of X-ray Fluorescence) examination of some low strength specimens revealed the presence of silicone at the adhesive/primer interface acting as a strength reducing factor.

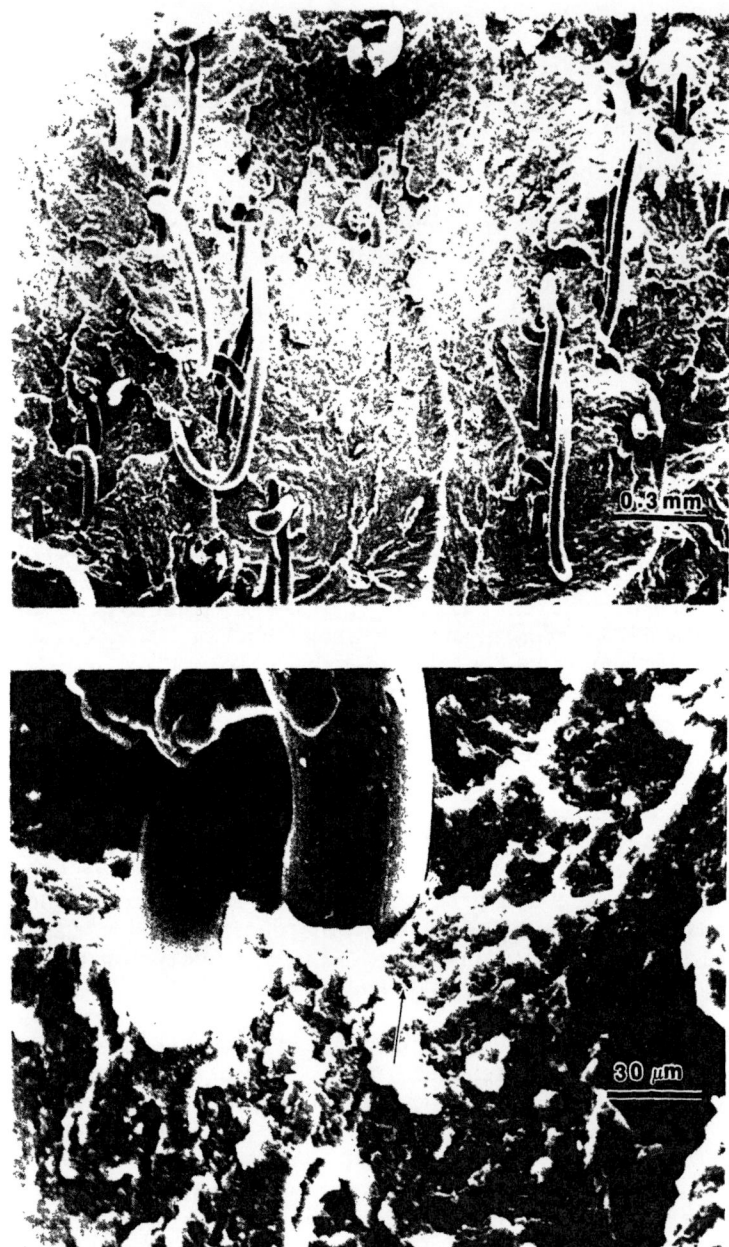

Fig. 9. Two magnification of scanning electron photo-
         micrographs of bulk fracture in Metlbond 1113.
         Tiny cavities (arrow) and larger "horseshoe"
         pattern derive from rubbery inclusions in the
         epoxy matrix.

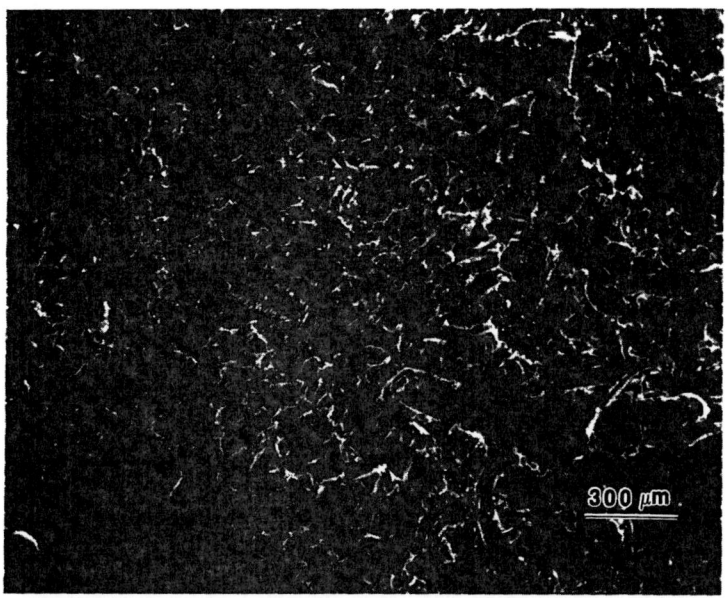

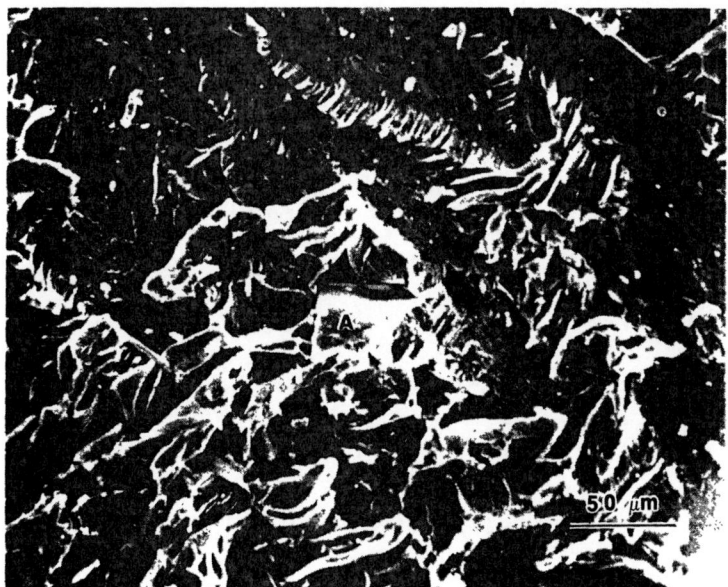

Fig. 10. Two magnifications of scanning electron photo-
          micrographs of lap shear specimen #17. Note
          the plastic deformation at point A at the
          lower left.

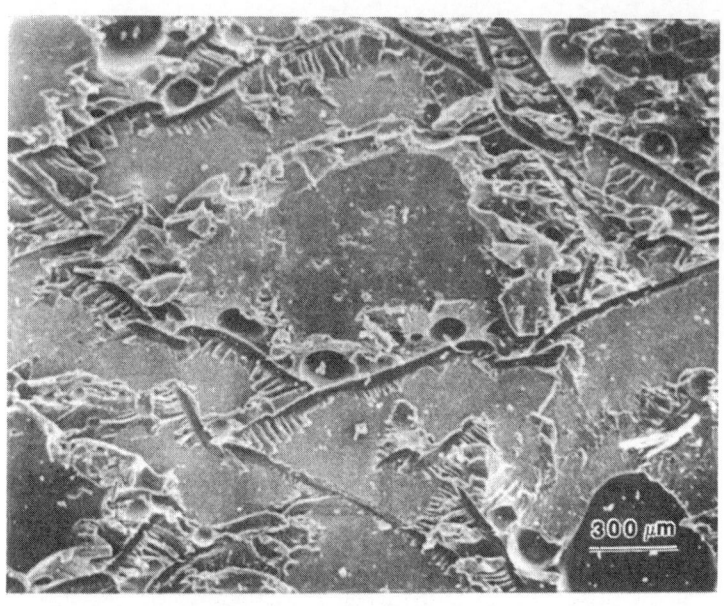

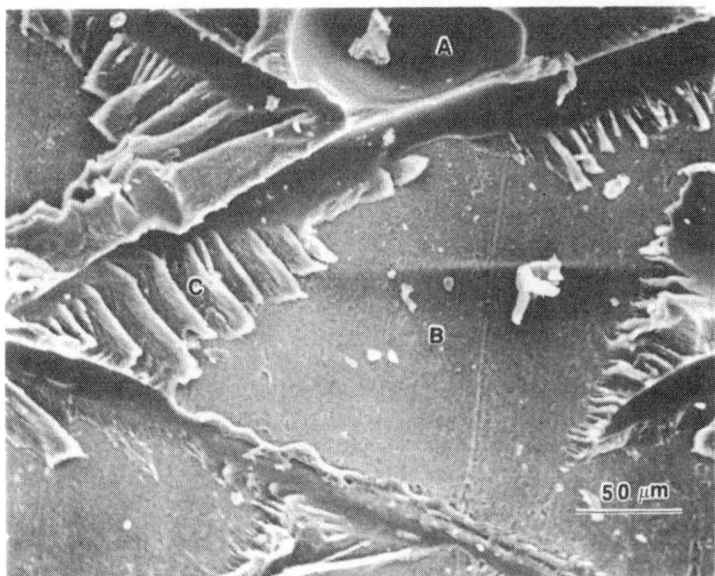

Fig. 11. Two magnifications of scanning electron
         photomicrographs of lap shear specimen #23.
         Note the voids (A), primer/matrix interfacial
         failure (B) and brittle cleavage cracks (C).

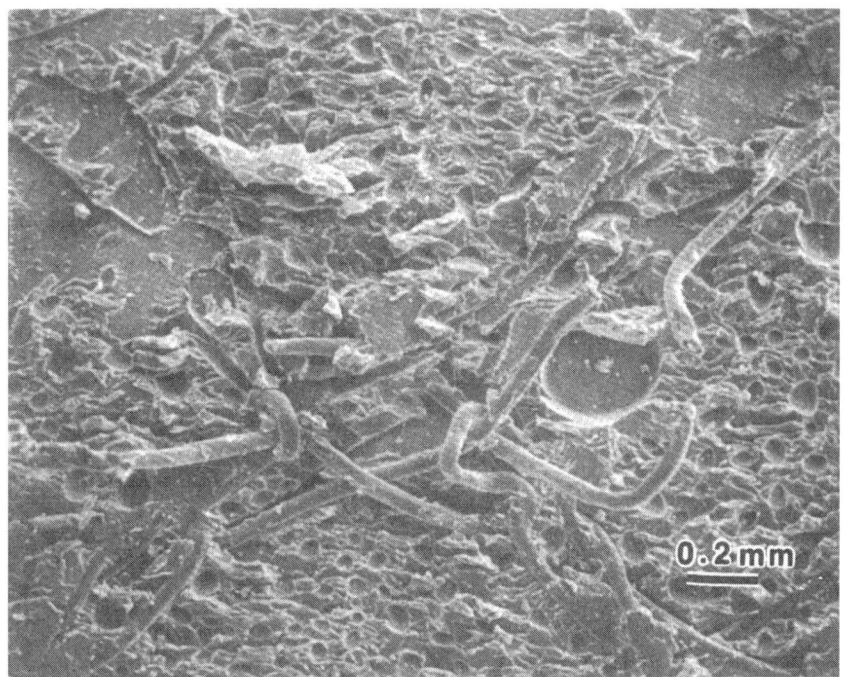

Fig. 12. Scanning electron photomicrograph of lap
          shear specimen #9. Notice how different
          is the void distribution from the previous
          two specimens.

     Above cases were chosen carefully for they represented the
whole fracture. Stress concentrations in crucial sections such as
overlap edges and in between voids act as strength reducing factors.
Much effort is needed to understand the process and materials
interactions that govern these inherent flaws.

CONCLUSIONS

     Strength vs. rate experiments on an epoxy adhesive, in bulk
(tensile and shear) produce accurate and predictable time dependence. However, the tensile data does not predict the highest
possible shear stresses and strains that can be obtained from the
bulk shear specimens. The bulk shear specimens constrained with
tabs produced about 60% higher yield stresses compared to the bulk
tensile predictions. The predictions by the data of the bulk shear
specimens with tabs, however, is believed to be valid for joints
with ideal bond conditions (i.e. no inherent voids or flaws, strong

adhesive/adherend interfacial forces without chemical impurities, uniform stress state, no residual stresses, and rate effects dominant over fracture effects). In fact, fractographic study with Neutron Radiography, optical and scanning electron microscopy (SEM), and surface profiles showed the bonded state of the adhesive was dominated by new constraints and random, inherent flaws and stress states, leading to scatter in the lap shear data.

Comparison of the bulk tensile, bulk shear and bonded shear data shows that only the elastic behavior is common to all three modes. In the bulk and especially bonded shear states, however, new stress states and fracture initiation processes appear to be the dominating factors.

Therefore, characterization of bonded joint should include a careful study of structure-process relationships to assess the presence and effects of inherent flaws and voids, interfacial chemical and physical state and possible residual curing stresses. This information, combined with quantitative fractography and viscoelastic modeling of the bulk adhesive, should make it possible to characterize the bonded state accurately.

ACKNOWLEDGEMENTS

This work was supported by the NASA Langley Research Center under NASA Grant NGR 47-004-090. A research associateship (E.S.) is gratefully acknowledged.

REFERENCES

1. D.W. Dwight, J. Colloid and Interface Sci. 59 (3), 447 (1977).
2. D.W. Dwight, M.E. Counts and J.P. Wightman, in "Colloid Interface Science, V. III. Adsorption, Catalysis, Solid Surfaces, Wetting, Surface Tension and Water", M. Kerker, ed., Academic Press, New York, p. 143 (1976).
3. H.F. Brinson, M.P. Renieri and C.T. Herakovich, in "Fracture Mechanics of Composites", ASTM, STP 593, p. 177 (1975).
4. M.J. Crochet, J. Appl. Mech. 33, 326 (1966).
5. P.M. Naghdi and S.A. Murch, J. Appl. Mech. 30, 321 (1963).
6. P.G. Ludwik, Elemente der Technologischen Mechanic, J. Springer, Berlin, p. 9, 1909.
7. C.J. Lin and J.P. Bell, J. Appl. Polym. Sci. 16, 1721 (1972).
8. R.J. Good, J. Adhesion 4, 133 (1972).
9. L.W. Jennings, in "Recent Advances in Adhesion", L.H. Lee, ed., Gordon and Breach, New York, p. 469 (1971).
10. W.D. Bascom and S. Mostovoy, in "Organic Coatings and Plastics Chemistry", S.S. Labana, ed., Vol. 38 (1), American Chemical Society, p. 152.

11. E.H. Andrews and A.J. Kinloch, J. Polym. Sci.: Symp. No. 46, 1
    (1974).
12. R.A. Gledhill and A.J. Kinloch, J. Adhesion 6, 315 (1974).
13. L.J. Hart-Smith, in "Air Force Conference on Fibrous Com-
    posites in Flight Vehicle Design", Dayton, Ohio, Sep-
    tember 1972.
14. D. Kutscha, Forest Products Laboratory Air Force Contract
    Report No. ML-TDR-64-298, December 1964.
15. W.J. Renton and J.R. Vinson, ASTM Composites Reliability Con-
    ference, STP-580, April 1974.
16. T.R. Guess, R.E. Alred and F.P. Gerstle, Jr., J. Test Eval. 5,
    84 (1977).
17. W.E. Dance, D.H. Petersen, J. Appl. Polym. Sci.: Appl. Polym.
    Symp. 32, 399 (1977).
18. J.M. Whitney, D.L. Stansbarger and H.B. Howell, J. Composite
    Mater. 5, 24 (1971).
19. E. Sancaktar and H.F. Brinson, This Volume.

Dielectric Relaxation Gradients in an Adhesive Bond

John L. Crowley and Arthur D. Jonath

Lockheed Palo Alto Research Laboratory

3251 Hanover Street, Palo Alto, California   94304

ABSTRACT

The dielectric relaxation of a filled epoxy adhesive bond has been measured as a function of the distance from one aluminum adherend by means of dielectric depolarization spectroscopy. Four thermally stimulated current peaks have been identified and characterized using for each a single activation energy and relaxation time approximation. A gradient in the magnitude of the peaks is observed as one approaches the adhesive-adherend interface. The difference in shape and magnitude of the curves can be explained in two ways: (1) differences in the chemical composition of the adhesive at the center of the bond and at the interface arise during the curing process; or (2) dipoles used in the bonding process near the interface are not free to respond to the electric field. The gradients occur over distances up to 2.5×10^{-4}m. This is contrasted with the typical range of tens of Angstroms for interfacial bond forces involving dipole or van der Waals interactions.

INTRODUCTION

Aluminum to aluminum epoxy bonds are in widespread use in many industries. The bond strength depends not only upon the mechanical properties of the resin but also upon the surface preparation of the adherend, the curing cycle of the resin, the ambient atmosphere and several poorly understood aging characteristics. In addition, it has been shown (1) that the locus of failure in an aluminum epoxy bond depends upon the state of stress within the bond. For failure in Mode I (opening) the fracture surface is most often observed to occur at the center of the bondline, while the fracture surface in

Mode II (shear) failure is characteristically found within the adhesive very near the adhesive-adherend interface. The observation of such failure led to the investigation of possible structural gradients in the cured resin within the interfacial region. Both mechanical and dielectric relaxation methods were used to interrogate the adhesive material properties within the adhesive-adherend interfacial region. The results of the thermally stimulated dielectric relaxation (TSDR) measurements will be discussed in this paper.

THERMALLY STIMULATED DIELECTRIC RELAXATION

The technique of TSDR, first used to study the impurity defects in alkali-halide crystal (2), has been used to probe the microstructural relaxations in an aluminum-epoxy bond. A recent review of its application to polymers has been given by Hedvig (3). The TSDR experiment measures a thermally stimulated current in a dielectric material which has been polarized by means of an externally applied d.c. electric field. The measured current can be due to the release of trapped mobile charge, relaxation of induced dipoles or to the movement of polar side groups or chain ends in the polymer.

To describe what is happening in the experiment, it is assumed that the thermally stimulated current is due to the reorientation of dipoles. The sample is cooled to immobilize the dipoles in the applied field orientation. The electric field is then removed and the sample is connected to a sensitive current measuring instrument. Next, the sample is heated and the dipoles begin to thermally randomize. This decay of the internal field due to oriented dipoles induces a current in the external circuit. The resultant current/temperature thermogram can be analyzed for dipole density, activation energy, and relaxation time. The current peaks recorded in this way are found to correlate well with the transition temperatures of the polymer as measured by mechanical relaxation or conventional (a.c.) dielectric spectroscopy (3).

The motion of dipoles is assumed to be thermally activated. The relaxation time(s) for dipole motion, $\tau(T)$, can therefore be described by an Arrhenius relation, $\tau(T) = \tau_o \exp(E/kT)$, where τ_o is the extrapolated relaxation time at infinite temperature, E is the activation energy for dipole motion and k is Boltzmann's constant. The reorientation is carried out at a temperature at which the relaxation time, $\tau(T)$, is much less than the time, t_p, for which the electric field is applied. After reorientation and immobilization of the dipoles, the induced current caused by their thermal randomization can be expressed by the following relationship (4):

$$i(T) = \frac{N_d F p^2}{kT_p} \left[\tau_o \exp(E/kT)\right]^{-1} \exp\left\{-\int_0^T \left[b\tau_o \exp(E/kT')\right]^{-1} dT'\right\}$$

where F is the applied electric field, p is the dipole moment, N_d is the density of dipoles, b is the heating rate, and T_p is the temperature at which the dipoles were polarized. These data were evaluated with this equation using the solution provided by Jenkins (5).

EXPERIMENTAL

The epoxy used in this experiment was Epoxylite 810 (Epoxylite Corporation, Anaheim, California), a silicate loaded, dianhydride-cured glycidyl ether of a phenolformaldehyde Novolac resin. TSDR and mixed-mode fracture samples were prepared with resin for bonding to aluminum surfaces which were first polished with 1-μm diamond polish and subjected to a common acid-dichromate etchant. After bonding, the resin was cured at 120°C for two hours. Fracture characteristically left a thin film of material or interfacial accommodation zone (IAZ) on one aluminum adherend and a bond thickness (bulk) of adhesive on the other. After fracture, additional TSDR samples were cut from the adherends. An aluminum electrical contact, 0.635 cm in diameter, was evaporated onto the adhesive surface of each specimen, while the aluminum substrate served as the second electrical contact. All samples discussed here were polarized at 25°C for 20 minutes.

Sample heating rates were varied from 0.12 deg K/s to 0.50 deg K/s. These rates, although higher than usual for tests on polymers, were still sufficiently slow to maintain thermal equilibrium because the thin film under measurement was bonded to a very good thermal conductor.

RESULTS AND DISCUSSION

The TSDR thermograms of the bulk and IAZ material appearing in Fig. 1 indicate that there are certain differences in the dielectric relaxation behavior of the adhesive depending upon location within the bond. First, there appears to be a general difference in the density of dipoles contained in the bulk and in the IAZ; i.e., the TSDR curve for the IAZ materials is lower-valued than that for the bulk material. Secondly, the data for the bulk material show peaks C and D appearing in the size ratio of 3:1, while in the IAZ data, this ratio is at best 1:1. Two explanations for the difference in shape and magnitude of the curves are offered: (1) differences in the chemical composition of the adhesive at the center of the bond

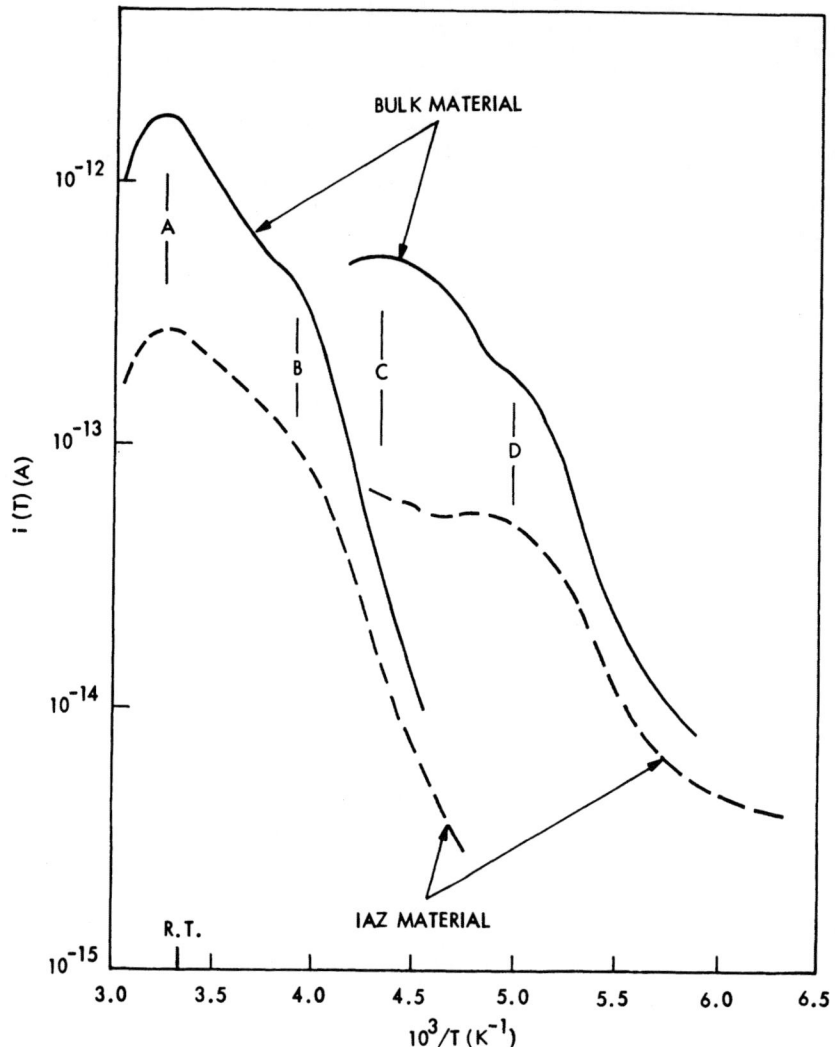

Fig. 1. The Thermally Stimulated Dielectric Current of
         Epoxylite 810 Bulk and IAZ Samples, Polarized
         with 9.55 x 10² V/cm at 25° C for 20 Minutes.
         Four resolvable peaks are indicated as A, B,
         C, and D.

and at the interface may arise during the curing process; or
(2) dipoles used in the bonding process near the interface are not
free to respond to the applied electric field.

    Of the four discernible peaks in Fig. 1, the magnitude of three
peaks, B, C, and D, were found to vary linearly with applied

electric field.   The current maximum for peak A increases at a sublinear rate with increase in field.   It is also observed that the peak A maximum shifts in temperature as the polarization temperature is changed.   Therefore, it does not appear that peak A is due to dipole reorientation.   Rheovibron measurements of the loss tangent made on sheet cast samples of the epoxy were able to identify only one peak in the temperature range of present interest. This peak appears to correspond to peak C in Fig. 1.

Figures 2, 3, and 4 show the results of the TSDR measurements on epoxy samples whose thicknesses were sequentially decreased by metallographic sectioning techniques.   The magnitude of the peak

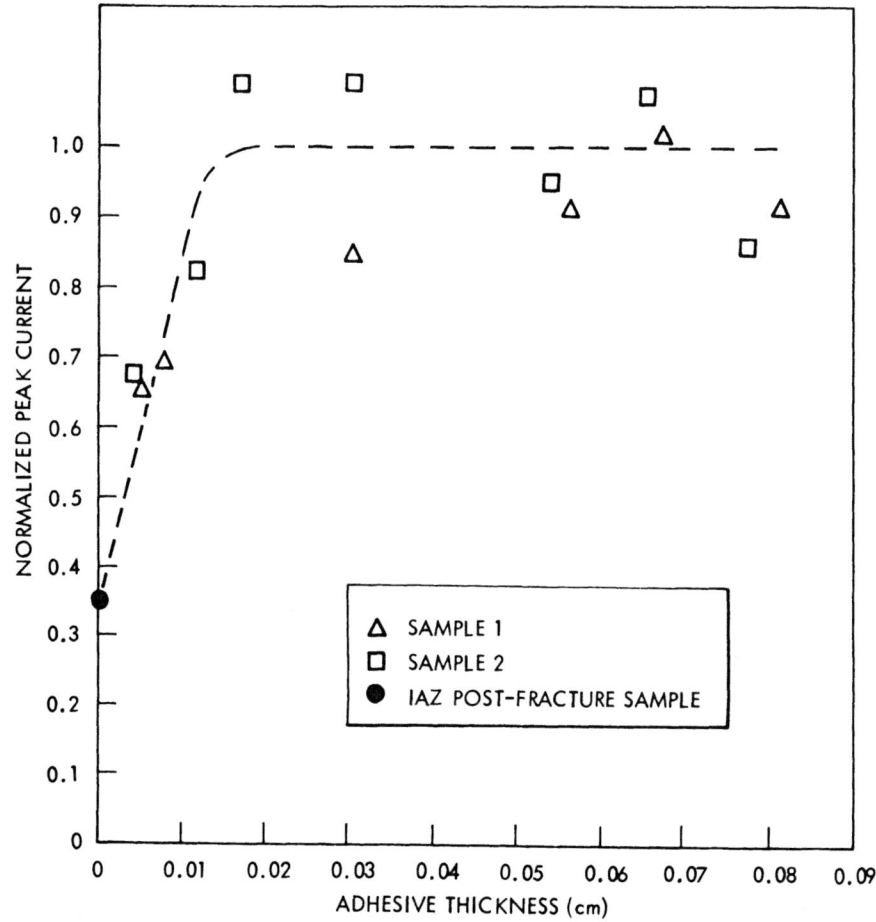

Fig. 2.   Magnitude of Peak B as a Function of Thickness
of Material Remaining on Adherend (Normalized to
a Peak Current of 4.75 x 10^{-13} A at a Field of
1.59 x 10^3 V/cm).

currents have been normalized to a constant field and average peak
magnitude. Because the polarization temperature and heating rate
were the same for all tests, the graphs essentially represent the
variation of the average dipole density in the epoxy as a function
of distance from the adherend. The density of dipoles responsible
for peaks B and C is seen to remain constant throughout the bulk of
the material and to decrease at distances less than 2×10^{-4} m. The
data for peak D do not justify a straight line approximation in the
bulk of the material, but its density is also seen to decrease as
the adhesive-adherend interface is approached. This result by
itself is not surprising; it has long been known that bulk properties differ from surface properties; i.e., any material must accom-

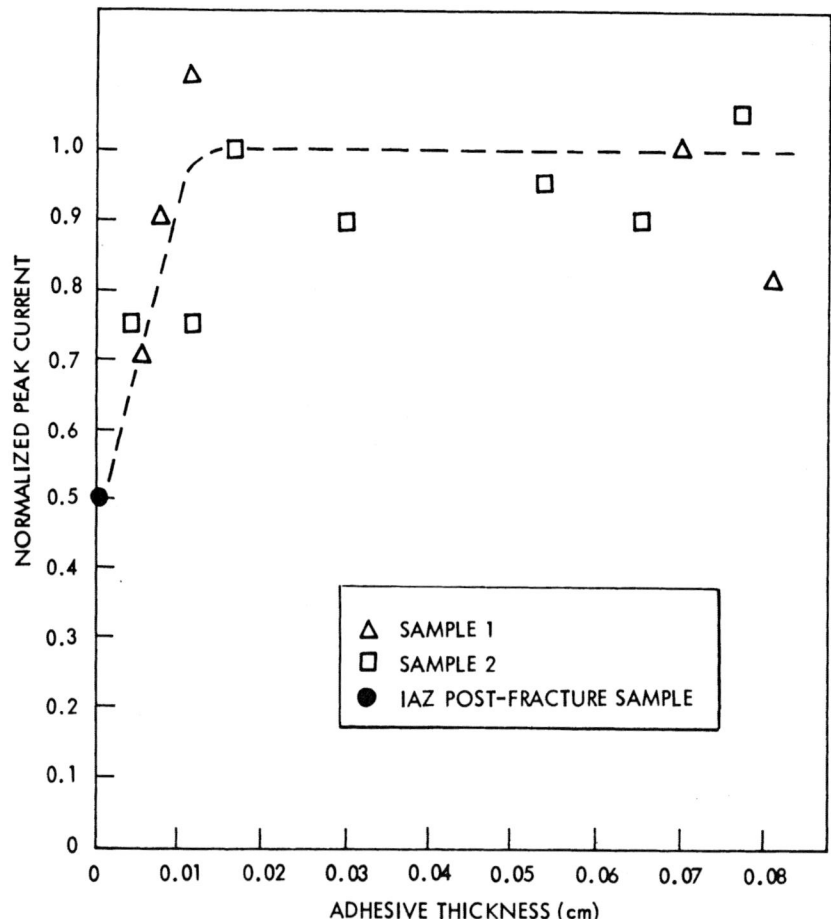

Fig. 3. Magnitude of Peak C as a Function of Thickness
         of Material Remaining on Adherend (Normalized to
         a Peak Current of 2.0×10^{-13} A at a Field of
         1.59×10^{3} V/cm).

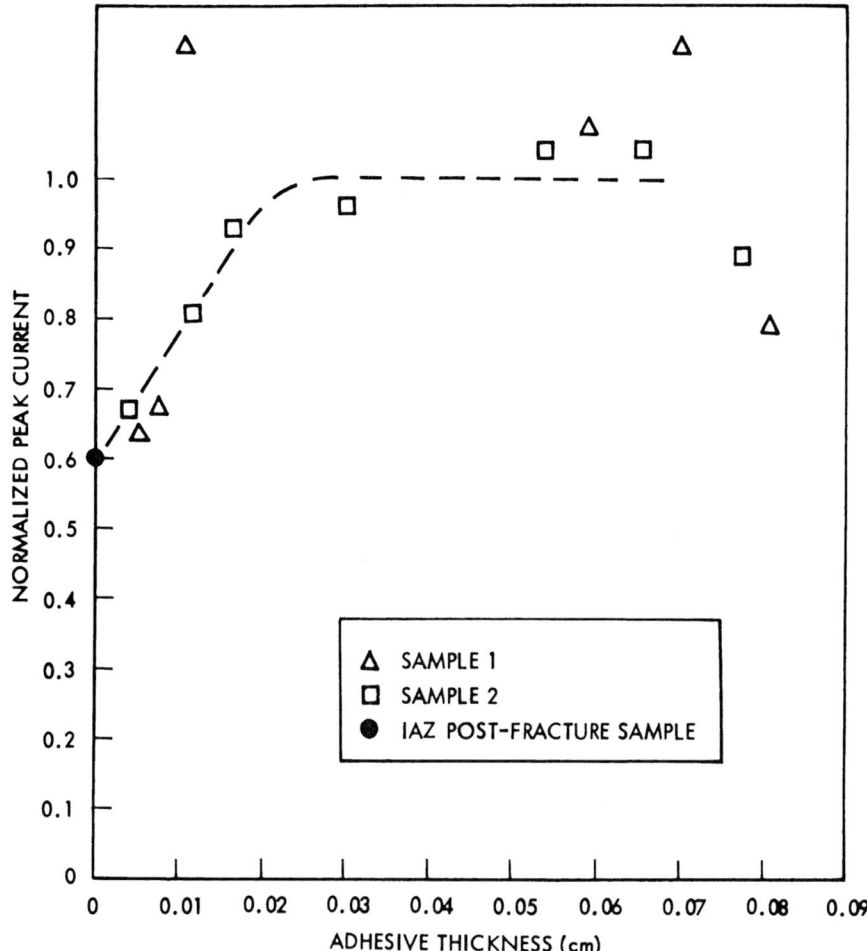

Fig. 4. Magnitude of Peak D as a Function of Thickness
of Material Remaining on Adherend (Normalized to
a Peak Current of 2.7 x 10^{-13} A at a Field of
1.59 x 10^3 V/cm).

modate itself to molecular (atomic) bond discontinuities caused by
surfaces.   Rather, it is the measure of the spatial range of the
accommodation zone, approximately 200μm, that is of interest. This
is contrasted with the tens of Angstroms typical range for interfacial bond forces involving dipole or van der Waals interactions.
A long-range mechanism, perhaps in conjunction with the short-range
phenomena, is needed to explain the result.   For example, if the
polymer contains interacting dipoles, ordering due to interfacial
electric fields might propagate a significant distance from the
interface.   Also, electrokinetic phenomena associated with col-
loids (6), postulated to explain the ordering of floccules seen in
two-phase epoxy polymers (7), might describe long-range effects
taking place in the epoxy cure process.

In addition to the TSDR measurements, an independent study was made of the shear wave speed, C_s, in the adhesive as a function of the bond thickness (8). Samples for this study were identical to those used in the TSDR measurements. The shear wave speed is given by the expression

$$C_s = \sqrt{G/\rho}$$

where G is the shear modulus, and ρ is the adhesive density. By differentiating the above expression, the change in shear speed can be expressed as

$$\frac{\Delta C_s}{C_s} = \frac{1}{2}\left(\frac{\Delta G}{G} - \frac{\Delta\rho}{\rho}\right)$$

The results of the shear wave speed measurements are shown in Fig. 5. The data show a variation in shear speed over the same magnitude of bond thickness as the variation in normalized peak current amplitudes shown in Figs. 2, 3, and 4. However, the variation in the shear speed is 10 percent of the original speed measured in the bulk, while the variation in the magnitude of peak B at comparable bond thickness is approximately 50 percent. If it is assumed that the variation in the magnitude of peak B is due entirely to a change in the density of the adhesive in the IAZ, then the change in shear speed represents a 70-percent decrease in the shear modulus. Although this reduction in shear modulus is consistent with the observed locus of failure for Mode II fracture, density gradients alone do not account for the change in ratio of peak magnitudes (see Fig. 1).

The normalized peak currents from the bulk materials shown in Fig. 1 are a factor of three greater than the asymptotic values shown in Figs. 2, 3, and 4. This anomaly remains unexplained. However, it is noted that the bulk fracture material underwent severe deformation and microcracking.

SUMMARY

The dielectric relaxation of a filled epoxy adhesive has been measured as a function of the distance from an aluminum adherend by means of thermally stimulated dielectric relaxation. Four thermally stimulated current peaks have been identified and characterized. A gradient in the magnitude of the peaks is observed as the adhesive-adherend interface is approached. The difference in shape and magnitude of the curves can be explained in two ways: (1) differences in chemical composition of the adhesive at the center of the bond and at the interface may arise during the curing process; or (2) dipoles used in the bonding process near the interface are not free to respond to the electric field. The gradients in the

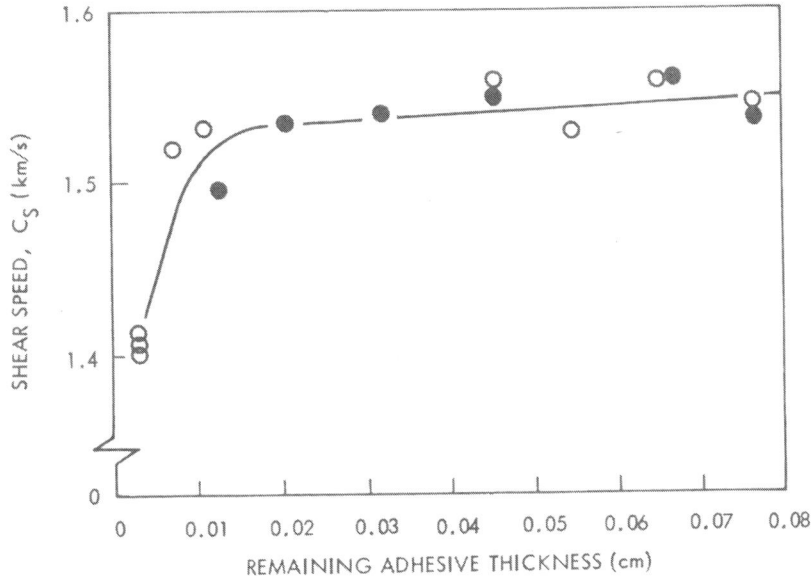

Fig. 5. Shear Wave Speed Variation Within an Adhesive
         Bond Thickness Measured Using Rayleigh Wave
         Method (20 MHz). (See Ref. 8.)

dipole density occur over distances up to 2×10^{-4} m.

ACKNOWLEDGEMENTS

        This work was supported by Lockheed Missiles & Space Company,
Inc. Independent Research funds. The authors wish to acknowledge
the support of R. Holsinger and T.J. Stultz in preparing the samples
for test.

REFERENCES

1.   W.D. Bascom, C.O. Timmons, and R.L. Jones, Apparent Inter-
       facial Failure in Mixed-Mode Adhesive Fracture, J. Mate-
       rials Sci. 10, 1037 (1975).
2.   C. Bucci, R. Fieschi, and G. Guidi, Ionic Thermocurrents in
       Dielectrics, Phys. Rev. 148, 816 (1966).
3.   P. Hedvig, "Dielectric Spectroscopy of Polymers", John Wiley
       and Sons, New York, pp. 148-155 (1977).
4.   C. Bucci, Ionic Thermocurrents in Alkali Halide Crystals Con-
       taining Substitutional Beryllium Ions, Phys. Rev. 164,
       1200 (1967).

5.   T.R. Jenkins, On Computing the Integral of Glow Curve Theory,
       J. Comp. Phys. 28, No. 2 (1978).
6.   A.W. Adamson, "Physical Chemistry of Surfaces", Interscience,
       New York, pp. 179-226 (1960).
7.   R.E. Cuthrell, Epoxy Polymers. II. Microstructure, J. App.
       Polym. Sci. 12, 1263 (1968).
8.   G.C. Knollman and J.J. Hartog, Lockheed Palo Alto Research
       Laboratory, Palo Alto, CA (private communication).

New Approach to the Understanding of Adhesive Interface Phenomena and Bond Strength

Arthur D. Jonath

Lockheed Palo Alto Research Laboratory

Palo Alto, California 94304

ABSTRACT

Microscopic and macroscopic aspects of adhesion science are currently under investigations in order to better understand the fundamentals of adhesive behavior from bond formation to bond deterioration/failure. Results of three novel experimental methods are described: a) Measurements using an improved shear specimen design indicate the absence of Mode II failure in a brittle epoxy system. b) Results of ultrasonic Rayleigh wave measurements indicate a possible mechanical properties gradient in the cured resin within the interfacial region of an epoxy-aluminum bond. c) Thermally stimulated dipole relaxation measurements indicate structural gradients within this region.

INTRODUCTION

Adhesive bond failure has been associated with missile malfunctions in various aerospace systems. In most instances, the contributing factors are not quantitatively identified. The problem is to determine the extent of failures resulting from improper

---

*This is an abridged version of the paper "Some Novel Approaches to the Study of Adhesive Interface Phenomena and Bond Strength" presented at the Structural Adhesives and Bonding Conference, Technology Conferences Associates, P.O. Box 842, El Segundo, Ca., March 13-15, 1979.

adhesive selection, interface preparation, adhesive application, or changes in adhesive characteristics due to age and environment. The empirically established procedures for surface preparation prior to application of adhesive do not guarantee reproducible quality bonds, and tests for the achieved bond quality are generally not related to the state of stress existing in an actual component. Understanding the aging processes in adhesives is especially critical to the aerospace industry because debonding of insulation and structural materials limit the reliability and lifetime of adhesive bonded systems such as solid propellant motors. Thus the overall objective of the research program discussed herein is to provide guidelines for selecting adhesive systems and surface conditioning requirements and to predict the age environment behavior of adhesive systems. To accomplish this, several novel experimental methods are being investigated for studying bonding/interface phenomena and bond-joint failure processes. They include

o   A novel shear test technique investigating assumptions used in adhesive fracture toughness measurements.

o   An ultrasonic surface wave method for measuring mechanical properties in thin adhesive layers.

o   A dipole relaxation method for experimental determination of adhesive structural changes at adhesive/adherend interfaces.

The purpose of this paper is to introduce the adhesives researcher and applications engineer to these experimental concepts, some of which are adapted from other fields, and to indicate where results of such measurements can be of practical use.

FRACTURE MECHANICS

Fracture toughness measurement is one of the materials concepts presently used to determine design data for adhesive bonded structures. This is considered to be a measure of the ultimate strength of a material in which the failure mode is by crack propagation, causing failure at engineering stresses less than the tensile or shear strengths. While fracture toughness is the appropriate failure criterion for linear elastic brittle failure, recent work at the Naval Research Laboratory and elsewhere has highlighted several pitfalls in applying these methods to the testing of adhesive joints (1). For example, joint geometry and bond thickness effects dominate the fracture toughness measurements in structural adhesive joints. Thus the approach for this aspect of the study was to explore, in selected adhesives, the limitations of fracture mechanics and to attempt to relate fracture energy measurements to the adhesive polymer structure.

Before discussing the novel adhesive shear test method, let us describe the fracture toughness measurement results which led to our need for measuring "pure" shear in an adhesive bond. Both silicate filled and unfilled versions of the Epoxylite 810, a dianhydride cured glycidyl ether of a phenolformaldehyde Novolac resin were tested in the tapered double cantilever beam (TDCB) and 45° scarf-joint configurations shown in Fig. 1 to measure Mode I (opening) and Mode II (shear) fracture toughness. The aluminum adherend surfaces were prepared with dichromate acid etch. Table 1 lists the results. The Mode I fracture energy values, G_{Ic}, are in the range expected for unmodified epoxies. G_{Ic} in the filled system is greater than that in the unfilled system implying that the silicate filler contributes somewhat to toughening this epoxy.

In either filled or unfilled versions of this adhesive, the opening-mode adhesive fracture energy G_{Ic} is nearly an order of magnitude greater than the combined-mode $G_{I,II}$ value. This is contrary to the results $(G_{I,II} > G_{Ic})$ generally found for the fracture of isotropic materials (2) and more like the results being found for elastomer-modified and commercial structural adhesives (3).

The locus of failure in the TDCB specimens precracked at the center-of-bond (COB) was observed consistently to follow the COB. In the scarf-joint samples, the failure occurred very near the adherend-adhesive interface whether the precrack was at either the COB or the interface. Although there is typically less than 50μm of adhesive remaining on the adherend nearest the failure locus, the failure is clearly cohesive in nature. In fact, adhesive failure (no detectable adhesive remaining on an adherend) was never observed in specimens where proper adherend surface preparation and handling procedures are followed.

The questions arising are thus: Why in the presence of shear stress does the failure locus follow a path so close to the interface? Is it simply the goemetry of the scarf-joint which drives the propagating crack to the interface? Is there an asymmetric stress concentration about the crack tip forcing this path? Or perhaps there are material properties changes (such as in shear modulus) occurring within the bondline thickness close and normal to the adherend surface? To answer these questions we sought to understand from both microscopic and molecular views what is happening in the interfacial accommodation zone (IAZ), this region very close to the adherend through which the crack propagates when significant shear stresses are present in the failing joint. More about this later.

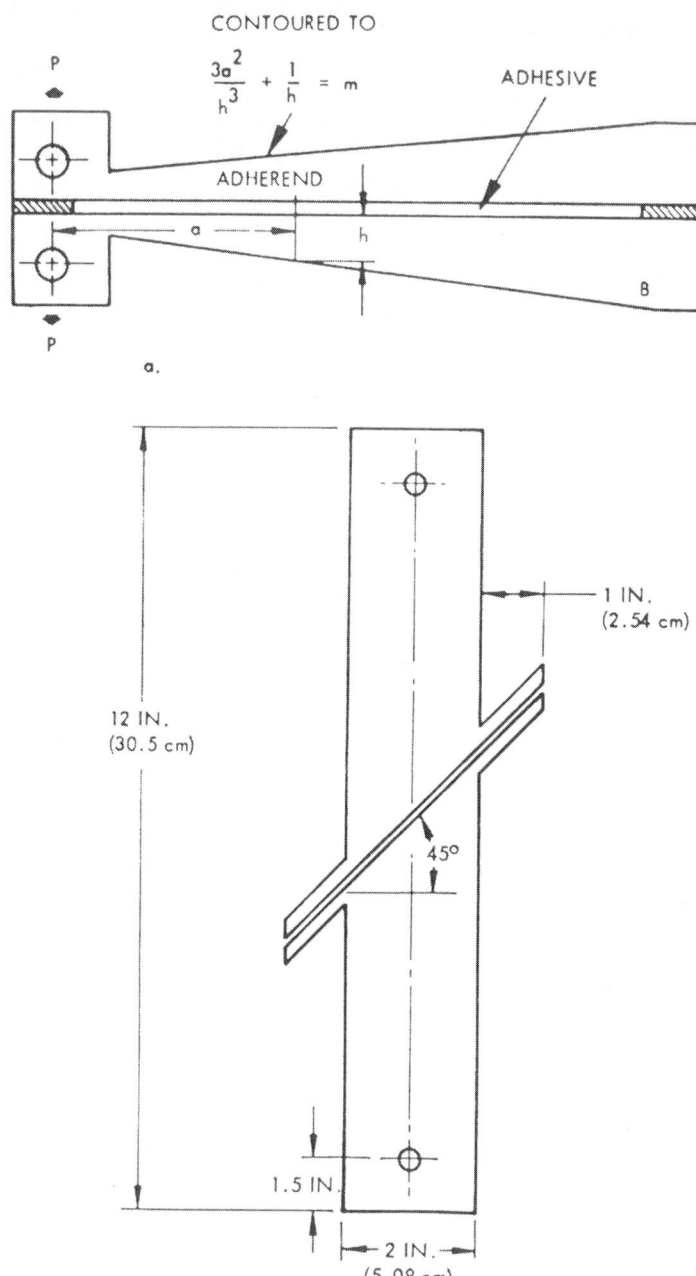

Fig. 1. Adhesive Fracture Test Specimens:
(a) Tapered Double Cantilever Beam for Mode I
and (b) 45° Scarf-Joint for Combined Modes I
and II.

Table 1

RESULTS OF FRACTURE TOUGHNESS MEASUREMENTS

	Relative Crack Length	G_{Total}		K_I	K_{II}
		Nm/m^2	$in-lb/in^2$	$N/m^{3/2}$	$N/m^{3/2}$

FILLED SYSTEM

$G_{I,II}$

SCARF	.1	93	0.53	2.08×10^6	1.54×10^6
	.2	100	0.57	2.32×10^6	1.35×10^6
	.4	110	0.63	2.47×10^6	1.28×10^6

G_{Ic}

TDCB	.022	700	4.0	7.1×10^6	
	.045	700	4.0	7.1×10^6	
	.1	700	4.0	7.1×10^6	

UNFILLED SYSTEM

$G_{I,II}$

SCARF	.2	51	0.29	1.66×10^6	9.63×10^5
	.3	49	0.28	1.66×10^6	8.96×10^5
	.4	39	0.22	1.49×10^6	7.75×10^5

G_{Ic}

TDCB	.05	400	2.3	5.4×10^6	
	.1	320	1.9	4.8×10^6	

First let us return to the fracture toughness results in Table 1 for both the filled and unfilled versions of the Epoxylite 810. The critical stress intensity factor K_{II}, calculated from the scarf-joint data using the finite element displacement method of Trantina (4), decreased as crack length increased. Since K_{II} is expected to vary with crack tip radius and not crack length, the existence of Mode II fracture therefore became suspect. The concern

is that failure is shear might be determined by the shear strength
in the remaining ligament material and not by propagation of a shear
mode crack.

To test this hypothesis a modification of the Iosipescu specimen (5) used to measure shear in metals was adapted (6) for
measuring "pure" shear in an adhesive bond. This specimen with
simplified shear and moment diagrams is shown in Fig. 2. The
computed results of a finite element analysis (7) of the shear and
peel stresses along the bond length for this shear specimen is shown
in Figs. 3A and 3B respectively. In this case, the applied load is
1.0 lb. and the total bond length is 1.0 in. Young's modulus and
Poisson's ratio for the aluminum adherends are 10.67×10^6 psi and
0.3368; 2.65×10^4 psi and 0.32 were used for the adhesive. Note
that for this specimen the shear stress exceeds its average value by
less than 3% in magnitude for less than 5% of the bond length. Peel
stresses vary more than this, but they are two orders of magnitude
less than the shear stress. A typical lap joint calculation is
included in Figs. 4A and 4B for comparison (8). The differences in
a) uniformity of shear stress along the bond and b) magnitude of
peel stresses between the modified Iosipescu sample and the lap
shear are striking.

In Fig. 5 the adhesive shear strength, as measured (6) in the
above described shear specimen, is plotted as a function of the
normalized crack length. The bondwidth is 1.0 in. and thickness is
0.020 in. The line in the figure represents the predicted behavior
if failure is determined by shear strength. Values above the line
at large crack lengths are probably volume effects and are being
investigated. Systematic deviation of the data to values below the
line implies the presence of Mode II. The data represented by open
triangles (center of bond moment = 0) are a strong indication that
Mode II is absent and that the failure is determined by the shear
strength of the remaining material. The measured shear strength
using this specimen is 4235 psi. The adhesive manufacturer reports
less than half this value measured in lap shear.

The solid triangles in Fig. 5 represent failure stresses
resulting when the zero moment position was deliberately moved
0.025 in. away from the centerline of the bond. The effect is
drastic--introduction of Mode I failure into the specimen. It is
evident that adhesive joint strength depends critically on the type
of loading experienced by the joint.

As in the scarf-joint speciments, the failure locus in this
shear specimen lies within the IAZ. The symmetry of the specimen
and its loading tends to rule out a geometrical argument for this
behavior. Crack tip asymmetries could not be detected micro-
scopically. Thus we turned our attention to search for variations

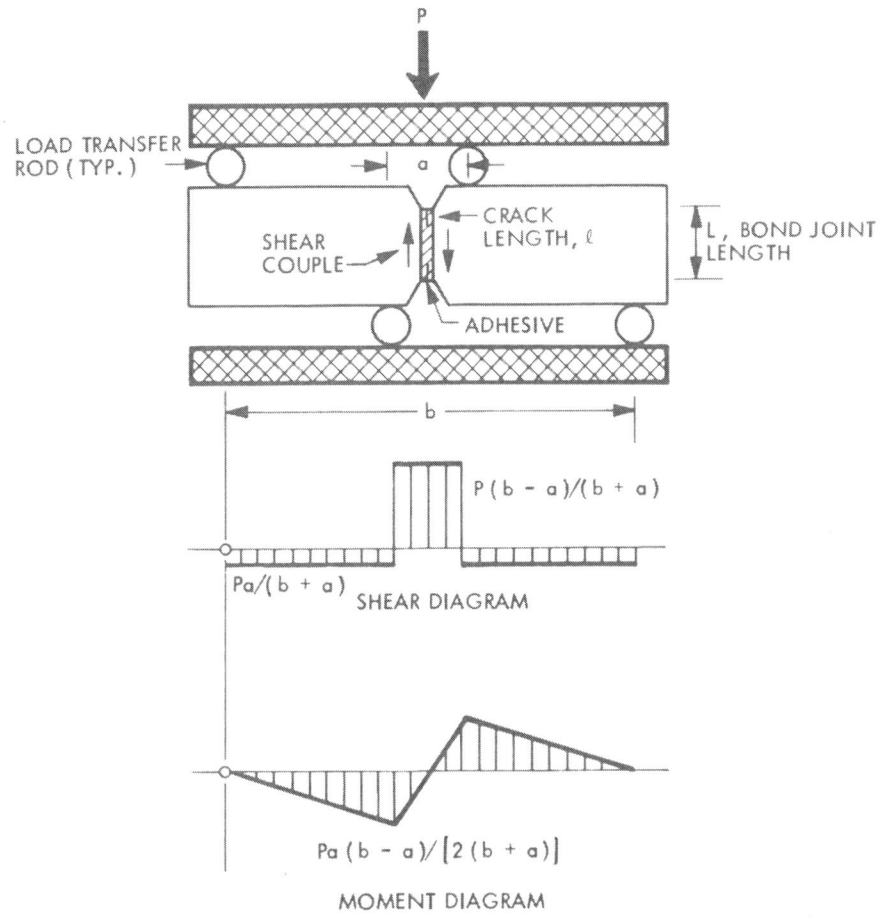

Fig. 2.   Adhesives Shear Test Specimen with Simplified
           Shear and Moment Diagrams.

in material properties and changes in material structure within the
IAZ.

MECHANICAL PROPERTIES OF THIN LAYERS BY ULTRASONIC RAYLEIGH WAVES

Elastic waves similar to surface gravitational waves can be
transmitted by an elastic medium, as was first shown by Lord Rayleigh. Their particle motions decrease rapidly with depth; their
speed, which depends on the value of Poisson's ratio, is always less
than that for a shear wave.

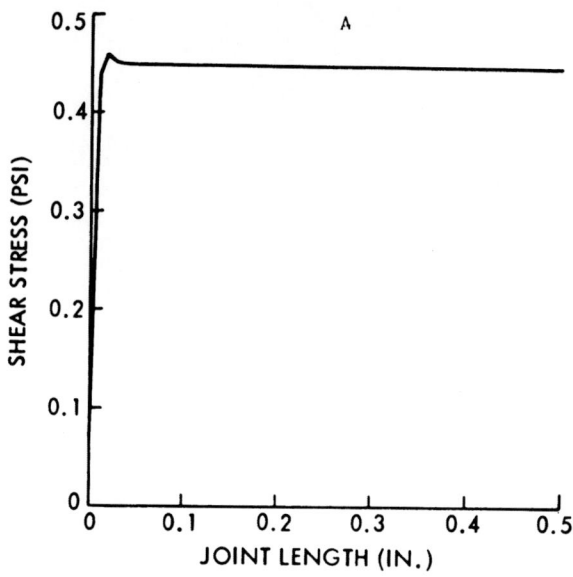

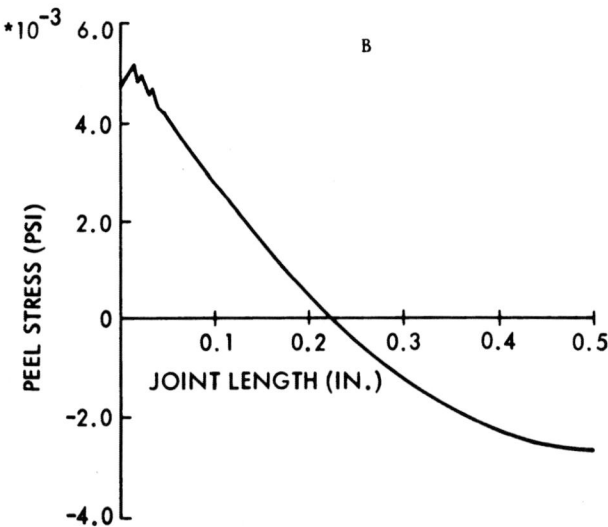

Fig. 3. Stress Distributions from Half-Length
         Finite-Element Analysis for A.) Shear
         and B.) Peel in Shear Specimen Shown
         in Fig. 2.

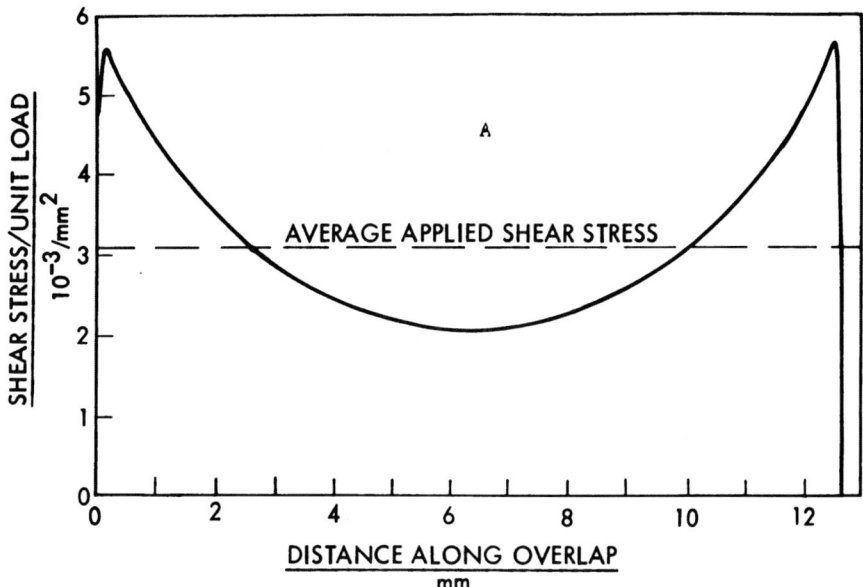

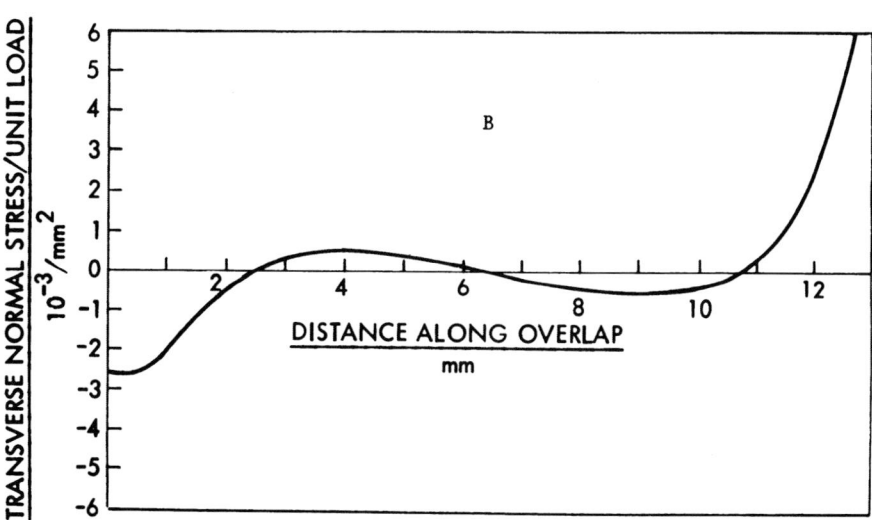

Fig. 4.   Stress Distribution from Full Length
          Double-Lap Finite-Element Analysis (8).
          A. Shear Stress; B. Transverse Normal
          Stress (Peel).

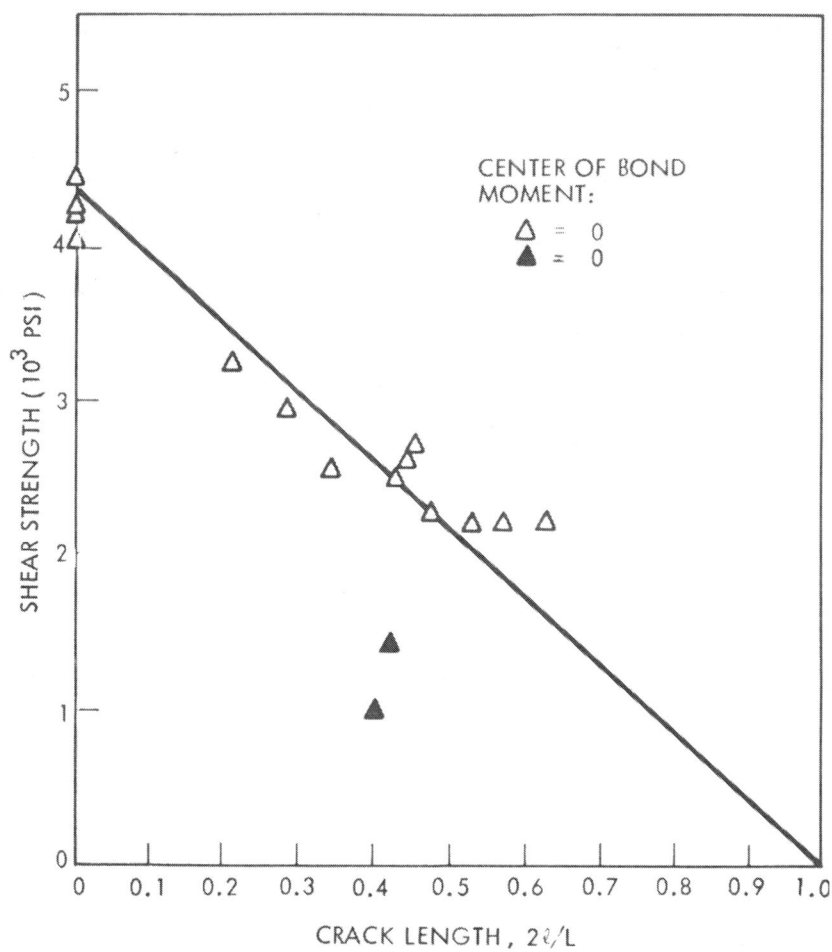

Fig. 5. Shear Strength Versus Crack Length for
Epoxylite Using Adhesive Shear Test
Specimen Shown in Fig. 2 (Ref. 6).

When a solid is immersed in a liquid and an ultrasonic wave
impinges at some angle onto the solid from the liquid, several wave
modes may be generated in the solid by mode conversion. That is,
energy in the incident mode converts to another mode at the expense
of the reflected wave. The critical angle for a surface mode is the
angle of incidence which produces an angle of refraction of 90° for
that mode. According to Snell's law, the refraction angle is
related to the angle of incidence and the speeds of sound in the
liquid and in the solid.

$$\frac{C_R}{\sin 90^\circ} = \frac{C_i}{\sin \theta_i^R} \tag{1}$$

where C_i is the incident sound speed in the liquid, C_R is the surface wave speed in the solid, and θ_i^R is the angle of incidence at which the surface mode conversion occurs. The shear wave speed, C_S, is related to C_R by

$$C_S = C_R/k \tag{2}$$

where k is a function of Poisson's ratio. The shear modulus, G, is a function of C_S and density ρ

$$G = \rho C_S^2 \tag{3}$$

Most of the energy in the Rayleigh wave is concentrated within a distance less than a wavelength from the solid surface and thus the shear wave speed in thin adhesive layers (such as that adjacent the cohesive fracture surface in the IAZ) can be determined.

A simple laboratory arrangement for measuring the Rayleigh angle appropriate for a thin layer of adhesive deposited on a substrate is shown in Fig. 6. A cylindrical glass vessel is mounted on a spectrometer table so as to be rotatable about its longitudinal axis (normal to the plane of the figure). This vessel is filled with a liquid, and the test specimen is positioned at the axial center of the vessel as shown. Thereby, the specimen can be rotated about its longitudinal axis by rotation of the spectrometer table. A 20 MHz ultrasonic transducer is positioned as shown in the figure. The stainless steel mirror allows the single transducer to be used as both acoustic transmitter and receiver. By adjustment of the sample and the transducer post the transmitted wave angle of incidence is kept equal to the angle of reflection as the incidence angle is varied. Then the received single displays an amplitude minimum and a phase reversal at θ_i^R, the Rayleigh angle.

The apparatus was evaluated by determining shear speed in various materials. For metals immersed in water the reflection minimum was used to determine θ_i^R and thus C_S. The agreement with published shear speeds was typically +3%, -0%. The relative error for repeating a measurement (sensitivity) was less than 0.15%. However, for typical polymer materials immersed in water. there is no surface generation mode, i.e. $C_i > C_R$ (see Eq. 1). Therefore, Dow Corning Fluid DC-200A (of 1cs. viscosity) was selected as an immersion medium and phase reversal was used to determine θ_i^R. Agreement was +8%, -0% suggesting a systematic experimental error, perhaps in the determination of the incident sound speed in the

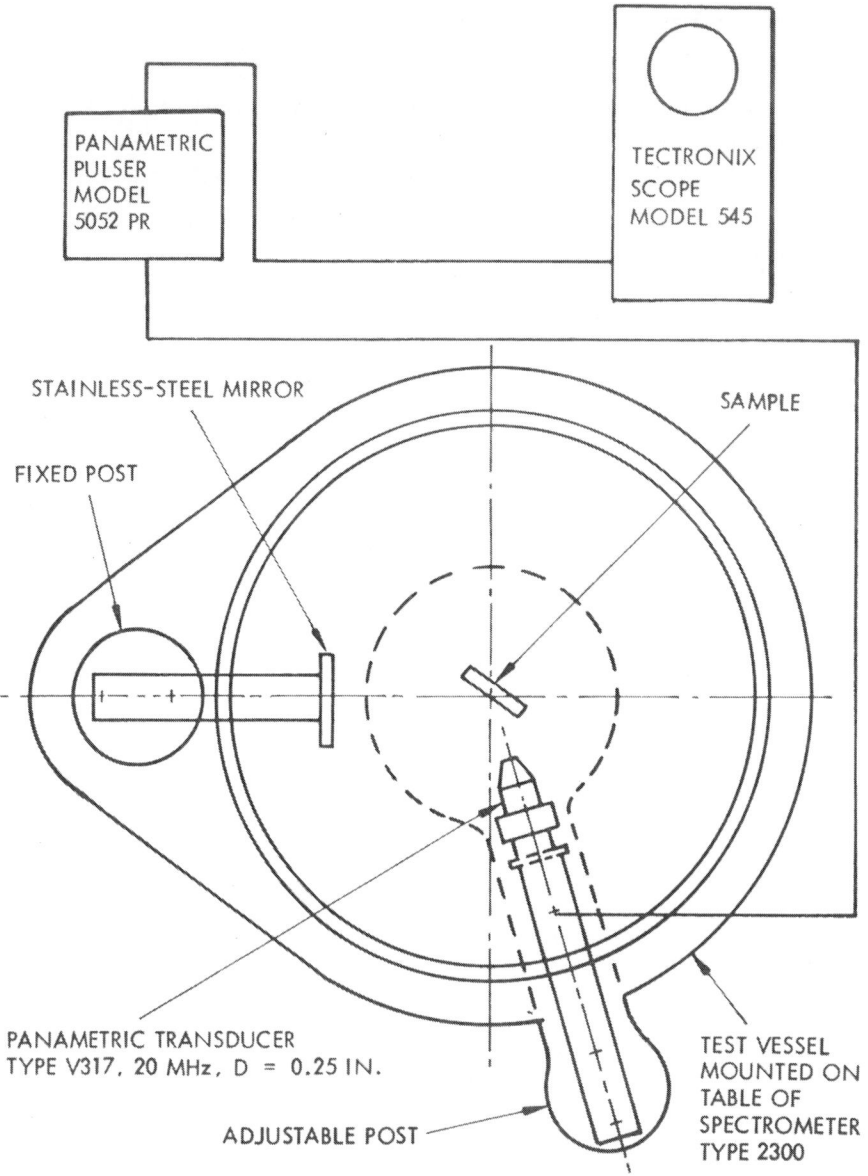

PANAMETRIC
PULSER
MODEL
5052 PR

TECTRONIX
SCOPE
MODEL 545

STAINLESS-STEEL MIRROR

SAMPLE

FIXED POST

PANAMETRIC TRANSDUCER
TYPE V317, 20 MHz, D = 0.25 IN.

ADJUSTABLE POST

TEST VESSEL
MOUNTED ON
TABLE OF
SPECTROMETER
TYPE 2300

Fig. 6.   Arrangement for Measurement of the Rayleigh
          Angle.

liquid. But the relative error of the measurement was found to be
approximately 1%.

Samples consisting of 0.03 in. thick Epoxylite 810 bonded to single aluminum adherends were prepared. Figure 7 shows the results of Rayleigh wave shear speed measurements (9) on these epoxy samples whose thicknesses were decreased by metallographic sectioning techniques. For adhesive thicknesses greater than 100μm, C_s is asymptotic (within 2 percent) to the bulk cast value. At 20μm from the aluminum adherend, C_s has decreased by about 11 percent, to 1.4 km/sec., the values measured in the 500μm and thicker postfracture specimens. The latter specimens were fashioned from scarf-joint test specimens whose fracture surfaces were located within the IAZ about 13μm from the adhesive/adherend interface. Since the surface wave displacement is significantly reduced at depths greater than 20μm it is not surprising that C_s values measured for the IAZ surface of these post-fracture specimens match those (points on extreme left in figure) for samples whose thickness was reduced to 25μm by polishing. Computational models for stress-strain distribution or crack propagation in adhesive bonds assume constant mechanical property values throughout the bond thickness. Yet density gradients are probable in polymers adjacent to interfaces.

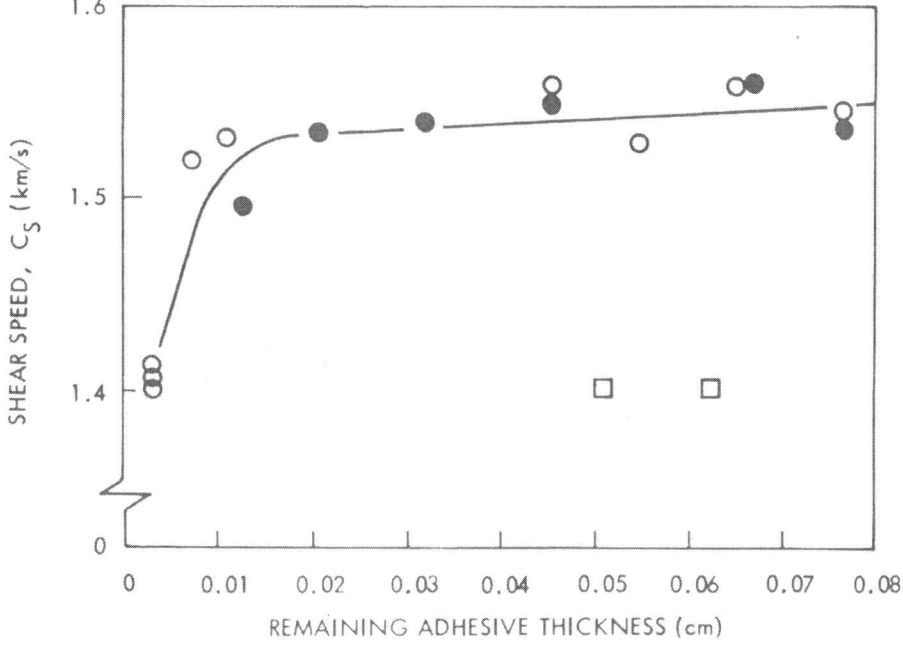

Fig. 7. Shear Wave Speed Variation Within an Adhesive Bond Thickness for Samples 2 (0) and 5 (●) and the Post-Fracture Sample (▢) Measured Using Rayleigh Wave Method (20MHz) (Ref. 9).

These are most likely caused by enhanced diffusion towards surfaces of the low molecular weight portion of the polymer size distribution. The data for this adhesive/adherend system indicate existence of a shear speed gradient extending 100 μm from the interface into the adhesive bond. Relative contributions of shear modulus and density to this property gradient (see Eq. (3)) have not been separated and must be determined before any conclusion can be drawn regarding modulus variation effect on adhesive joint failure locus. However, that this gradient region contains the failure surface when shear is present indicates a necessity for reexamination of concepts based upon homogeneity of adhesive properties throughout a bond.

THERMALLY STIMULATED DIPOLE RELAXATION

In order to afford a molecular view of behavior in the interfacial region, the experimental technique of thermally stimulated dipole relaxation (TSDR), originally used to study defects in alkali halide crystals (10), was applied to probe the microstructural relaxation in the aluminum-epoxy system (11). The TSDR experiment involves the reorientation of dipoles in a dielectric material by means of an externally applied electric field. Details of the experiment are given by Crowley and Jonath elsewhere in this volume.

Typical TSDR thermograms for IAZ and bulk samples are shown in Fig. 8. In obtaining these data, the samples were heated to 235°K yielding peaks D and C and then quickly recooled to 125°K. Peaks B and A were then obtained by reheating at the prior rate. In this manner, individual dipole responses can be separately observed.

The TSDR response caused by any interaction between the dipoles and the silicate particles can be seen by comparing the control (unloaded) sample thermogram with that of the loaded sample. Peaks A and B are virtually unchanged in magnitude, while the presence of the filler decreases the relaxation current due to reordering of the dipoles related to peaks C and D. It is concluded that these dipoles play a role in bonding the filler particles and thus fewer are available to respond to the polarization field.

Bulk and IAZ thermograms for the loaded material in the figure differ in two ways: (1) The magnitude of the current is less in the IAZ material than in the bulk. (2) The ratios of peak magnitudes in the IAZ material differs from those in the bulk material.

It was proposed (11) that dipoles taking part in the bonding process near the interface, and thus not free to respond to the polarizing electric field, would explain the first observation.

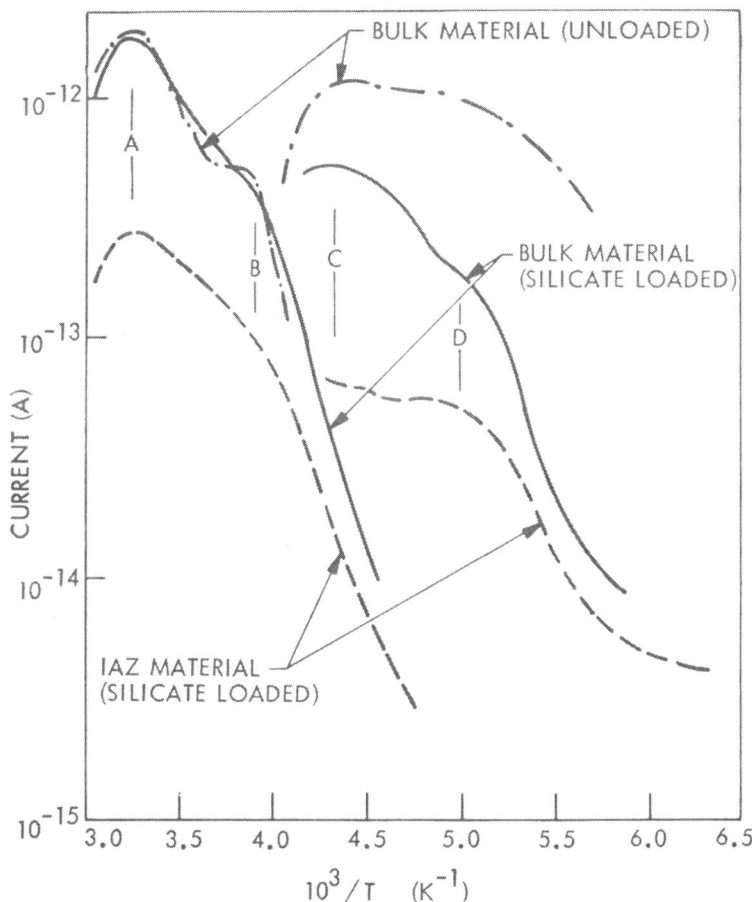

Fig. 8. The Thermally Stimulated Dipole Relaxation
Current of Epoxylite 810 Bulk and IAZ Samples,
Polarized with 9.55×10^2 V/cm at 25°C for
20 Min. Four resolvable peaks are indicated
as A, B, C and D.

The same authors postulate chemical composition changes, perhaps
due to inhomogeneities in the cure process, to explain the second.

    At the present time, the behavior of peak A presents a dilemma.
Contrary to expected dipole response, its magnitude does not vary
linearly with applied field (see Fig. 9) and its peak maximum shifts
with polarization temperature (temperature at which field is
applied). Yet its magnitude does change with location in the
adhesive. The implications of these results are unclear.

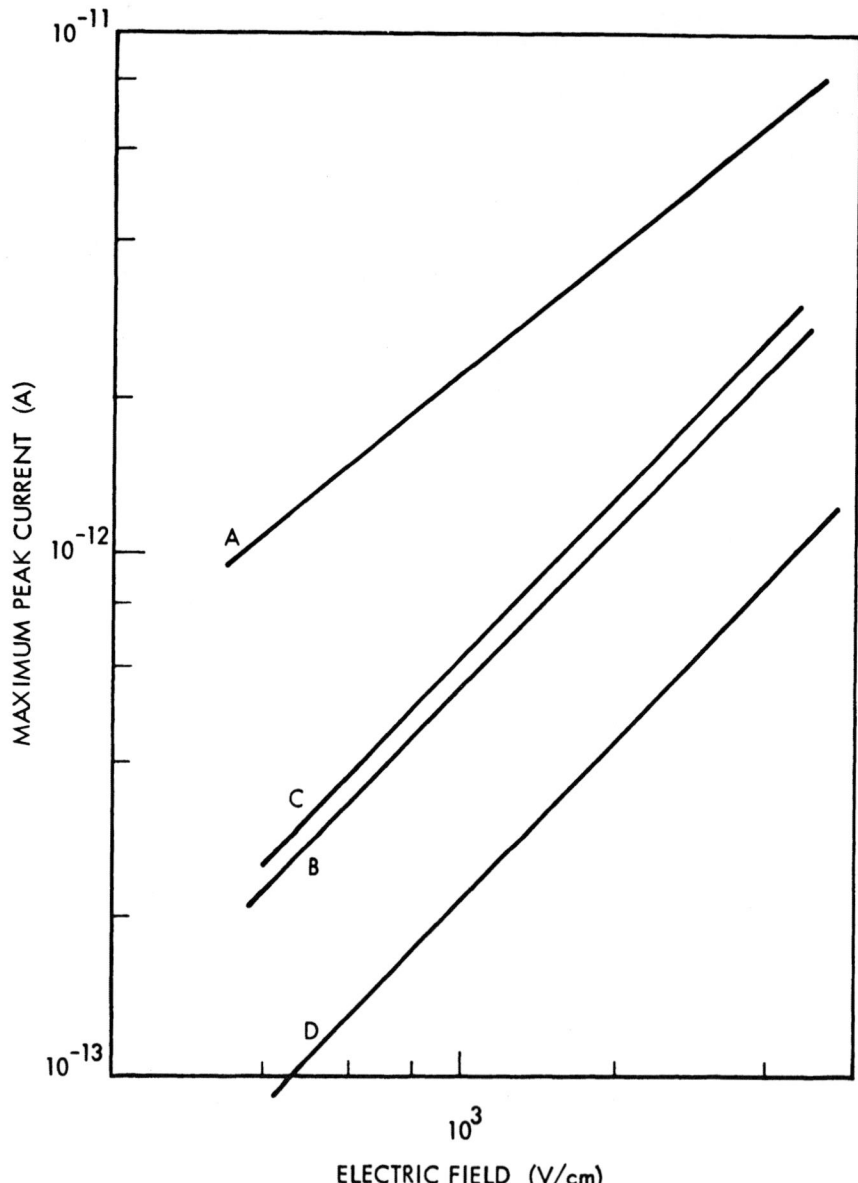

Fig. 9. The Dependence of the Peak Current on the
Strength of the Polarizing Electric Field
for Peaks A, B, C and D. Note sublinear
behavior of A.

From the temperature shift in the peak maximum with heating rate and from the slope of the low temperature tail, an activation energy of 0.71 eV and a relaxation time constant of 0.1 sec. at room temperature was calculated for dipole B. Using these values and assuming a value for the dipole moment, an estimate of the dipole density can be computed (11). The identity of the molecular species responsible for this dipole is now known at this time but typical values of dipole strength for most organic molecules lie between 1 and 2 debye. For a dipole moment of 1 debye, a dipole density of 2×10^{20} cm^{-3} was calculated for the IAZ and 1.2×10^{21} cm^{-3} for the bulk material.

Experiments to determine spatial extent of dipole gradients in the IAZ are reported elsewhere (11). Although there are differences among the peaks, the results are similar to that of the Rayleigh wave experiment reported in the previous section. An example of these results is shown for peak B in Fig. 10. The scatter in the data is considerably greater than in the acoustic experiment but it is evident that the percent change in TSDR current (i.e. dipole density to first approximation) is greater than the percent change in shear speed. (cf. Fig. 7).

SUMMARY AND CONCLUSIONS

We have observed interfacial cohesive failure using a symmetric shear specimen design newly adapted for testing shear strength in adhesives. A finite element analysis of the Epoxylite 810/aluminum system shows shear stress nearly constant along the bond length and peel stress two orders of magnitude less than the shear stress. Using this specimen we have concluded that failure in shear is determined by the shear strength of the epoxy and is not caused by stress concentration at a crack tip (Mode II). A detailed analysis of the limitations of this modified Iosipescu specimen and comparison to the symmetric lap shear specimen analysis (12) has yet to be made. Such an analysis might include effects due to load rate, bond thickness, adherend surface preparation and less brittle adhesives.

Work in progress to determine the extent and nature of the IAZ, a region over which adhesive microstructure and properties adapt from surface or interface to bulk conditions, was presented. Using an ultrasonic Rayleigh wave generation method, we have observed a shear speed gradient extending approximately 100 µm from the adhesive-adherend interface. Measurement of the density gradient in this region is needed to complement the shear speed data for determination of shear modulus gradient. Then we will understand more about the nature of the interfacial cohesive failure which occurs when significant shear stresses are present.

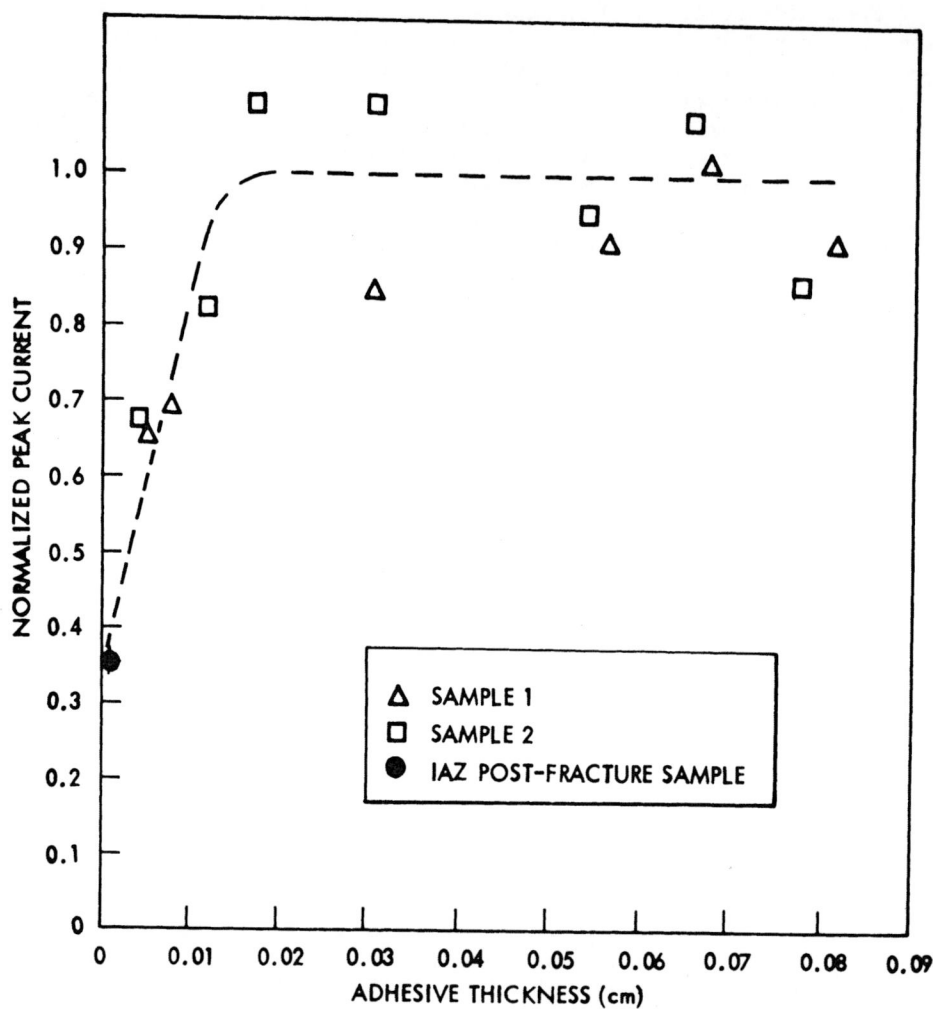

Fig. 10.   Magnitude of Peak B as a Function of
           Thickness of Material Remaining on
           Adherend (Normalized to a Peak Current
           of 4.75 x 10^{-13}A at a Field of
           1.59 x 10^3 V/cm) (Ref. 11).

With a sensitive thermally stimulated relaxation technique, we have observed the response of several distinct dipoles. The activation energy and room temperature relaxation time deduced for one (0.71 eV and 0.1 sec.) are reasonable for dipoles associated with molecular side groups in polymers. TSDR measured dipole structure gradients are seen to extend up to 200μm from the interface. Because these gradients and that of the shear speed differ both in magnitude and spatial extent, it seems unlikely that a single underlying cause explains the behavior. From the TSDR results to date we infer a long-range dipole contribution to the bonding process and chemical composition changes within the interfacial region. Fourier transform infrared (FTIR) and electron spectroscopy for chemical analysis (ESCA) techniques are being used to identify the dipoles.

Oxidation and moisture diffusion processes associated with adhesives aging/environmental deterioration often are found to be interfacial phenomena. Therefore details of the molecular structure in the IAZ will provide insight into these mechanisms.

ACKNOWLEDGEMENTS

Throughout the various phases of this work, the author has received help and encouragement from discussions with John Crowley, John Bjeletich and Gil Knollman. This work was supported by the Lockheed Missiles & Space Co. Independent Research Fund.

REFERENCES

1. W. D. Bascom, private communication.
2. J. G. Williams and P. D. Ewing, Int. J. Fracture Mech., 8, 441 (1972).
3. W. D. Bascom, C. O. Timmons, and R. L. Jones, J. Mat. Sci., 10, 1037 (1975).
4. G. G. Trantina, J. Composite Materials, 6, 371 (1972).
5. N. Iosipescu, J. of Materials, 2, #3, 537 (1967).
6. J. G. Bjeletich and A. D. Jonath, to be published.
7. D. L. Flaggs, private communication.
8. R. D. Adams and N. A. Peppiatt, J. of Strain Analysis, 9, #3, 185 (1974).
9. G. C. Knollman, J. J. Hartog and A. D. Jonath, to be published.
10. C. Bucci, R. Fieschi, and G. Guidi, Phys. Rev., 148, 816 (1966).
11. J. L. Crowley and A. D. Jonath, Adhesion and Absorption of Polymers, ed. by L. H. Lee, Plenum Press, New York, New York (1979).
12. W. J. Renton, Experimental Mechanics, Nov. 1976, p 409.

Discussion

A. Silberberg (Weizmann Institute, Israel): Have you any idea about the thinnest layer that could be scanned by this technique, particularly if there are no strongly absorbing chromophores present. Could one detect the presence of solution adsorbed polymer layer which typically are present in the amounts of 2-10 mg/M^2 of surface?

L.H. Lee (Xerox Corporation): No, we are just beginning to work on adsorption and do not have any data on the thinnest layer which could be scanned. The polymers without chromophores could present problems.

A. Jonath (Lockheed Palo Alto Research Laboratory): Have you measured PAS on indirect vs. direct semiconductors; i.e., how does the phonon structure (for indirect semiconductors) appear in the PAS spectra?

L.H. Lee: No. We have not measured the semiconductors you mentioned.

A Jonath: What experimental "tricks" did you use to make measurements in the U.V. region?

L.H. Lee: We used ozone-generating Xenon lamp for getting good PAS spectra in the U.V. region. In this region, we also have to be careful about false signals caused by the instability of the system. It is important to repeat several measurements, particularly with fresh samples, to confirm the spectra. The U.V. radiation tends to degrade some polymers or dyes after repeating scans.

B.D. Ratner (University of Washington): In order to study the nature of adhesive bonding on a molecular level, a higher level of resolution than that seen in your spectra would probably be necessary. Is there any prospect for increasing the resolution of this photoacoustic technique or for meaningfully deconvoluting the broad peaks observed in the U.V. region?

L.H. Lee: So far, resolution appears to be a problem, especially in the U.V. region. We have tried to reduce the amount of sample, or to use a diluent for the same purpose. I have not seen any work on deconvolution.   However, I have seen a paper on the Fourier transformation*. It is likely that the latter approach can improve the resolution in the future.

On the Paper by J.S. Solomon

D.W. Dwight (Virginia Polytechnic Institute): Why does aluminum drop at the metal-oxide interface?

J.S. Solomon (University of Dayton Research Institute): The apparent decrease in the aluminum at the oxide-metal interface is an artifact of the derivative recording mode which is related to the peak shape change and shift of $AL_{K\ell\ell}$.

D.W. Dwight: Why is there no oxygen in the epoxy?

J.S. Solomon: Some of the sputter profiles shown were constructed using normalized signals which means that the minimum signal levels were arbitrarily assigned a value of zero. Oxygen was always detected in the bulk epoxy adhesive.

On the Paper by D.W. Dwight

B. Ratner (University of Washington): Infrared analysis of the etched Teflon surface could not detect chemical changes. Could you identify by ESCA what chemical structural changes occurred on the surfaces to impart the darkened color which was observed?

D.W. Dwight (Virginia Polytechnic Institute): Yes, we discovered evidence for unsaturation as well as carbonyl and carboxyl groups. The details are in J. Colloid and Interface Sci. <u>47</u>, 650 (1974).

---

*   R.K. Burnham, M.M. Farrow and E.M. Eyring, paper presented to the Symposium on Photoacoustic Spectroscopy, FACSS-V, Boston, October 1978.

P. Datta (RCA Laboratories): Viscoelastic properties of the surface of the polymer contribute to the strength of adhesive bond. Do you have any method to measure modulus of the polymer surface?

D.W. Dwight: I believe this variable is the most important parameter that still eludes even qualitative evaluation. Two approaches have promise to help here: 1. tribological-friction and wear and 2. surface wave method.

On the Paper by J.L. Crowley and A.D. Jonath

D.W. Dwight (Virginia Polytechnic Institute): Why do there appear to be two spectra, each with two peaks?

J.L. Crowley (Lockheed Palo Alto Research Laboratory): There appear to be two spectra in Fig. 1 because peak cleaning techniques were used to better observe peak B. Specifically the measurement of the thermally stimulated current was stopped after peaks C & D had been discharged and the sample was recooled. When the measurement was resumed there was no longer a contribution to the current from peaks C & D and peak B could be observed more distinctly.

·D.W. Dwight: Can you interpret the peaks in terms of molecular motions?

J.L. Crowley: We are not able to correlate specific peaks with known molecular motions in the adhesive at this time. We hope to make some determinations based on a comparison with FTIR spectrum.

D.W. Dwight: What has been published on the surface wave technique for these materials?

J.L. Crowley: See Sound Waves in Solids, H.F. Pollord, Pion Ltd., 1977, pp. 65-77.

On the Paper by A.D. Jonath

L.H. Lee (Xerox Corporation): How do you foresee the shear wave analysis as a complementary device in determining shear strength?

A.D. Jonath (Lockheed Palo Alto Research Laboratory): In the approximation that shear strength is proportional to shear modulus, a shear wave analysis might give the relative strength of an adjoint. This could be especially useful in determining adhesive aging characteristics. However, the strength of an adjoint is often flaw-determined and the shear wave analysis would only be useful in predicting locus-of-failure.

D.W. Dwight (Virginia Polytechnic Institute):   Has the G of
steel or other model materials been determined by surface waves?

A.D. Jonath: We have determined the shear wave speed of model
materials (as is mentioned in the written version of the paper)
using the surface wave method. Some of these measured values can be
compared with published data as follows:

Material	Shear Wave Speed (km/sec)	
	Measured	Published
304 Stainless	3.14	3.13 (Ref. A1)
6061 Aluminum	3.18	3.10 (Ref. A2)
Glass Prism	2.60	2.56 " "
Microscope Glass Cover Slip	2.93	2.84 (Ref. A3)
Plexiglas	1.44 - 1.55	1.32 (Ref. A2)
Polystyrene	1.19 - 1.20	1.12 " "

Note that all surface wave (measured) values are greater than
corresponding published values. The latter are bulk values, usually determined by an ultrasonic transit time technique. The difference in the two might be explained by density variations between
bulk and surface materials. Thus to a first order approximation,
the shear modulus of model materials has been inferred from surface
wave measurements. But the technique is sensitive to mechanical
properties very near the surface (to a depth of one wavelength of
the incident acoustic wave) and so a density measurement in a very
thin surface layer is required to accurately measure the shear
modulus in the Rayleigh wave experiment.

REFERENCES

A1. R.S. Sharp, Research Techniques in Nondestructive Testing,
        Academic Press, New York, 1970.
A2. D. Ensminger, Ultrasonics, The Low- and High-Intensity Appli-
        cations, Marcel Dekker, Inc., New York, 1973.
A3. W.P. Mason, Physical Acoustics and The Properties of Solids,
        D. Van Nostrand Co., Inc., Princeton, 1958, p. 20.

D.W. Dwight: What is the effect of roughness on this technique?

A.D. Jonath: Surface roughness, if small compared to the
incident acoustic wavelength (as in these experiments), has no
effect.

PART THREE

POLYMERIC STRUCTURAL ADHESIVES

Introductory Remarks

D.H. Kaelble

Rockwell International

Thousand Oaks, California   91360

Polymeric structural adhesives now find a wide range of applications in mobile structures where high structural reliability with reduced weight and cost result by proper adhesive bonding. Component mechanical design, materials selection and manufacturing process control each play a critical role in achieving proper adhesive bonding. In this session the papers are arranged so as to discuss several aspects of fracture mechanics and rheology, followed by discussions of surface chemistry and adhesive chemistry, aspects of joint performance and reliability. Summary discussions on crack healing and UV curing deal with important issues relating to the microstructural aspects of adhesion intermediate between the chemical and continuum mechanical levels of analysis. The evolution of the technical viewpoint which intimately dovetails chemical, physical and mechanical viewpoints is reflected in this session which is a small but important part of this international conference.

(Editor's Note: The paper on "Adhesive Characteristics of Ultraviolet Radiation Curable Resins" originally presented to this Session by T. Kobayashi and K. Nate is included in Part 5 of this book.)

Fracture Mechanics and Adherence of Viscoelastic Solids

D. Maugis and M. Barquins

Equipe de Recherche de Mecanique des Surfaces

CNRS 1 Place A. Briand 92190 Meudon, France

ABSTRACT

Contact of two elastic solids is treated as a thermodynamic problem. It is shown that $U = U_E + U_S$ and $U_T = U_E + U_P + U_S$ are thermodynamic potentials respectively for transformations at fixed grips and at fixed load conditions (U_E, U_P, U_S are the elastic, potential, interface energies). Equations giving the displacement δ and the strain energy release rate G as a function of the contact area A and the load P appear to be the equations of state of the system. Two bodies in contact on an area A are in equilibrium if G=w, where w is the thermodynamic (or Dupre's) work of adhesion. This equilibrium is stable if $\partial G/\partial A$ is positive, unstable if negative. The quasi-static force of adherence is the load corresponding to $\partial G/\partial A = 0$. But equilibrium may be stable at fixed grips and unstable at fixed load, so that the quasistatic force of adherence may depend on the stiffness of the measuring apparatus. When G>w, the separation of the two bodies starts, and can be seen as the propagation of a crack in mode I. G-w is the force applied to unit length of crack; under this force, the crack takes a limiting speed v, which is a function of the temperature, and one can write

$$G - w = w\phi(a_T v).$$

The second term is the drag due to viscoelastic losses at the crack tip and is proportional to w as proposed by Gent and Schultz, and Andrews and Kinloch. The function ϕ is a characteristic of the material (most probably linked to the frequency dependence of E' and E", the real and imaginary part of the Young modulus) and is independent of the geometry and loading system. In this proposed formula surface properties and viscoelastic losses are clearly decoupled from elastic properties and loading conditions that appear in G. If ϕ is known, this equation allows one to predict any feature such as kinetics of detachment at fixed load, fixed grips or fixed cross-head velocity $\dot{\delta}$. (This last point completely solves the problem of tackiness). The only hypotheses are that failure is an adhesive failure and that viscoelastic losses are limited to the crack tip; this last condition means that gross displacements must be elastic for G to be valid in kinetic phenomena.

Three geometries are investigated: adherence of punches, adherence of spheres and peeling. The variation of energies with area of contact is given, and the kinetics of crack propagation under various conditions is studied. Experiments on the adherence of polyurethane to glass confirm the theoretical predictions with a high precision.

1.   INTRODUCTION

Historically, studies of adherence have approached the subject from two viewpoints: (i) the one of surface chemistry, trying to correlate adhesive strengths to van der Waals forces or to the thermodynamic work of adhesion; (ii) and the one of fracture mechanics in an attempt to predict the debonding threshold from analysis of stresses around preexisting flaws. Neither of these approaches was successful in studying problems as simple as peeling, or adherence of spheres or punches on elastic half spaces.

Our understanding of the adherence of elastic solids made a marked advance with the introduction of the energy balance concept by Kendall (1) and Johnson et al (2). This concept was successfully applied to a number of cases by Kendall: adherence of punches (Kendall, 1); peeling (Kendall, 2,3,4,5), lap shear joints (Kendall, 5-8), composites (Kendall, 9-12), laminates (Kendall, 5,7), and fibre pull-out (Kendall, 13).

This energy balance theory is based on optimizing the total energy of the system at equilibrium, and the adherence force is derived by equating the first derivative of the total energy to zero. But this procedure cannot give information about the stability of the system, that depends on the second derivative of the energy. Thus, one is led to reintroduce the concepts of fracture mechanics such as the strain energy release rate G, or the stress intensity factor K, and to study the stability by sign of the derivative of G or K as proposed in fracture mechanics by Rice (14) and Huet (15). This approach enables one to study the kinetics of crack propagation, and provides a synthesis that reconciliates surface chemistry and fracture mechanics (Maugis and Barquins, 16).

2.   GENERAL THEORY

2.1 The Thermodynamic Point of View

Let us consider the system of two elastic solids in contact over an area A. This system can exchange work and heat with outer systems, but no matter (closed systems). A force P (either compressive or tensile) can be applied on the two elastic bodies, either by a fixed load, as in Fig. 1a, or by a more complex system as in Fig. 1b.

At this stage two points must be emphasized:

a)   The area of contact A is allowed to vary, so that the geometry of the system can change, and linear elasticity must be excluded.

b)   This variation of A may be done independently of the load P, or of the elastic displacement δ, so that the state of the system depends in general on two independent variables P,A or δ,A. As an example the area of contact can decrease at fixed P or fixed δ, until the separation of the two bodies occurs, corresponding to the rupture of a joint under fixed load or fixed grips conditions.

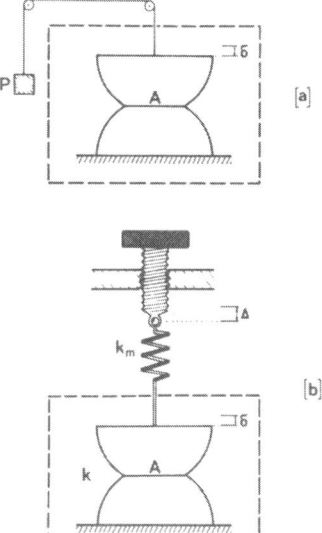

Fig. 1.   Study of the state of equilibrium at fixed load condition (a), and with a testing machine with a finite stiffness k_m (b). Fixed grips condition corresponds to $k_m = \infty$.

We will consider the reduction of the area of contact as the propagation of a crack toward the center of the contact, the crack extension being in mode I (opening mode, with rims moving perpendicularly to the crack plane). Our purpose is to study the equilibrium and stability of the system under various internal constraints. We will follow the general method of thermodynamics (see in particular the illuminating book by Callen (17)). The general method is as follows.

The state of a thermodynamic system is defined by a number of independent extensive variables X_i, as S entropy, V volume, N_i number of moles of constituent i. The system is in equilibrium for virtual variations of these extensive variables if the energy U is minimum. The equation

$$U = U (S,V,X_i \ldots)$$

is called the fundamental equation and its first differential

$$dU = (\frac{\partial U}{\partial S})_{V,X_i} \, dS + (\frac{\partial U}{\partial V})_{S,X_i} \, dV + (\frac{\partial U}{\partial X_i})_{S,V} \, dX_i + \ldots$$

can be written

$$dU = TdS - pdV + P_i dX_i$$

with the following intensive parameters:

$$(\frac{\partial U}{\partial S})_{V,X_i} \equiv T \quad \text{the temperature}$$

$$-(\frac{\partial U}{\partial V})_{S,X_i} \equiv p \quad \text{the pressure}$$

$$(\frac{\partial U}{\partial X_i})_{S,V} \equiv P_i$$

These intensive parameters are functions of $S,V,X_i \ldots$ and the knowledge of all the equations of state:

$$T = T (S,V,X_i \ldots)$$

$$p = p (S,V,X_i \ldots)$$

$$P_i = P_i (S,V,X_i \ldots)$$

is equivalent to the knowledge of the fundamental equation.

But, in general, the equilibrium must be studied with various constraints, as no heat exchange, constant temperature, constant pressure, constant volume, etc. The elegant solution used in thermodynamics is to use Legendre transformations of the energy U that exchange any variable X_j with the partial derivative $P_j = (\frac{\partial U}{\partial X_j})$. The equilibrium of the system at constant P_j corresponds to the minimum of the Legendre transform

$$\psi(P_j, X_i) = U - P_j X_j$$

These Legendre transformed functions are called thermodynamic potentials. So, the Helmholtz free energy F replaces the entropy S by the temperature T

$$F = U - TS$$

$$dF = -SdT -pdV + P_i dX_i \ldots$$

the enthalpy H replaces the volume V by the pressure p

$$H = U + pV$$

$$dH = TdS + Vdp + P_i dX_i$$

and the Gibbs free energy $\mathcal{G}$ replaces the entropy by the temperature, and the volume by the pressure

$$\mathcal{G} = U - TS + pV$$

$$d\mathcal{G} = -SdT + Vdp + P_i dX_i$$

Thus, any partial derivative P_i can be defined as

$$P_i = (\frac{\partial U}{\partial X_i})_{S,V,X_j} = (\frac{\partial F}{\partial X_i})_{T,V,X_j} = (\frac{\partial H}{\partial X_i})_{S,P,X_j} = (\frac{\partial \mathcal{G}}{\partial X_i})_{T,P,X_j} \quad (1)$$

2.2 Equilibrium of the System. Griffith's Criterion

Let us now return to our particular system. Its energy

$$U = U(S,\delta,A)$$

is a function, besides entropy S, of δ and A, and can be divided in elastic energy U_E and interface energy U_S. This interface energy is solely function of A and will be written

$$U_S = -(\gamma_1 + \gamma_2 - \gamma_{12})A$$

$$= - wA$$

where γ_1 and γ_2 are the surface energies of the two bodies, γ_{12} their interfacial energy, and w the Dupre's energy of adhesion, or thermodynamic work of adhesion. The first differential of energy

$$dU = (\frac{\partial U}{\partial S})_{\delta,A}dS + (\frac{\partial U}{\partial \delta})_{S,A}d\delta + (\frac{\partial U}{\partial A})_{S,\delta}dA$$

can be written

$$dU = TdS + Pd\delta + (G-w)dA \qquad\qquad (2)$$

with the following parameters

$$(\frac{\partial U}{\partial \delta})_{S,A} = (\frac{\partial U_E}{\partial \delta})_{S,A} = P$$

$$(\frac{\partial U}{\partial A})_{S,\delta} = (\frac{\partial U_E}{\partial A})_{S,\delta} + (\frac{\partial U_S}{\partial A})_{S,\delta} = G-w$$

G that describes the variation of elastic energy with A at constant δ is called "strain energy release rate".

The three equations of state[*]

$$S = S(\delta,A)$$

$$P = P(\delta,A) \qquad\qquad (3)$$

$$G = G(\delta,A)$$

carry on the same information as the fundamental equation $U = U(S,\delta,A)$.

For equilibrium under various constraints, the following thermodynamic potentials are to be used:

- at constant temperature, the Helmholtz free energy

$$F = U-TS$$

$$dF = -SdT + Pd\delta + (G-w)dA \qquad\qquad (4)$$

---

[*] Note that the implicit equation (3) implies

$$(\frac{\partial P}{\partial A})_{\delta} (\frac{\partial A}{\partial \delta})_{P} (\frac{\partial \delta}{\partial P})_{A} = -1 \qquad\qquad (3\ bis)$$

- at fixed load, the enthalpy

$$H = U - P\delta$$

$$dH = TdS - \delta dP + (G-w)dA \qquad (5)$$

- at fixed load and temperature, the Gibbs free energy

$$\mathcal{G} = U - TS - P\delta$$

$$d\mathcal{G} = -SdT - \delta dP + (G-w)dA \qquad (6)$$

Let us now simplify the problem by assuming that thermal effects are negligible (T=o, S=o). One has

$$dF = dU = d(U_E + U_S) = Pd\,\delta + (G-w)dA \qquad (7)$$

$$d\mathcal{G} = dH = d(U_E + U_S + U_P) = -\delta dP + (G-w)dA \qquad (8)$$

noting that $U_P = -P\delta$ in the potential energy of the load P. Relation similar to eq. (1) can be written for G-w, leading to

$$G = (\frac{\partial U_E}{\partial A})_\delta = (\frac{\partial U_E}{\partial A})_P + (\frac{\partial U_P}{\partial A})_P \qquad (9)$$

Note also that the crossed derivatives (Maxwell relations) of eqs. (7) and (8) give

$$(\frac{\partial G}{\partial \delta})_A = (\frac{\partial P}{\partial A})_\delta \qquad (10)$$

$$(\frac{\partial G}{\partial P})_A = -(\frac{\partial \delta}{\partial A})_P \qquad (11)$$

Equilibrium at fixed load conditions (dP=o) corresponds to an extremum of F or U (eq. 8)), and equilibrium at fixed grips conditions (dδ=o) to an extremum of $\mathcal{G}$ or H (eq. (7)). In either case, equilibrium is given by

$$G = w \qquad (12)$$

This equilibrium relation may be called Griffith's criterion and links two of the three variables δ, P, A of the equation of state (3), so that the equilibrium curves $\delta(A)$, $A(P)$, $P(\delta)$ are function of w.

If G≠w, the area of contact will spontaneously change so as to decrease thermodynamic potentials. (In spontaneous evolution they cannot increase). If G<w, eq. (7) and (8) show that A must

increase, and the crack recedes.   Conversely, if G>w the area of
contact must decrease to have dF<o or dq<o, and the crack extends.
GdA is the mechanical energy released when the crack extends by dA.
The breaking of interfacial bonds requires an amount of energy wdA,
and the excess (G-w)dA is changed in kinetic energy if there is no
dissipative factor. G is often called "crack extension force" but
properly speaking the crack extension force is G-w, which is zero
when G=w at equilibrium.

2.3 Stability of Equilibrium, and Stiffness of the Testing Machine

The equilibrium defined by eqs. (12) and (9) can be stable,
unstable or indifferent.   A thermodynamic system under a given
constraint is stable if the corresponding thermodynamic potential
is minimum, i.e. if its second derivative is positive.

Stability at fixed grips conditions ($d\delta$=o) as deduced from eq.
(7) corresponds to

$$(\frac{\partial G}{\partial A})_\delta > 0 \tag{13}$$

Stability at fixed load conditions (dP=o) corresponds to

$$(\frac{\partial G}{\partial A})_P > 0 \tag{14}$$

But stability can be studied with a testing machine with a finite
stiffness k_m.   Figure 1b is a schematic representation of such a
machine.   The stiffness k_m is that of the spring.   The crosshead
displacement Δ can be obtained by turning the screw (that is
thermodynamically equivalent to place a load P onto the spring, and
then to clamp it without external work), and is divided in elastic
displacement δ_m of the spring, and elastic displacement δ of the two
solids in contact.   The spring exerts a force $P = k_m \delta_m$ on the two
bodies and one has

$$\Delta = \delta + \delta_m = \delta + \frac{P}{k_m} \tag{15}$$

Let us study the stability of the system involving the two elastic
bodies and the spring at fixed crosshead displacement Δ.   The energy
of the system includes now the elastic energy U_m of the springs

$$U = U_E(A, \delta) + U_m(\delta_m) + U_S(A)$$

and its first differential is

$$dU = (\frac{\partial U_E}{\partial A})_\delta dA + (\frac{\partial U_E}{\partial \delta})_A d\delta + \frac{dU_m}{d\delta_m} d\delta_m + \frac{dU_S}{dA} dA$$

$$= GdA + Pd\delta + Pd\delta_m - wdA$$

$$= Pd\Delta + (G-w) dA$$

Equilibrium at fixed Δ is still given by $G=w$, and the stability by

$$(\frac{\partial G}{\partial A})_\Delta > 0 \qquad\qquad (16)$$

but this stability depends on the stiffness k_m of the apparatus. Intuitively, one can see the spring as a reservoir that provides energy for crack propagation at constant Δ.   Let us compute $(\frac{\partial G}{\partial A})_\Delta$ as a function of $(\frac{\partial G}{\partial A})_\delta$ by considering $G[A, \Delta(\delta, A)]$ as a function $G[A, \delta(\Delta, A)]$

$$(\frac{\partial G}{\partial A})_\Delta = (\frac{\partial G}{\partial A})_\delta + (\frac{\partial G}{\partial \delta})_A (\frac{\partial \delta}{\partial A})_\Delta \qquad\qquad (17)$$

Differentiating eq. (15) gives

$$d\Delta = d\delta + \frac{1}{k_m} \left[(\frac{\partial P}{\partial A})_\delta dA + (\frac{\partial P}{\partial \delta})_A d\delta \right]$$

hence

$$(\frac{\partial \delta}{\partial A})_\Delta = -(\frac{\partial P}{\partial A})_\delta \frac{1}{k_m + (\frac{\partial P}{\partial \delta})_A} \qquad\qquad (18)$$

Taking into account eqs. (18) and (10), eq. (17) becomes

$$(\frac{\partial G}{\partial A})_\Delta = (\frac{\partial G}{\partial A})_\delta - (\frac{\partial P}{\partial A})_\delta^2 \frac{1}{k_m + (\frac{\partial P}{\partial \delta})_A} \qquad\qquad (19)$$

The quantity $(\frac{\partial P}{\partial \delta})_A = k$ is the stiffness of the two elastic solids, and is positive. So, $(\frac{\partial G}{\partial A})_\Delta$ can be zero, whereas $(\frac{\partial G}{\partial A})_\delta$ is still positive.   Eq. (19) given by Courtel et al. (18) shows that the stability range monotonically increases with the stiffness, from the fixed load case ($k_m=o$) to the fixed grips case ($k_m \infty$).   When $k_m \to o$ one approaches the fixed load condition and eq. (19) becomes

$$(\frac{\partial G}{\partial A})_P = (\frac{\partial G}{\partial A})_\delta - (\frac{\partial P}{\partial A})_\delta^2 (\frac{\partial \delta}{\partial P})_A$$

This last equation is more easily found by writing

$$dG = (\frac{\partial G}{\partial A})_P dA + (\frac{\partial G}{\partial P})_A dP = (\frac{\partial G}{\partial A})_\delta dA + (\frac{\partial G}{\partial \delta})_A d\delta = o$$

and using eq. (3 bis) and Maxwell relations.

2.4 Quasistatic Adherence Force

The more general equation for stable equilibrium is thus

$$G = w$$

$$(\frac{\partial G}{\partial A})_\Delta > o$$

If a fluctuation decreases A (dA<o), G incrementally decreases and one has G<w: the crack recedes to its equilibrium position. It can only advance if the load P or the displacement δ is varied, bringing back G to the value w: one is dealing with controlled rupture of an adhesive joint. In this case one has

$$dG = (\frac{\partial G}{\partial A})_\Delta dA + (\frac{\partial G}{\partial \Delta})_A d\Delta = o \qquad\qquad (20)$$

with the two particular cases

$$dG = (\frac{\partial G}{\partial A})_\delta dA + (\frac{\partial G}{\partial \delta})_A d\delta = o$$

$$dG = (\frac{\partial G}{\partial A})_P dA + (\frac{\partial G}{\partial P})_A dP = o$$

Starting from a stable equilibrium with the two bodies compressed (P>o, δ>o) let us decrease the cross-head displacement Δ: one generally encounters a progressive reduction of area of contact, i.e. a controlled rupture with $(\frac{\partial G}{\partial A})_\Delta$>o up to a point where $(\frac{\partial G}{\partial A})_\Delta$=o; the equilibrium becomes unstable and the crack spontaneously extends toward the rupture under this given cross-head displacement. It extends with acceleration, for the crack extension force G-w increases as A decreases.

The load P_c corresponding to the limit of stability

$$G = w$$

$$(\frac{\partial G}{\partial A}) = o$$

will be called the __quasistatic adherence force__ of the two bodies. As the limit of stability is a function of the stiffness k_m, the measured adherence force can depend on the stiffness of the apparatus.

The Fig. 2 outlines relations between the thermodynamic poten-tial $U_T = U_E + U_P + U_S$ (for experiments at fixed load) and area of contact A. (This figure displays the case of the contact of two spheres). The equilibrium curve is the locus of points with $(\frac{\partial U_T}{\partial A})_P = o$ (i.e. G = w). By decreasing the area of contact, one encounters stable equilibrium (heavy line) and controlled rupture up to the point C where

$$(\frac{\partial^2 U_T}{\partial A^2})_P \equiv (\frac{\partial G}{\partial A})_P = o$$

At this point spontaneous propagation occurs at fixed load, (dP=o) and total energy decreases at constant load (quasistatic adherence force) along the curve $(U_T)_P$.

From eq. (8), one has along equilibrium curve

$$\frac{dU_T}{dA} = - \delta\frac{dP}{dA}$$

and hence the point C corresponds to a horizontal tangent on the equilibrium curve. It must be emphasized that neither $(\frac{\partial U_T}{\partial A}) = o$ nor $\frac{dU_T}{dA} = o$ (see point B) are suitable criteria for spontaneous crack extension.

Similar curve can be drawn for experiments at fixed grips conditions, with $U_E + U_S$ as the thermodynamic potential. The rupture corresponds to an extremum of the equilibrium curve given by

$$\frac{d(U_E + U_S)}{dA} = P\frac{d\delta}{dA} = o$$

2.5 Further Considerations About Forces and Displacements

Let us return to eq. (8); $U_T = U_E + U_P + U_S$ is the total energy of a system like that of Fig. 1a and one can write:

$$\delta = -(\frac{\partial U_T}{\partial P})_A = -\frac{dU_T}{dP} \tag{21}$$

where total derivative holds for displacement along equilibrium curve (G = w). One can also eliminate $dU_S = -wdA$ and write

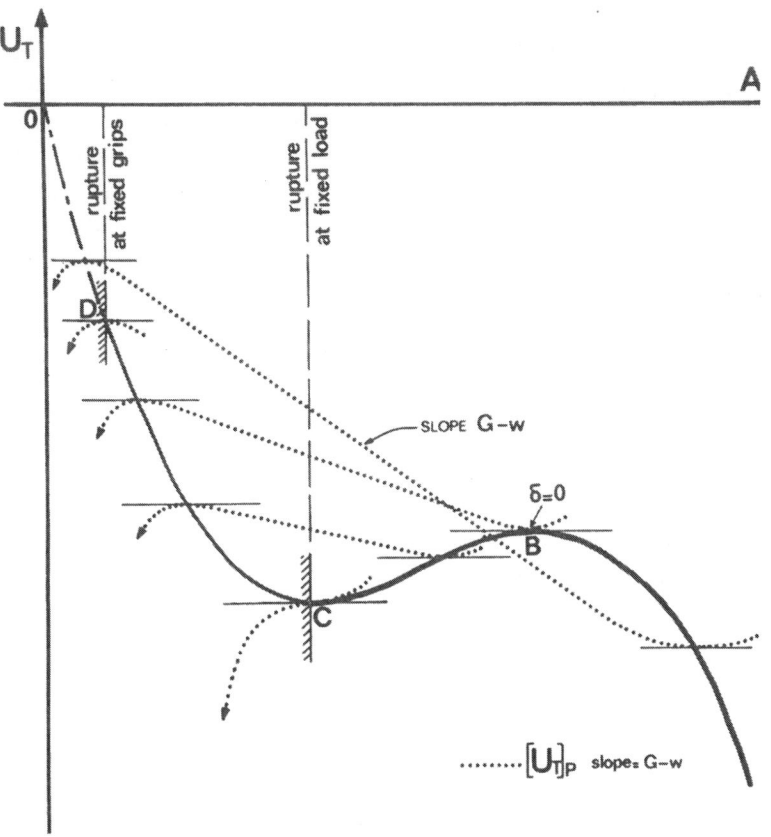

Fig. 2.   Variation of the total energy U_T with the
          area of contact A in the case of the contact
          of two spheres. By decreasing the area of
          contact, equilibrium is stable (at fixed load
          or fixed grips) up to the point C (heavy line);
          between the points C and D, the equilibrium
          is unstable at fixed load but stable at fixed
          grips conditions; beyond the point D (dashed
          line) the equilibrium is always unstable.

$$dU_C = \delta dP - GdA \tag{22}$$

with $U_C = -(U_E+U_P) = -U_T+U_S$, a term we will call the __complementary__
__energy__. One has:

$$G = -(\frac{\partial U_C}{\partial A})_P$$

$$\delta = (\frac{\partial U_C}{\partial P})_A$$

This last expression is known in classical elasticity as the generalized Castigliano theorem, that gives the displacements in nonlinear elasticity, when Castigliano's theorem $\delta = (\frac{\partial U_E}{\partial P})_A$ is irrelevant. The complementary energy is thus the thermodynamic potential used in classical elasticity for transformation at fixed load. In classical elasticity $(U_S = o)$

$$U_C = P\delta - \int Pd\delta = \int \delta dP$$

but, in problems with surface energy, the elastic energy is no longer given by $\int Pd\delta$ as we will see below.

Eq. (7) can be rewritten as

$$dU_E = Pd\delta + GdA \tag{23}$$

that shows that evaluation of elastic energy in systems with surface energy needs special care. Besides external forces, there are forces due to molecular attraction that cause elastic deformations and lead to elastic energy storage. In classical mechanics the derivative of the elastic energy gives the applied load. Here, three derivatives can be defined:

$$(\frac{\partial U_E}{\partial \delta})_A = P$$

$$(\frac{\partial U_E}{\partial \delta})_P = P + G (\frac{\partial A}{\partial \delta})_P = P + \Pi \tag{24}$$

$$\frac{dU_E}{d\delta} = P + w \frac{dA}{d\delta} = P + F_S \tag{25}$$

The force $F_S = -\frac{dU_S}{d\delta}$ is associated with displacements along the equilibrium curve; the force Π, also given from eq. (22) by

$$\Pi = -(\frac{\partial U_C}{\partial \delta})_P$$

acts in addition to applied load P in a displacement δ at constant P.

2.6 The Stress Intensity Approach

It is shown in fracture mechanics that stresses vary as $\frac{1}{\sqrt{\rho}}$ with the distance ρ near a crack tip, and the stress intensity factor K, for crack propagation in mode I is defined by

$$\sigma_z = \frac{K_I}{\sqrt{2\pi\rho}} \qquad\qquad (26)$$

where σ_z is the normal stress in the crack plane. This stress intensity factor is related to strain energy release rate by

$$G = \frac{K_I^2}{E} \qquad\text{for plane stress} \qquad (27)$$

$$G = K_I^2 \frac{1-\nu^2}{E} \qquad\text{for plane strain} \qquad (28)$$

This intensity factor can also be seen as a discontinuity of displacement factor (17), for discontinuity of displacement $[u_z]$ can be written as

$$[u_z] = \frac{K_I}{\mu} (k+1)\sqrt{\frac{\rho}{2\pi}} \qquad\qquad (29)$$

with μ the shear modulus and

$$k = \frac{3-\nu}{1+\nu} \qquad\text{for plane stress} \qquad (30a)$$

$$k = 3-4\nu \quad\text{for plane strain} \qquad\qquad (30b)$$

2.7 Viscoelastic Losses

The force P_o corresponding to G=w and $(\partial G/\partial A)=0$ is the quasi-static or thermodynamic force of adherence (not to be confused with the thermodynamic work of adhesion w). When G>w, the force G-w is applied to the unit length of crack that accelerates, until a limit speed v imposed by viscoelastic losses is reached. In the case of spontaneous crack extension, G-w increases with time under the constant load P_o so that the crack velocity increases with time. But there are cases (e.g. peeling) where G is independent of the area of contact, so that the crack does not move under the force P_o. A force $P > P_o$ must be applied to move the crack at detectable velocity, and the value P_o can be obtained only by extrapolation towards zero crack speed.

Viscoelastic losses vary with deformation rate and temperature, but results at various temperatures can be shifted to a
reference temperature $T_s=T_g+50$ (where T_g is the glassy transition
temperature) by using the WLF shift factor a_T given (Ferry, 20) by

$$\log a_T = - \frac{8.86 \ (T-T_S)}{101.6 + T-T_S} \ . \tag{31}$$

So, for a given geometry, adherence forces may be studied as a
function of the reduced parameter a_Tv, where v is the crack speed: a
master curve is thus obtained (Kaelble, 21,22), (Gent and Petrich,
23), (Gent, 24), (Gent and Kinloch, 25). If one assumes that
viscoelastic losses are localized at the crack tip and proportional
to the thermodynamic work of adhesion* as suggested by peeling
experiments in various liquids (Gent and Schultz, 26) or on various
substrates (Andrews and Kinloch, 27) one can write

$$G-w = w \, \phi(a_Tv) \tag{32}$$

where ϕ is a dimensionless function of crack speed and temperature,
independent of the geometry of the system. At high crack speed one
has G>>w, and the variation of crack speed with w, arises from the
multiplication term w in the right hand side of eq. (32).

Equation (32) can be rewritten as

$$G = w \, [1+\phi(a_Tv)] = w_f \tag{33}$$

where w_f may be called apparent work of adhesion, or fracture work
of adhesion by analogy with fracture surface energy γ_f. This
apparent work of adhesion is thus the product of two terms: the
thermodynamic work of adhesion w and a function of velocity and
temperature, as proposed by Gent and Schultz (26) and Andrews and
Kinloch (27). Equation (33) shows that influence of geometric
parameters entering in G can be tested by working at constant
temperature and crack velocity, i.e. working at constant w_f.
Measuring adherence force with a non-zero crack speed, can lead to
very high values, since viscoelastic losses, and hence w_f, can be
enormous, even for 1 μm S^{-1}. Inspection of the WLF shift factor
shows that it is beneficial to work at high temperature to reduce
w_f.

---

* This means that viscoelastic losses can only arise if the interface is itself capable of withstanding stress (Andrews and Kinloch, 27).

Application of eq. (32) implies that gross displacements are purely elastic with G computed from the relaxed elastic modulus E_o, and that the frequency dependence of E only appears at the crack tip where deformation velocities $\dot{\varepsilon}$ are high. The loss factor $\frac{E''}{E'}$ (where E' and E" are the real and imaginary parts of E) and its frequency dependence are taken into account in the function $\phi(a_T v)$.

Note that decreasing w at constant bulk properties and loading system increases the crack velocity, so that the viscoelastic solid seems to be less "viscous". At very low w values, the interface can appear "brittle". Conversely, inserting a very thin film of glue between the two solids considerably increases the work of adhesion but does not change the variation of elastic potential energies, so that the above treatment may be used for glued joints.

3.   EXPERIMENTAL METHOD

3.1 Apparatus

The apparatus used (Fig. 3) to study the adherence of glass spheres and punches is a modified form of the apparatus described elsewhere (Barquins and Courtel, 28) for the study of rubber friction.   The area of contact, illuminated by reflection of monochromatic light, is observed through the punch or the spherical cap with a microscope.   Figure 4 is a diagrammatic view of the mechanical system.   The hemisphere or the punch is supported by a balance.   This balance has an axis of rotation given by two micro-ball bearings; its sensitivity is about 10 dyn.   The vertical displacement of the arm can be recorded by a displacement transducer with a precision of about 0.1 μm.   The whole mechanical system can be laterally shifted to adjust the centre of the contact with the microscope axis.

A 16 mm camera (25 or 50 frames per second) is located at the top of the microscope, and the contact areas are recorded at a magnification of about x10.   Any sequence begins with the photograph of a micrometer scale; the areas of contact are such that the view through the glass ball does not introduce optical distortion.   The contact radii are thus obtained with a precision of about 1 μm. Experimental results of sections 4.4 and 5.5 correspond to about 160,000 frames, of which about 60,000 have been measured.

3.2 Material

The viscoelastic material chosen was an optically smooth poly-urethane, recommended for dynamic studies in photoelasticity (PSM 4 Vishay) and delivered as plates of thickness h=3.175 mm (1/8 inch).

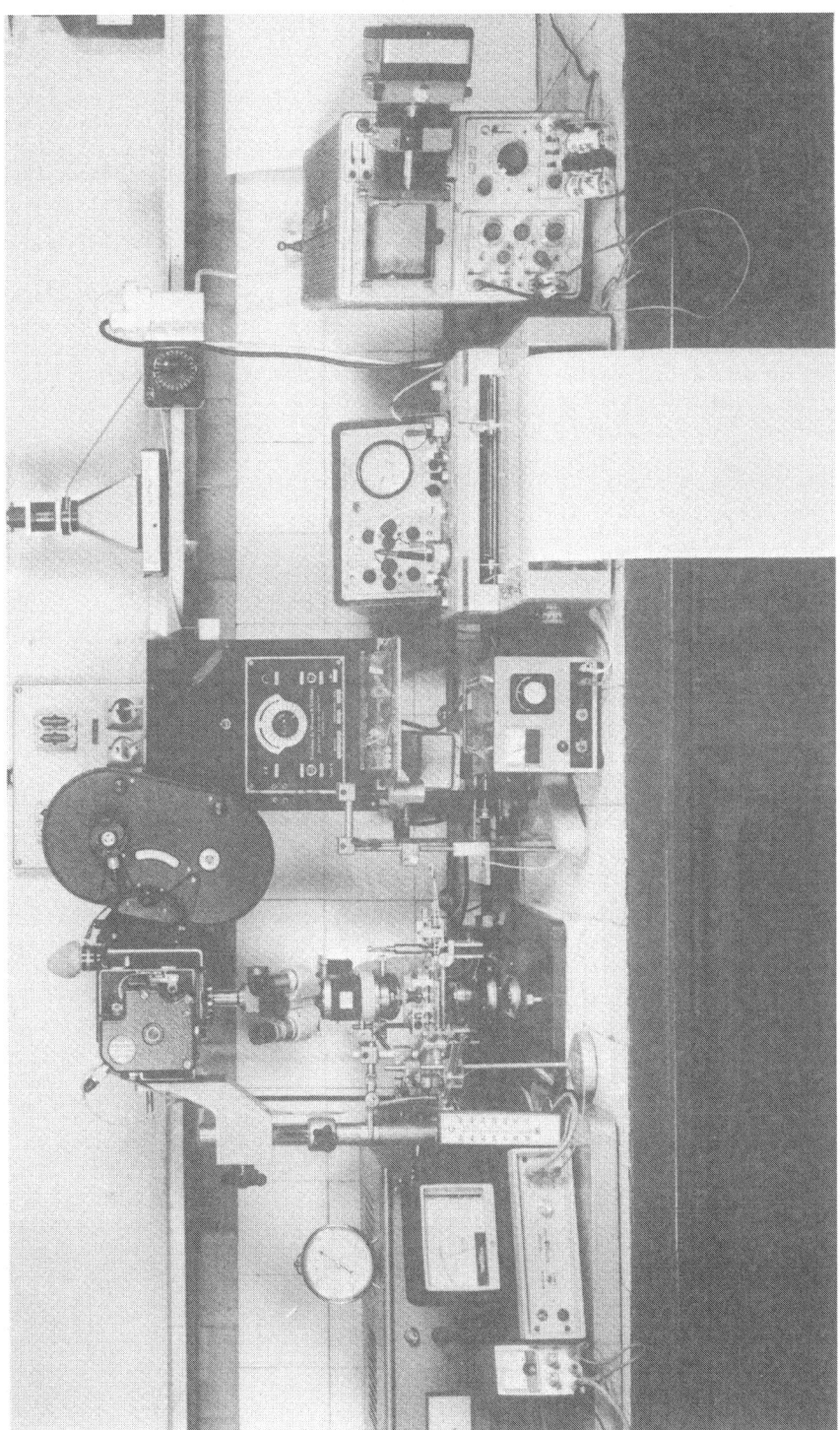

Fig. 3. General view of the apparatus used to study the adherence of glass spheres and punches to viscoelastic materials.

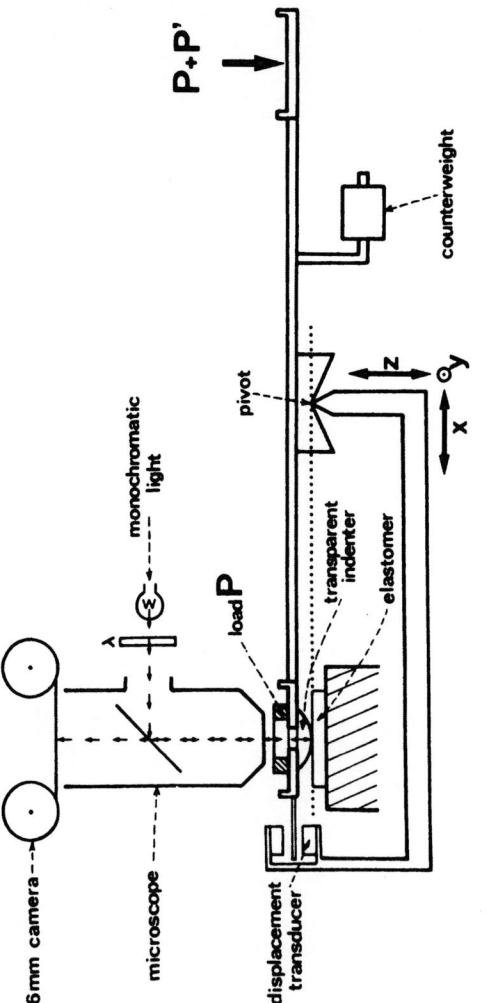

Fig. 4.   Schematic arrangement of the apparatus shown in Fig. 3.

The surfaces were cleaned with alcohol, dried with warm air, and left during 30 min for the equilibrium with room atmosphere to be reached. Microscopic examination fails to reveal any surface feature such as the bloom crystals observed by Roberts and Othman (28 bis) on rubber surfaces. The Young's modulus E and the thermo-dynamic work of adhesion w to glass were measured by the method proposed elsewhere (Barquins and Courtel (28), i.e. the verifica-tion of eq. (72) for the contact of a small glass ball. This method was previously used to study the variation with temperature of E and w for rubbers (Barquins et al., 29). The Young's modulus of this polyurethane, so measured, is $E=5\times10^7$ dyn cm^{-2}, the Poisson's ratio is taken as $\nu=\frac{1}{2}$, and w can vary from day to day, most probably with the relative humidity of the atmosphere from $w=50$ erg cm^{-2} to $w=85$ erg cm^{-2} (high values corresponding to fine weather). So w was always measured before any set of experiments, and the contact during any set of experiment was always made at the same point. In these conditions, the reproducibility is excellent.

4. ADHERENCE OF A RIGID PUNCH

4.1 Indentation By a Punch. The Boussinesq's Problem

Stresses and deformations provoked by a rigid circular flat punch indenting an elastic half-space (Fig. 5a) were given by Boussinesq (30). Under the punch (radius a) the elastic displace-ment and the pressure are

$$\delta = \frac{1-\nu^2}{2E} \frac{P}{a} \tag{34}$$

$$\sigma_z = \frac{P}{2\pi a^2 \sqrt{1 - \frac{r^2}{a^2}}} \tag{35}$$

The vertical displacement of the surface* for $r>a$ is

$$u_z = \frac{1-\nu^2}{\pi E} \frac{P}{a} \text{ Arc sin } \frac{a}{r} \tag{36}$$

---

* From a theorem due to Way (31) the horizontal displacement in the surface, outside the contact is independent of the load distribu-tion inside the area of contact. It is thus the same as for a concentrated force

$$u_z = - \frac{(1-2\nu)(1+\nu)}{2\pi E} \frac{P}{r} \text{ for } r > a$$

The asymptotic value at large distances, $\frac{r}{a} \gg 1$, is

$$u_z = \frac{1-\nu^2}{\pi E} \frac{P}{r} \tag{37}$$

as for a concentrated force; the asymptotic value near the edge

$$u_z = \delta - \frac{2\delta}{\pi} \sqrt{\frac{2(r-a)}{a}} \tag{38}$$

thus, $\frac{du_z}{dr} = \infty$ at the edge of the contact. The same equations hold for an adhesive contact, when P is a tensile load (Fig. 5b).

4.2   Strain Energy Release Rate

The variation of energies in a fluctuation of area of contact at fixed load or fixed displacement is difficult to "see" in Fig. 5b, and it is more convenient to consider Fig. 5c, a rigid plate in contact with a viscoelastic half space on an area $A=\pi a^2$, under a tensile load P.

In fact, it is not necessary to restrict ourselves to a rigid plate, and the stresses developed at the interface will develop the same kind of elastic deformation in the "indenter" (Fig. 5d). So, as pointed out by Kendall (31 bis) adherence of a punch is representative of fracture experiment on a deeply notched cylindrical bar. In Fig. 5d, the elastic displacement δ is the sum of the displacement of the two bodies and will be written

$$\delta = \frac{2P}{3aK} \tag{39}$$

with

$$\frac{1}{K} = \frac{3}{4} \left(\frac{1-\nu_1^2}{E_1} + \frac{1-\nu_2^2}{E_2} \right) \tag{40}$$

The elastic displacement varies with P at fixed a in Fig. 5a, but can also vary with a at fixed P as in Figs. 5c and 5d. Equation (39) is the equation of state (3).

The elastic energy is computed from the observed geometry of the system, and from given values of δ and a it is the same as in Fig. 5c, i.e.

$$U_E = \frac{1}{2} P\delta = \frac{3aK\delta^2}{4} = \frac{P^2}{3aK}$$

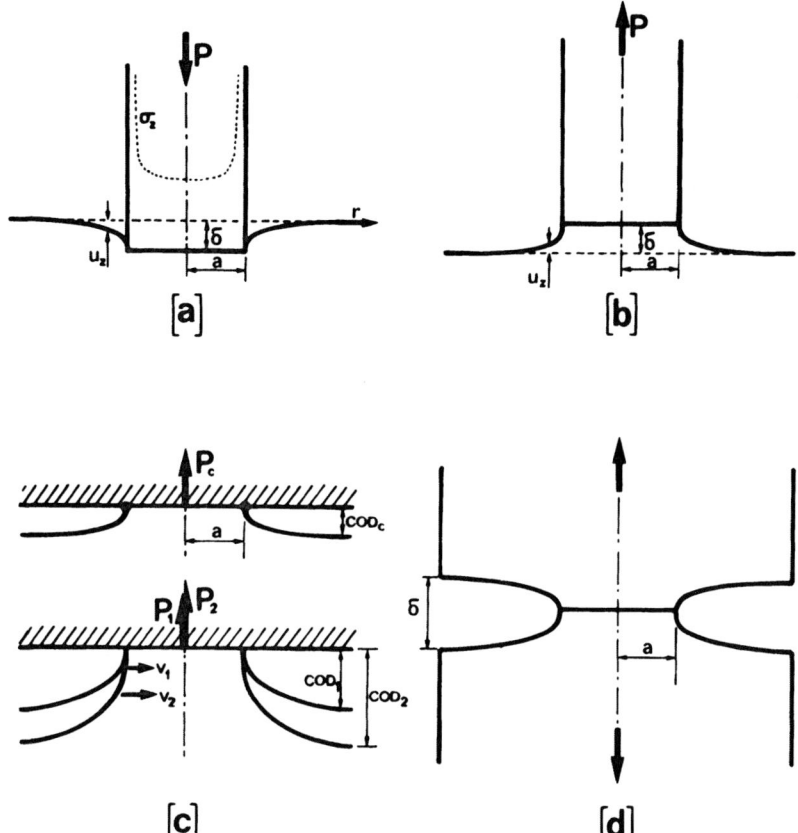

Fig. 5. Analogy between the adherence of a rigid
circular flat punch on elastic half-space
and a fracture experiment on a deeply
notched cylindrical bar.

With the potential energy, $U_p = -P\delta$, one can compute G from eq.
(9):

$$G = \frac{P^2}{6\pi a^3 K} = \frac{3K\delta^2}{8\pi a} \tag{41}$$

This is the second equation of state.

The equilibrium given by G=w is always unstable for

$$\left(\frac{\partial G}{\partial A}\right)_P = 3\left(\frac{\partial G}{\partial A}\right)_\delta = -\frac{P^2}{4\pi^2 a^5 K} = -\frac{9K\delta^2}{16\pi^2 a^3} = -\frac{3}{2}\frac{G}{A} \qquad (42)$$

is negative. If, from the equilibrium point, A increases, it will indefinitely increase; if it decreases, it will decrease until the rupture is reached. If the punch is of finite size (radius a_o) the area of contact, of course, cannot exceed $A_o = \pi a_o^2$; and the only possibility when P reaches the critical value

$$P_c = +(6\pi a_o^3 Kw)^{\frac{1}{2}} \qquad (43)$$

given by G=w, is a spontaneous reduction of area of contact down to rupture. This value P_c, first given by Kendall (1) is the quasi-static adherence force of this punch. Similarly, if displacement δ is imposed, spontaneous rupture occurs as soon as G=w, corresponding to a critical displacement*

$$\delta_c = \left(\frac{8\pi aw}{3K}\right)^{\frac{1}{2}} \qquad (44)$$

The Maxwell relations (10) and (11) becomes

$$\left(\frac{\partial G}{\partial \delta}\right)_A = \left(\frac{\partial P}{\partial A}\right)_\delta = \frac{P}{2\pi a^2} \qquad (45)$$

$$\left(\frac{\partial G}{\partial P}\right)_A = -\left(\frac{\partial \delta}{\partial A}\right)_P = \frac{P}{3\pi a^3 K} \qquad (46)$$

Note also the relation

$$\left(\frac{\partial G}{\partial A}\right)_\delta - \left(\frac{\partial G}{\partial A}\right)_P = \frac{G}{A} \qquad (47)$$

4.3 Energies of the System; Berry's Representation

As said in sec. 2.2, the "enthalpy" $U_T = U_E + U_P + U_S$ is the thermodynamic potential for transformations at constant load (see eq. (8)). Figure 6 shows the variations of this "enthalpy" along the equilibrium point, or during transformations at constant load

---

* As δ may be seen as a crack opening displacement (COD), it appears that a critical COD must be reached before the crack begins to start.

P.   At equilibrium points one has

$$U_E = -2U_S = 2Aw \qquad (48)$$

so that

$$U_T = -\frac{3}{2} U_E = \frac{3}{4} U_P = 3U_S = -3Aw \qquad (49)$$

and the curve $U_T(A)$ is a straight line with $(\frac{\partial U_T}{\partial A})_P = o$ everywhere and $\frac{dU_T}{dA} = o$ nowhere.

For transformations at fixed displacement, the internal energy $U = U_E + U_S$ is the thermodynamic potential and the Fig. 7 gives its variations along equilibrium points with

$$U = \tfrac{1}{2}U_E = -U_S = Aw.$$

For spontaneous transformation these potentials cannot increase.

Berry (32) has shown that plotting the locus of the Griffith criterion, and the loading curve on a stress-strain plot enables the various energies involved to have a geometric interpretation. Especially the surface energy of the system is the area between the loading curve and the locus of the Griffith criterion.   This locus is the cubic

$$P = \frac{9K^2}{16\pi w} \delta^3 \qquad (50)$$

and is drawn on Fig. 8.   Points such as B or B_1 on this locus have the area A as parameter.   For a punch of area A_o, the loading curve is the straight line OB (constant A_o).   At the equilibrium point B, the elastic energy $U_E = \tfrac{1}{2}P_c\delta_c$ is given by the triangle OBI, and the surface energy $|U_S| = \tfrac{1}{2}|U_E|$ by the double hatched zone (for a cubic $y=kx^3$, the area under the curve is $\tfrac{1}{4}yx$).   During crack propagation at fixed load (Fig. 8b) or at fixed displacement (Fig. 8c) the area of contact and the surface energy U_S decrease, and the elastic energy given by the triangle* OB_1I increases toward infinity or decreases to zero.

---

* Note that the elastic energy at the point B_1 in Fig. 8b is not given by

$$\int P d\delta = \tfrac{1}{2}P_c\delta_c + P_c(\delta - \delta_c)$$

(footnote continued on next page)

During crack propagation at fixed load, the variation of U_E is given by eq. (24) with $\Gamma = -\frac{P}{2}$. Hence $(\frac{\partial U_E}{\partial \delta})_P = \frac{P}{2}$ and

$$U_E = \frac{1}{2}P_c\delta_c + \int_{\delta_c}^{\delta} (P+\Gamma)d\delta = \frac{1}{2}P_c\delta.$$

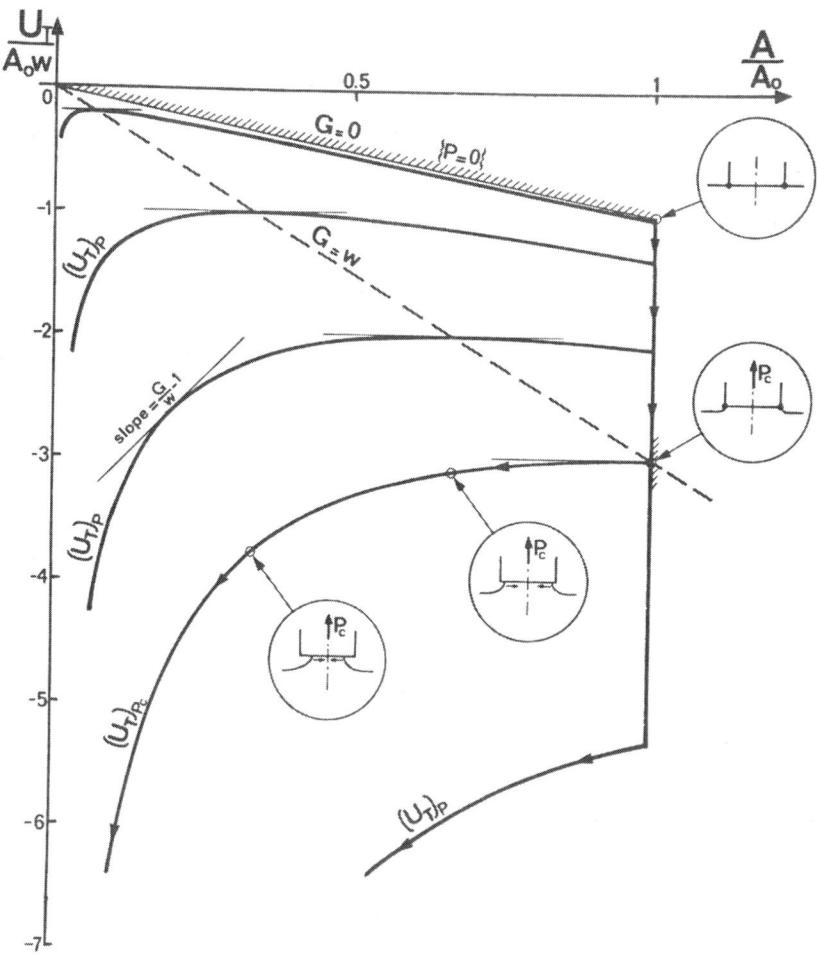

Fig. 6.   Total energy versus area of contact in reduced
coordinates. The dashed line is for equilibrium points, and heavy curves are for variations at constant load. A schematic evolution
of the contact under the load P_c is also shown.

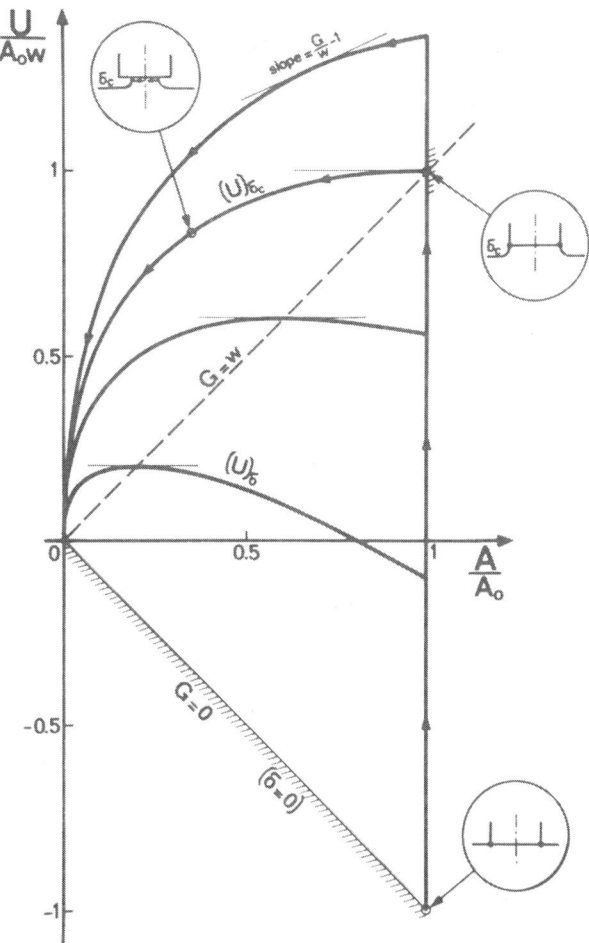

Fig. 7.   Internal energy versus area of contact in
          reduced coordinates. The dashed line is for
          equilibrium points and heavy curves are for
          variations at fixed grips condition. A schematic
          evolution of the contact under the displace-
          ment δ_c is also shown.

In Fig. 8b the thermodynamic potential $U_T = U_E + U_P + U_S$ at point B_1 is the negative area $\overset{\frown}{HB_1B_2}0$ and its variation BB_1B_2 (dotted zone) during crack propagation is the work $\int_{A_o}^{A}(G-w)dA$ done by the system: it corresponds to the kinetic energy of the crack and/or to heat dissipated in viscous drag. The work done at rupture is infinite, for rupture corresponds to δ infinite.

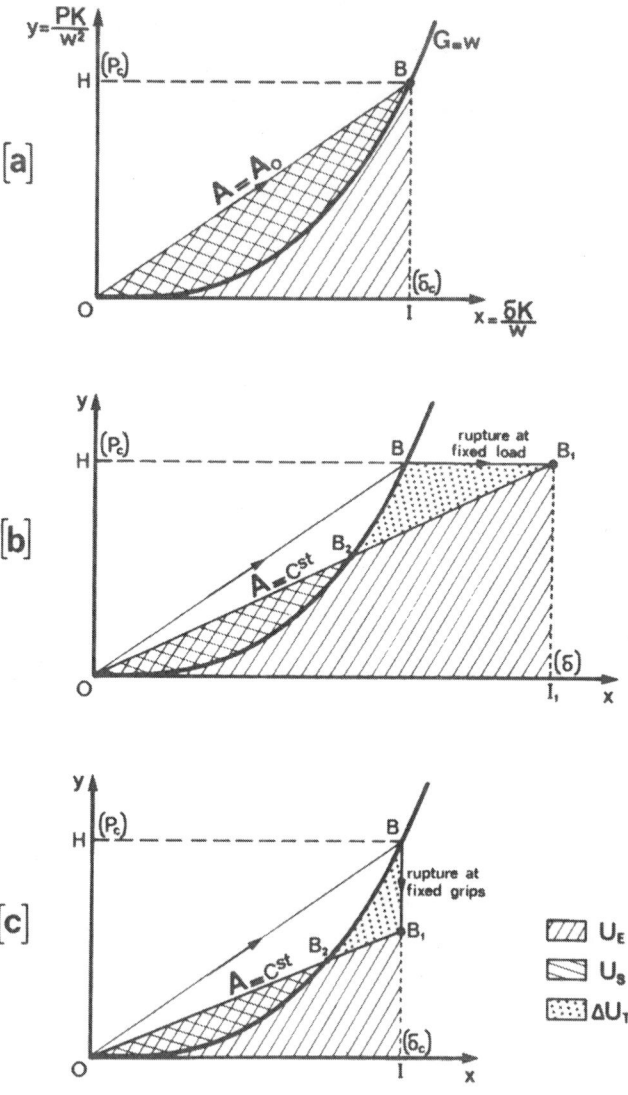

Fig. 8.   Berry's representation for the contact of
a rigid flat punch.   Stored elastic and
surface energies are given for equilibrium
condition (a) and for crack propagation at
fixed load (b) and at fixed grip condition
(c).   The variation of total energy during
propagation is also given.

In Fig. 8c the thermodynamic potential $U = U_E + U_S$ is the positive area $\widehat{OB_2B_1I}$ and the work $\int_{A_0}^{A}$ (G-w)dA done by the system is still BB_1B_2. At rupture this work is $\frac{1}{2}U_E = -U_S = A_0w$.

4.4   Crack Propagation at Fixed Load

Starting with a punch at $P=0$, $\delta=0$, let us suddenly apply a tensile load $P>P_c$ and study the kinetics of crack propagation by measuring the variation of radius of contact with time.

Figure 9 is a photograph of the area of contact during this crack propagation. Whatever the shape and size of the initial contact, the final shape is that of a circle. The grey border around the contact is due to light reflection along the polyurethane profile shown in Fig. 5c. This profile varies with P and a.

Figure 10 shows the variation with time of the radius of contact under loads varying from $P = -3000$ to -15000 dyn. The origin of times is that of rupture, so that all times are negative. For a given tensile load, the crack speed increases as the radius a decreases, and for a given radius a, the crack speed increases as the tensile load P increases. These results can be analyzed by the eq. (32) with G given by the eq. (41). On Fig. 11 the normalized "motive" $\frac{G-w}{w}$ computed from the value of P,w and the measured a is plotted as a function of the crack speed $v = \frac{da}{dt}$. The results obtained at $21°C$ were shifted to $23°C$ by the WLF method (eq. (31) with a glassy transition taken as $T_g = -50°C$) to compare with the peeling results (heavy curve, see sec. 6). This curve is the dissipative function $\phi(a_Tv)$ that appears to vary as $(a_Tv)^{0.6}$.

Once the dissipative function

$$\phi(a_Tv) = \alpha(a_Tv)^{0.6} \tag{51}$$

known, the differential equation (32), together with the eqs. (39) and (41) allows for the prediction of any evolution of the system. The kinetic is independent of a_0 and only function of the instantaneous radius of contact a. It is the kinetic of the rupture of plane/plane contact, under the action of a traction perpendicular to the interface. In the particular case where the initial crack speed is such that $G \gg w$, it comes

$$t = \frac{1}{6} (6\pi Kw\alpha)^{\frac{1}{0.6}} \frac{a^6}{P^{10/3}}$$

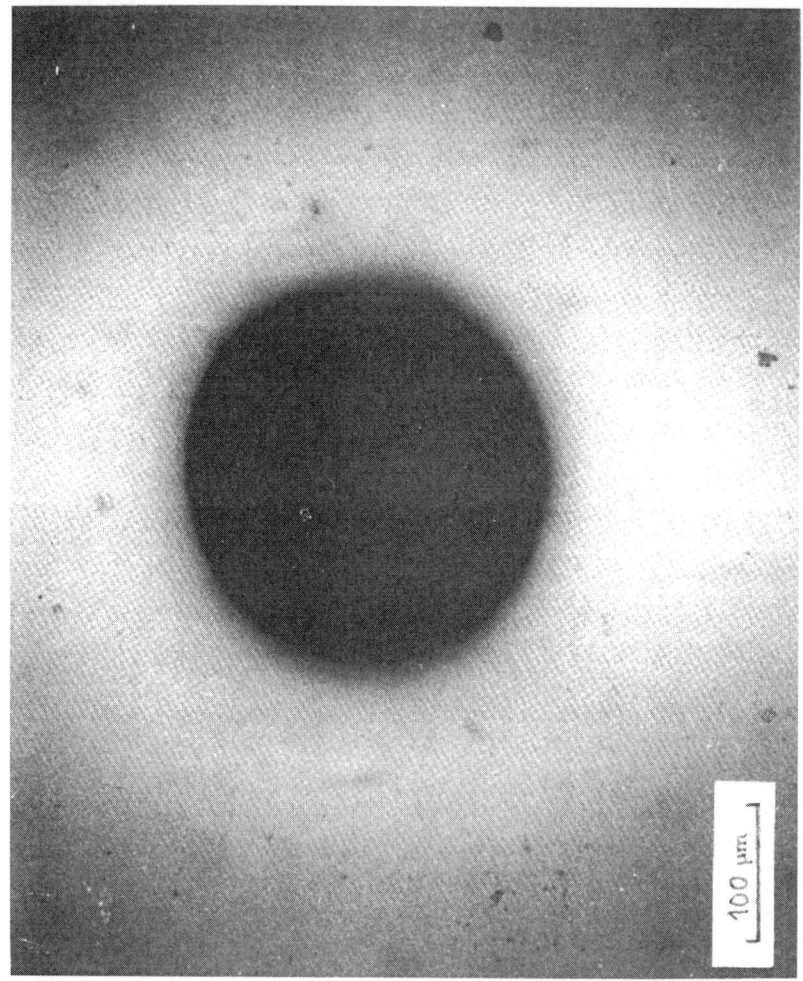

Fig. 9. View of the area of contact during the unloading of a glass flat punch from a polyurethane surface. The corresponding profile of contact is shown in Fig. 5c.

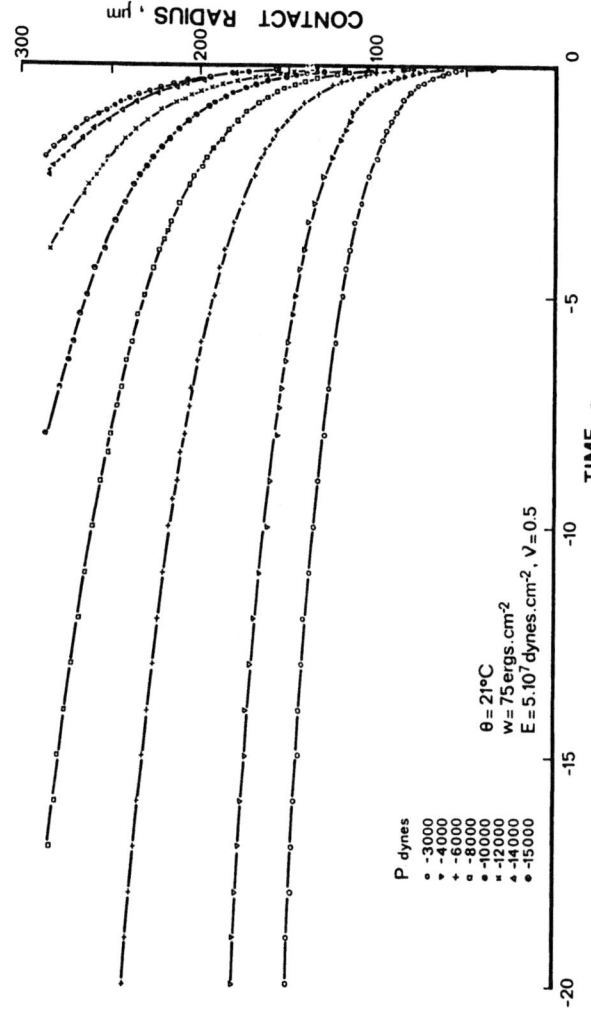

Fig. 10.   Adherence of a glass flat punch to polyurethane: radius of contact
versus time, for various tensile loads P. The origin of time is
that of rupture so that all times are negative. For a given tensile
load P, the crack speed increases as the radius a decreases, and
for a given radius $\underline{a}$ the crack speed increases as the tensile load
P increases.

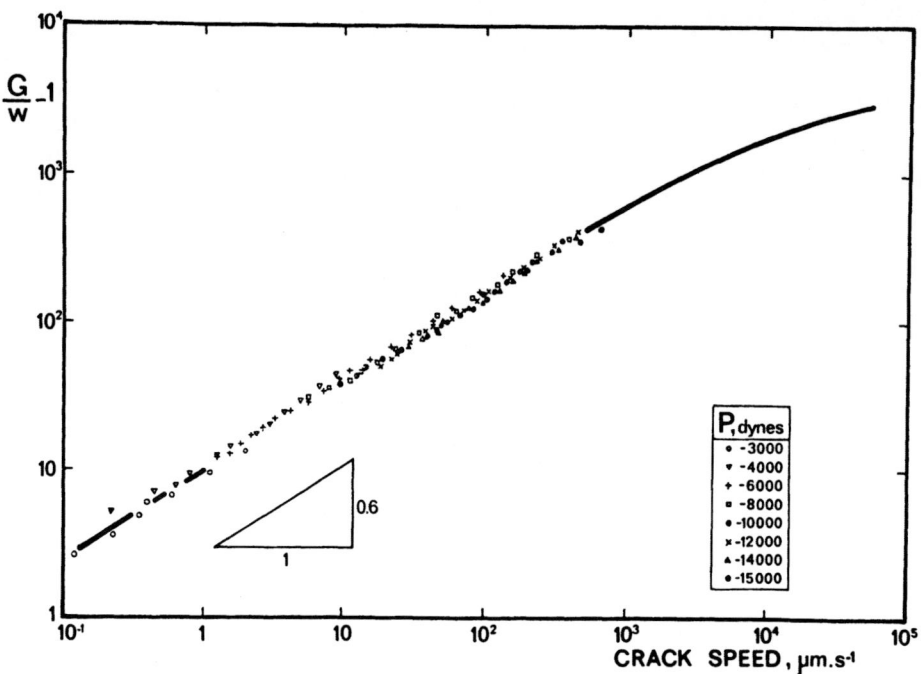

Fig. 11.   "Normalized motive" versus crack speed for
            adherence of a glass flat punch on polyure-
            thane surface. The same master curve as for
            peeling (heavy line) is obtained.

and the curves of Fig. 10 may be normalized by $\dfrac{a^6}{p^{10/3}}$ and not by $\dfrac{a^6}{P^4}$ as

erroneously inferred by Maugis and Barquins (33).

        The whole analysis is based on the hypothesis that eq. (41) is
still valid during crack propagation, i.e. that displacements δ are
purely elastic, and related to P by eq. (39) with relaxed elastic
moduli. This was checked. Figure 12 shows the theoretical relations between δ and a at equilibrium points, or during rupture at
constant load in the elastic hypothesis. The experimental displacements are given on Fig. 13. As the origin of the displacement
δ was difficult to obtain, the points (δ,a) corresponding to the
last frame (1/8 S before the rupture) of every unloading were put on
the theoretical curves. It can be seen that all the other points
fall on these theoretical curves. Thus δ varies with a, at constant
P, in agreement with eq. (39), i.e. the displacements are purely
elastic, so that viscoelastic losses are limited to the crack tip.
This is why the expression for strain energy release rate is still
valid far from equilibrium.

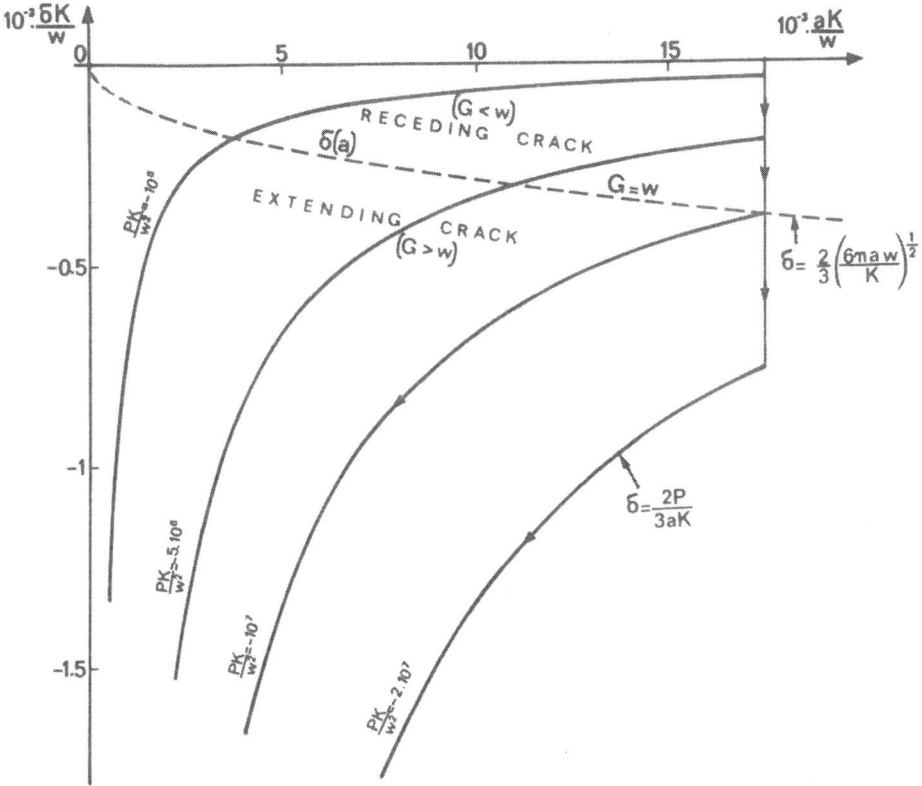

Fig. 12. Relations, in reduced coordinates, between
          elastic displacements δ and radius a of contact
          area of a flat punch with a viscoelastic
          half-space. The dashed curve is for equilibrium
          points and heavy lines are for variations
          at constant load.

4.4 Crack Propagation at Fixed Grips

Starting with a punch at P=o, δ=o one can suddenly apply a
displacement $\delta > \delta_c$ and record the variation of the force P during
crack propagation at fixed δ. The variation ABO of the force with
a, is depicted on Fig. 14 (eqs.(39) and (43)), but no experimental
verification was attempted.

4.5 Crack Propagation at Fixed Cross-head Velocity. Tackiness

Starting with a punch (radius a_o) with an initial displacement
corresponding to unstable equilibrium (eq. (43)) one can impose a

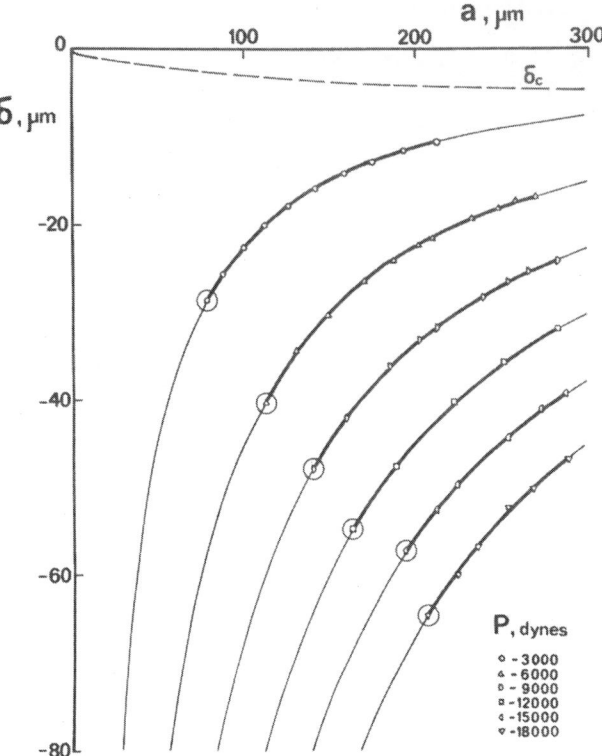

Fig. 13. Displacement versus radius of area of contact
          for adherence of a glass flat punch on a
          polyurethane surface. Experimental and theo-
          retical curves exactly coincide. The dis-
          placements are thus purely elastic, so that
          viscoelastic losses are limited to the crack
          tip.

fixed cross-head velocity $\dot{\delta} = \dfrac{d\delta}{dt}$ and compute the variation $P(t)$
from eqs. (32), (39), (41) and (51).

The crack velocity is given by

$$\frac{da}{dt} = \left(\frac{3K\delta^2}{8\pi a w \alpha} - \frac{1}{\alpha} \right)^{\frac{1}{0.6}}$$

and its variation by

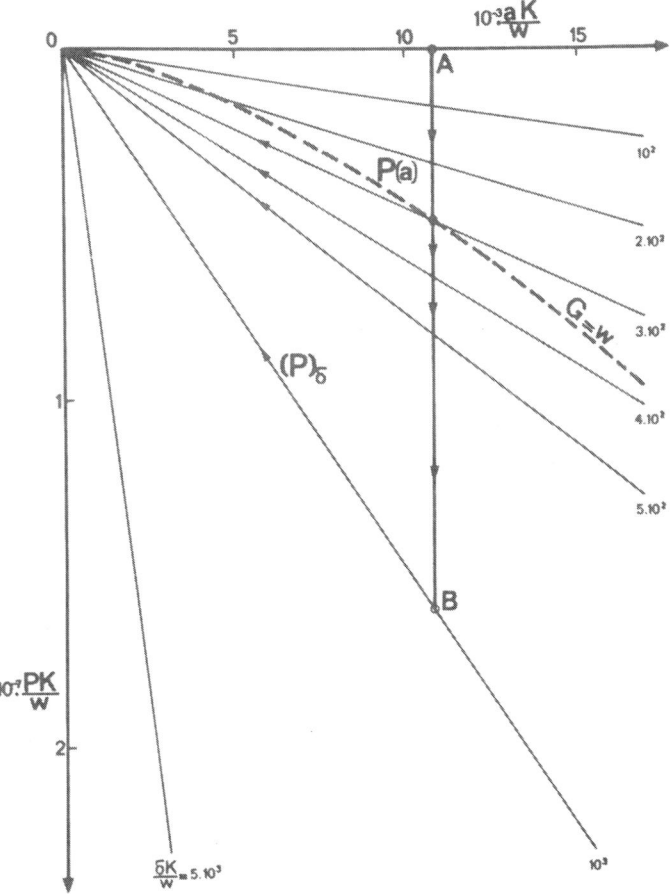

Fig. 14.   Relations, in reduced coordinates, between
          the load $\underline{P}$ and the radius $\underline{a}$ of contact area
          of a flat punch with a viscoelastic half-space.
          The dashed curve is for equilibrium points
          and heavy lines are for variation at constant
          displacement δ.

$$\frac{d^2a}{dt^2} = \frac{1}{0.6} \left(\frac{da}{dt}\right)^{0.4} \left(\frac{3K\delta^2}{8\pi wa^2\alpha}\right)\left(-\frac{da}{dt} + \frac{2a}{\delta}\frac{d\delta}{dt}\right)$$

From equilibrium values a_o and δ_o, one increments δ to δ_1 by
$\delta_1 = \delta_o + \dot{\delta}\Delta t$, and $\left|\frac{da}{dt}\right|_1$ and $\left|\frac{d^2a}{dt^2}\right|_1$ are computed by above equations
with a_o and δ_1 values.   An approximate value a_1^{*} is obtained by:

$$a_1^* = a_o - \left|\frac{da}{dt}\right|_1 \Delta t - \left|\frac{d^2a}{dt^2}\right|_1 \frac{\Delta t^2}{2}$$

from which the correct values $\dfrac{da_1}{dt}$ and $\dfrac{d^2a_1}{dt^2}$ are calculated, hence

the true a_1, by above equations with a_1^* and δ_1 instead of a_o, δ_1.

By such incrementations one can compute a_i and the force

$$P_i = \frac{3}{2} a_i K \delta_i.$$

Figure 15 displays such computed curves for various $\dot{\delta}$. The force goes through a maximum P_{max} that is often taken as a definition of the <u>tackiness</u> of the viscoelastic solid. It must be pointed out that P_{max} has no physical signification: for a given punch of radius a_o it increases with the cross-head velocity $\dot{\delta}$. From eq. (41) P_{max} corresponds to

$$3\frac{\dot{a}}{a} + \frac{\dot{G}}{G} = \frac{\dot{a}}{a} + \frac{\dot{\delta}}{\delta} = 0 \tag{52}$$

The increase of G with time is due to simultaneous increasing δ and decreasing a:

$$\frac{dG}{dt} = \left(\frac{\partial G}{\partial t}\right)_\delta \dot{a} + \left(\frac{\partial G}{\partial t}\right)_a \dot{\delta}$$

and eq. (52) leads to

$$2\left(\frac{\partial G}{\partial t}\right)_\delta = \left(\frac{\partial G}{\partial t}\right)_a$$

for P_{max}. The left hand side is a material (and geometric) characteristic but the right hand side depends upon $\dot{\delta}$.

Experimental verifications are in progress.

5.   ADHERENCE OF TWO SPHERES

5.1   The Hertz Problem

Stresses and deformation produced by the contact of two elastic spheres (radius R_1 and R_2, Young moduli E_1 and E_2, Poisson's ratio v_1 and v_2) were given by Hertz (1880). Under a load P, the area of contact has a radius a_H given by

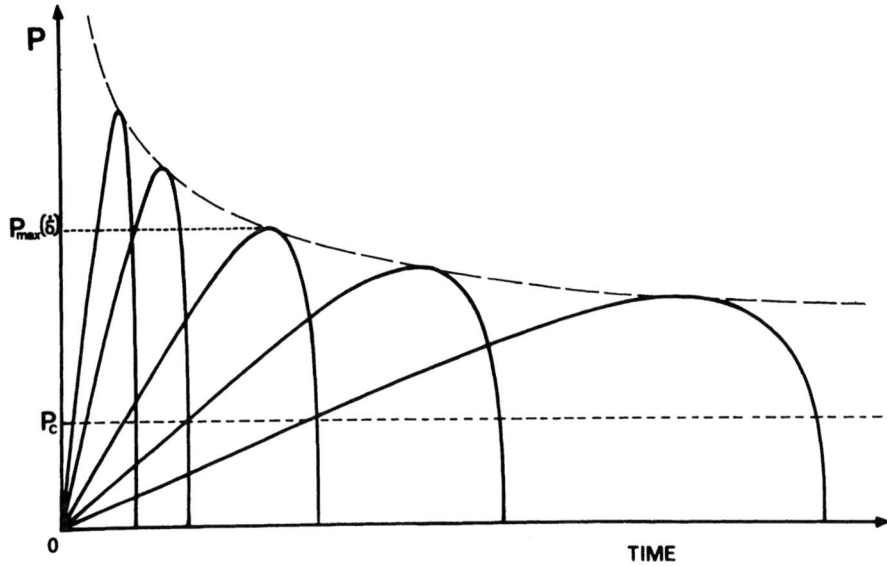

Fig. 15.   Computed curves, for various fixed cross-
head velocities, of the variation with time
of the adherence force of a flat punch in
contact with a viscoelastic surface.

$$a_H^3 = \frac{PR}{K} \tag{53}$$

with

$$\frac{1}{R} = \frac{1}{R_1} + \frac{1}{R_2}$$

and K given by eq. (40). The elastic displacement and the pressure
are

$$\delta_H = \frac{a_H^2}{R} \tag{54}$$

$$\sigma_z = \frac{3}{2} \frac{P}{\pi a_H^2} \sqrt{1 - \frac{r^2}{a^2}} \tag{55}$$

Note that the relation between P and δ is not linear (Fig.
16), so that the elastic energy is not $\frac{1}{2}P\delta_H$ but

$$U_E = \frac{2}{5} P\delta_H \tag{56}$$

and that elastic displacement is not given by the Castigliano's
theorem, but by the generalized Castigliano's theorem (eq. (21)):

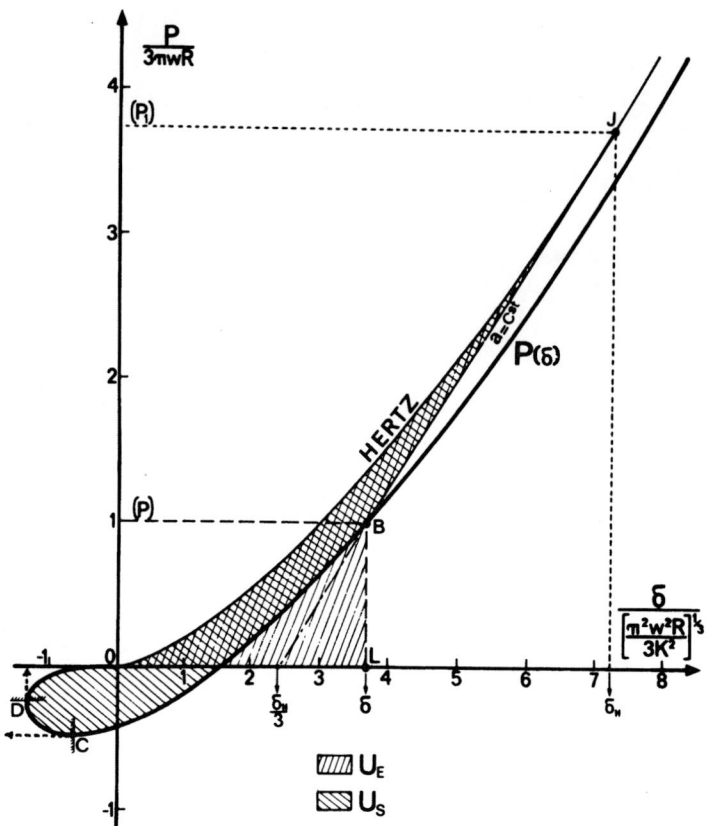

Fig. 16. Equilibrium curve for Hertzian (w=o) and
          adhesive (w≠o) contacts of two spheres.
          Berry's representation with stored elastic
          and surface energies at the equilibrium
          point B.

$$\delta_H = - \frac{dU_T}{dP}$$

using the thermodynamic potential U_T given by

$$U_T = - \frac{3}{2} U_E \qquad\qquad (57)$$

(as for punches, eq. (49)). Note also that the relation

$$\frac{dU_E}{d\delta} = P$$

is still valid.

The tangent to the Hertzian curve $P(\delta)$ at a point where the radius of contact is a, has a slope (the stiffness):

$$k = \frac{dP}{d\delta} = \frac{3aK}{2} \tag{58}$$

and intersect the δ axis in $\frac{\delta_H}{3}$ (the Legendre transform $\psi(\lambda) = \delta - \lambda P$ that exchange P and the compliance $\lambda = \frac{d\delta}{dP}$ in the (P,δ) relation) and the P axis in $-\frac{P}{2}$ (the Legendre transform $\psi(k) = P - k\delta$ that exchange δ and k). Note that compliance and stiffness are that of a punch of radius a.

5.2 The Strain Energy Release Rate

The Hertzian results do not take into account the molecular attraction between the two bodies leading to a larger radius of contact $a > a_H$ and elastic displacement $\delta > \delta_H$. The correct calculation has been performed by Johnson et al.(2).

The actual radius of contact under a load P can be expressed by an equation similar to eq. (53):

$$a^3 = \frac{P_1 R}{K} \tag{59}$$

where $P_1 > P$ is an apparent Hertz load. The actual elastic displacement δ and elastic energy U_E are found as following (Fig. 16). Assuming no surface energy, let us apply the load P_1 (point J): the displacement is $\frac{a^2}{R}$ (eq. (54)) and the elastic energy is $\frac{2}{5}\frac{P_1 a^2}{R}$ (Eq. (56)). Then, at constant radius of contact let us decrease the load from P_1 to P by increasing surface energy. The variation of displacement (J to B) and elastic energy are that of a punch of radius a:

$$\Delta\delta = \frac{2(P_1 - P)}{3aK}$$

$$\Delta U_E = \frac{P_1^2}{3aK} - \frac{P^2}{3aK}$$

Thus, at point B one has

$$\delta = \frac{a^2}{3R} + \frac{2P}{3aK} \tag{60}$$

$$U_E = \frac{P^2}{3aK} + \frac{a^5 K}{15R^2} \tag{61a}$$

$$= \frac{3aK}{4} (\delta - \frac{a^2}{3R})^2 + \frac{a^5 K}{15R^2} \tag{61b}$$

From eq. (9) it comes, with $U_P = -P\delta$

$$G = \frac{(P_1 - P)^2}{6\pi R P_1} \tag{62a}$$

$$= \frac{3a^3 K}{8\pi R^2} (1 - \frac{R\delta}{a^2})^2 \tag{62b}$$

Equations (60) and (62) are the equations of state of the system. Note that eq. (62) gives eq. (41) when $R \to \infty$. The derivatives of G are

$$(\frac{\partial G}{\partial A})_P = \frac{P_1^2 - P^2}{4\pi^2 a^2 R P_1} \tag{63}$$

$$(\frac{\partial G}{\partial A})_\delta = \frac{3K}{16\pi^2 R^2 a^3} (a^2 - \delta R)(3a^2 + \delta R) \tag{64}$$

Note that eq. (47) is still valid, and that U_E can be written

$$U_E = \frac{2}{5} (P\delta + GA) \tag{65}$$

The Maxwell equations (10) and (11) becomes

$$(\frac{\partial G}{\partial \delta})_A = (\frac{\partial P}{\partial A})_\delta = - \frac{P_1 - P}{2\pi a^2} \tag{66}$$

$$(\frac{\partial G}{\partial P})_A = -(\frac{\partial \delta}{\partial A})_P = - \frac{P_1 - P}{3\pi a^3 K} \tag{67}$$

5.3 Equilibrium Relations and Quasistatic Forces of Adherence

The equilibrium relation G=w links now P_1 to P and two of the three variatbles of the equation of state (60).

Figure 17 shows the relation between P_1 and P from eq. (62a); in reduced coordinates it gives a parabola inclined by $45°$ to the axis, the top of which corresponds to $P = -5P_1$, the slope being

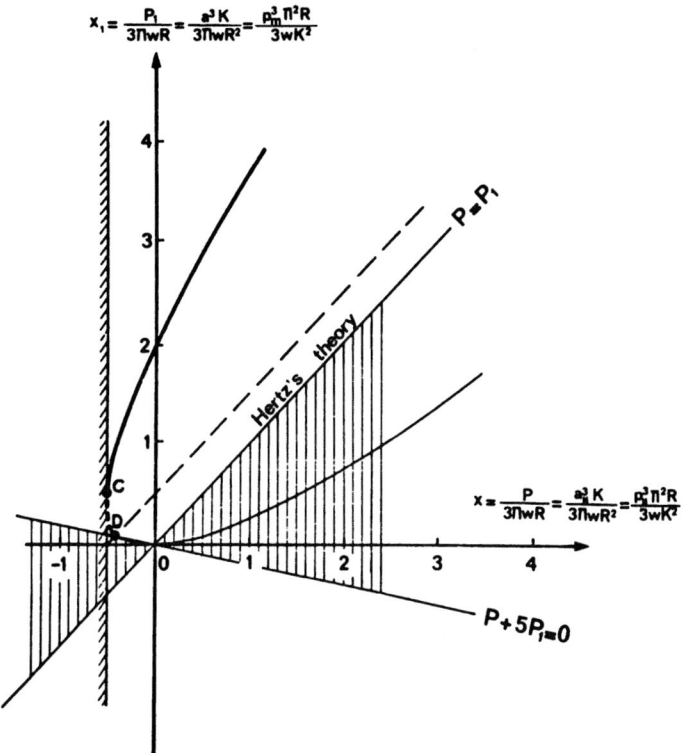

Fig. 17.   Elastic contact of two spheres: apparent
            Hertzian load P_1 versus normal load P in
            reduced coordinates.   This is a parabola
            inclinated at $45°$ to the axis.   The stability
            with fixed load conditions is shown by
            heavy curve.   Fixed grips conditions allow
            to increase the range of stability (dotted
            curve).

$$\frac{dP_1}{dP} = \frac{P_1-P}{P_1-P-3\pi wR} = \frac{2P_1}{P_1+P} \tag{68}$$

This figure thus shows the relation between $a^3 = \dfrac{P_1 R}{K}$ and $a_H^3 = \dfrac{PR}{K}$.

Figure 18 gives the same information for P(a) together with
variations at constant δ.   These last curves have their slope given
by eq. (66) and have their maximum on the Hertz curve (P_1=P).

Figure 19 shows the relation between δ and a from eq. (62b) in
reduced coordinates together with variations at constant load.   The

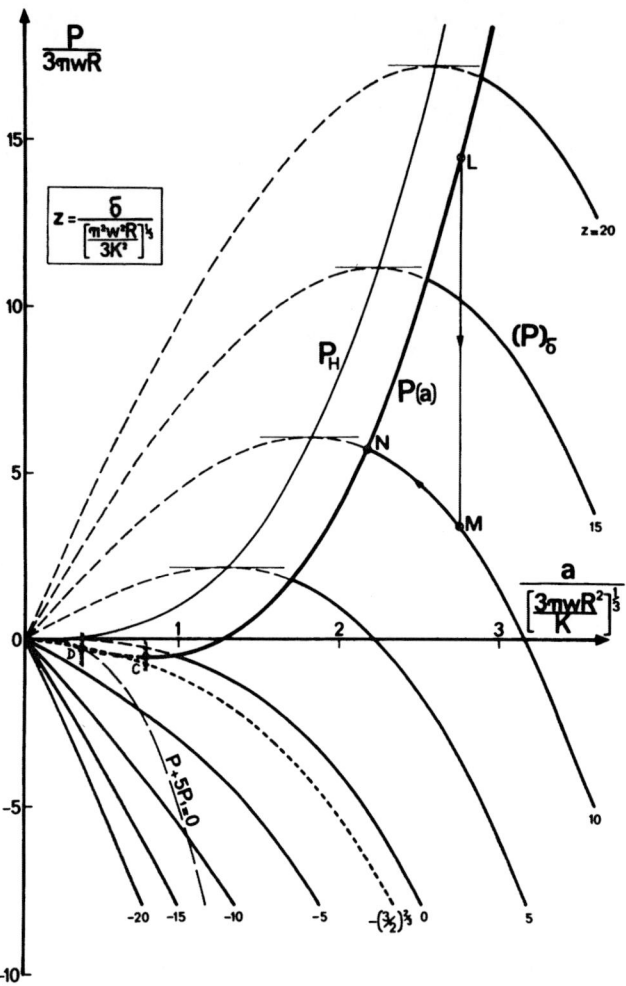

Fig. 18.   Relations between normal load P and radius
of area of contact a of two spheres, in
reduced coordinates. The equilibrium curve
is P(a); curves $(P)_\delta$ show the variation of
P with a at fixed grips condition. The curve
P_H is given by Hertz' theory. Displacement
from equilibrium value δ to δ' leads to an
instantaneous unloading at constant a (branch
LM) followed by an evolution along the curve
$(P)_\delta$, towards a new equilibrium if $\delta' >$

$$- (\frac{3\pi^2 w^2 R}{4K^2})^{1/3}.$$

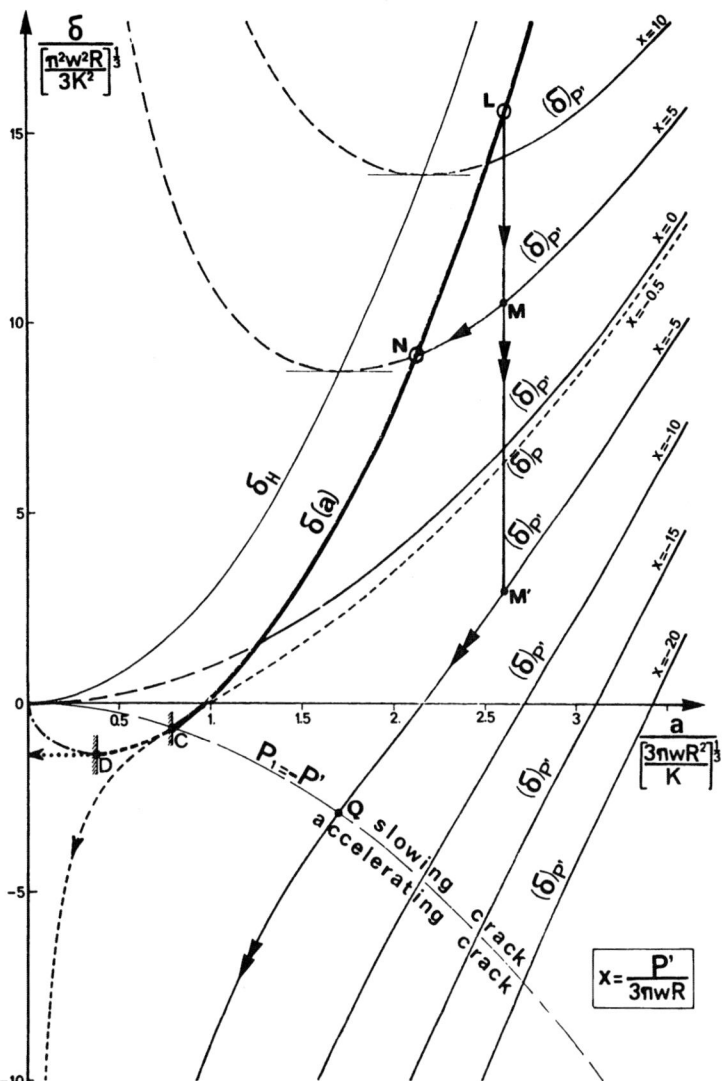

Fig. 19. Relations between elastic displacement δ and radius of area of contact a of two spheres, in reduced coordinates. The equilibrium curve is δ(a); curves $(\delta)_{P'}$ show the variation of δ with a at fixed load. The curve δ_H is given by the Hertz' theory. Unloading from P to P' leads to an instantaneous displacement at constant a (branch LM or LM'), followed by an evolution along the curve $(\delta)_{P'}$ towards a new equilibrium if $P' > -\frac{3}{2}\pi wR$ (branch MN), or towards rupture if $P' < -\frac{3}{2}\pi wR$ (branch M'Q).

$\delta(a)$ curve, first given by Johnson (34), has a slope

$$\frac{d\delta}{dA} = \frac{5P_1+P}{6\pi a^3 K} \tag{69}$$

The $(\delta)_P$ curves have their slope given by eq. (67) and have their minimum on the Hertz curve $(P_1=P)$.

Note that in Figs. 18 and 19 the curves $(P)_\delta$ and $(\delta)_P$ are independent of w.

The equilibrium curve $P(\delta)$, first given by Johnson (34) is displayed on Fig. 16. Its slope is

$$\frac{dP}{d\delta} = \frac{9Ka}{2}\left[\frac{P_1+P}{5P_1+P}\right] \tag{70}$$

The curves $(P)_a$ giving the variations of P with δ at constant a, are straight lines such as BJ, independent of w. Their slope

$$\left(\frac{\partial P}{\partial \delta}\right)_a = \frac{3aK}{2} = k$$

is the stiffness of the contact. They intersect the δ axis at $\frac{\delta_H}{3} = \frac{a^2}{3R}$ (the partial Legendre transform $\psi(\lambda,a) = \delta - \lambda P$ that exchanges P and the compliance $\lambda = \left(\frac{\partial \delta}{\partial P}\right)_a$ in the P,δ relationship) and the P axis at $-\frac{P_1}{2}$ (the partial Legendre transform $\psi(k,a) = P-k\delta$ that exchanges δ and k).

The stable equilibrium under the load P is defined by the positive value of $\left(\frac{\partial G}{\partial A}\right)_P$, eq. (63). The branches corresponding to stable equilibrium are the heavy curves in Figs. 17, 18, 19. They correspond to

$$P_1 = P + 3\pi wR + [6\pi wRP + (3\pi wR)^2]^{\frac{1}{2}} \tag{71}$$

$$a^3 = \frac{PR}{K}\left\{1 + \frac{3\pi wR}{P} + \left[2\left(\frac{3\pi wR}{P}\right) + \left(\frac{3\pi wR}{P}\right)^2\right]^{\frac{1}{2}}\right\} \tag{72}$$

given by Johnson et al. (2). The term between brackets is the correction to the Hertz's theory. The equilibrium is unstable and the crack spontaneously extends at fixed load (adherence force) for

$$P = -P_1 = -\frac{3}{2}\pi wR \tag{73}$$

$$a^3{}_{min} = 3\pi wR^2/ 2K$$

$$\delta = a^2/3R.$$

At this point (point C) one has $dP/da=0$, $d\delta/dP=\infty$, and the curves δ (a) and $(\delta)_P$ are tangents as expected above.

The equilibrium is stable at fixed grips conditions if $(\frac{\partial G}{\partial A})_\delta > 0$. The two terms between brackets in eq. (64) vanish respectively for $P=P_1$ and $P+5P_1=0$. The hatched zone of Fig. 17 are for instability. It can be observed that, at fixed δ, P_1 can decrease down to the top of the parabola. At this point, corresponding to the spontaneous rupture, one has

$$P = -5P_1 = - \frac{5}{6}\pi wR$$

$$a^3 = \pi wR^2/6K$$

$$\delta = -3a^2/R.$$

At this point $d\delta/dP=0$, and $d\delta/dA=0$ as expected. The equilibrium curves are thus more extended at fixed grips (up to the point D) as expected in sec. 2.3. Beyond the point C eqs.(71) and (72) are not valid, for the sign must be changed before the square root. However, the relation

$$P_1 - P = (6\pi wRP_1)^{\frac{1}{2}}$$

remains valid along the whole branch of equilibrium at fixed grips.

Figure 20 shows the geometry of an adhesive equilibrium contact under various loads, for a hard sphere on an elastic half-space, and Fig. 20a compares Hertzian contact and adhesive contact under the same load. In a Hertzian contact, the displacement δ_H is twice the height of the spherical cap $h=a^2|2R$. In an adhesive contact, one has $\delta < 2h$. The contact angle θ arises from the existence of surface and interface energies* (for Hertzian contact $w=0$ and $\theta =\pi$). Such an angle was effectively measured (Barquins et al, 35) in the case of glass-rubber contact ($\theta < 160^\circ$). Whereas in a Hertzian contact, the elastic material leaves the contact facing upward, in adhesive contact it leaves it facing downward, as shown in Fig. 20, unless high loads are applied, corresponding to $a|R > \sin \theta$.

* By analogy with Wenzel's relation for the wetting of rough
surfaces, it may be anticipated that the apparent contact angle
can vary with the roughness r of the surface. Reasoning as in the
thermodynamic derivation of the Wenzel's relation (Good, 36), one
can see that a variation dA in the apparent area of contact leads
to a variation dU_S = $-wrdA$=$w_a dA$ in the energy of surface, while U_P
and U_E are at first approximation not modified for slight roughness.   Equation G=w becomes G=w_a=rw, indicating that slight
roughness can increase the adhesion as compared with an ideally
smooth surface, as experimentally shown by Briggs and Briscoe
(37).

_____

The load P_1 is the apparent Hertz load (for w=0) that would
give the same area of contact, but the displacement would be
different.   To have the same displacement δ as the one calculated
for w≠0, the apparent Hertz load P' given by

$$P'^2 = P_1^{\,2} \left(\frac{P_1+2P}{3P_1}\right)^3$$ (74)

should be applied, but the area of contact would be different.

5.4   Energies of the System.   Berry's Representation

Figure 21 shows the variations of the thermodynamic potential
$U_T = U_E+U_P+U_S$ during transformations at constant load or along the
equilibrium curve.   The second branch of unstable equilibrium is
given by

$$P_1-P = -(6\pi wRP_1)^{\frac{1}{2}}$$

The variation of the total energy in the Hertz theory (w=0) is also
given for comparison.   The relation

$$U_E = -\frac{2}{3}U_T$$

for variations along the equilibrium curve, is the same as for
Hertzian contacts (eq (56)), or for punches (eq. (49)).

When eqs. (67) and (69) are taken into account, the forces due
to molecular interactions, eqs. (24) and (25) become

$$F_S = w\,\frac{6\pi RP_1}{5P_1+P}$$ (75)

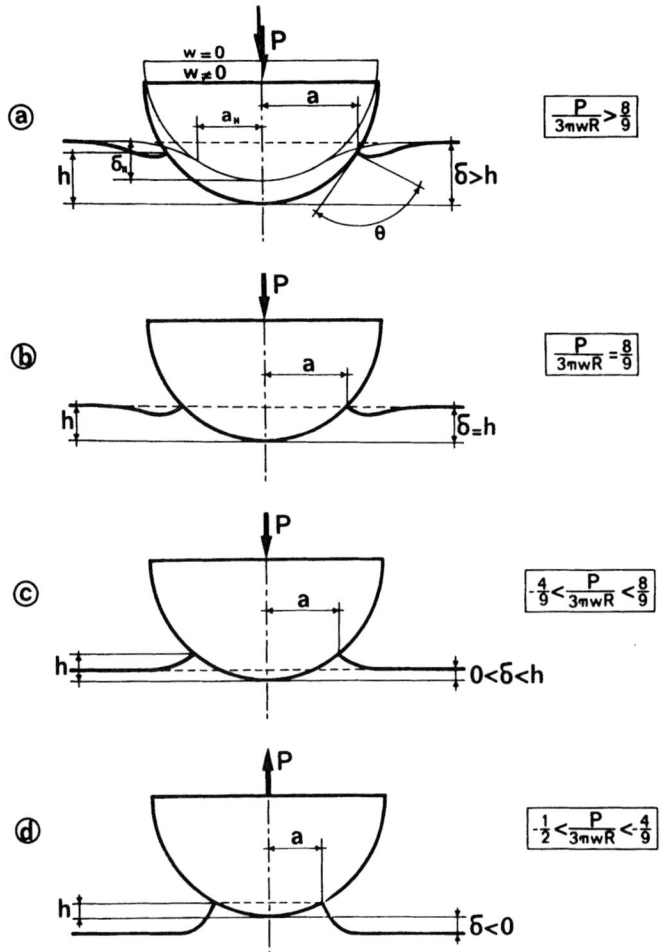

Fig. 20.   Geometry of an adhesive contact in equilibrium
           under various loads, for a hard hemisphere on
           an elastic half-space. The view a compares
           Hertzian contact and adhesive contact under
           the same load. Whereas in Hertzian contact
           the elastic material leaves the contact upwards,
           in adhesive contact it leaves it downwards.

$$\Pi = \frac{P_1 - P}{2} \tag{76}$$

and eq. (21) may be verified. The force F_S corresponds to a virtual
displacement along the equilibrium curve at an equilibrium point
such as N in Fig. 19.

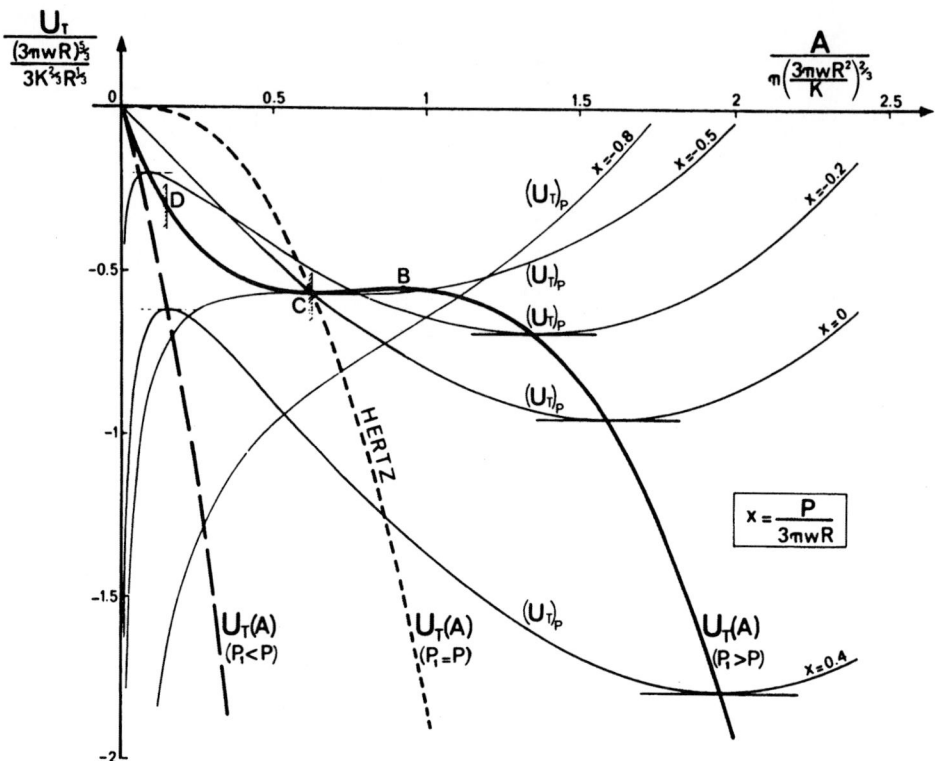

Fig. 21. Relations between total energy U_T and area
of contact A as outlined in Fig. 2, with
five curves $(U_T)_p$ for negative, positive
and zero load. The two branches of equilibrium
corresponding to the two determinations

$P_1-P = + \sqrt{6\pi wRP_1}$ (heavy line) and $P_1-P =$

$- \sqrt{6\pi wRP_1}$ (dashed line) are shown. Variation
of total energy in the Hertz' theory (w=o) is
also given for comparison (dotted line).
The Point B corresponds to δ=o.

The force Π corresponds to a virtual displacement at constant
load for any point such as N (equilibrium) or M (non-equilibrium).
At the rupture at fixed load $\Pi=F_S=\frac{3}{2}\pi wR$ is the adherence force. For
rupture at fixed grips F_S becomes infinite.

For transformations at fixed displacement, the internal energy
$U = U_E+U_S$ is the thermodynamic potential, and the Fig. 22 displays
its variations at fixed δ or along equilibrium points.

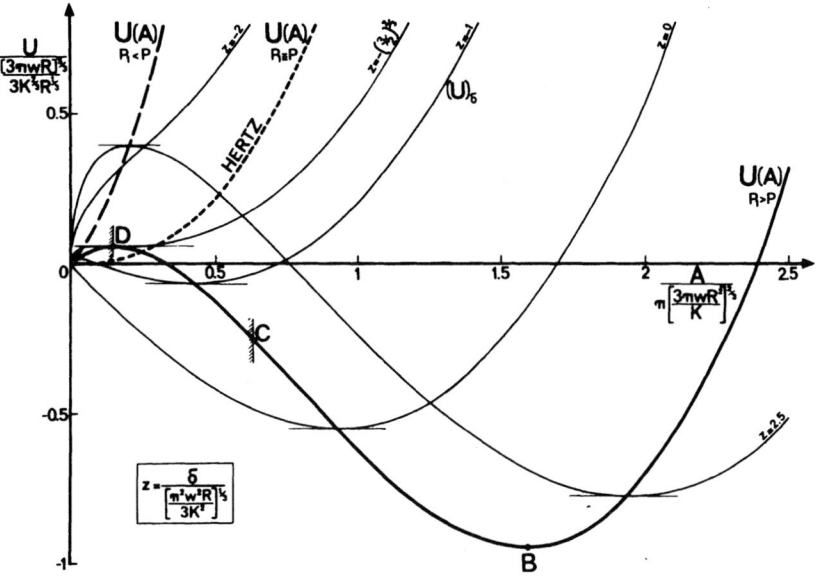

Fig. 22.   Relations between internal energy U and area
           of contact A with five curves (U)$_\delta$ for negative,
           positive and zero displacement. As in Fig. 21,
           the two branches of equilibrium (heavy and
           dashed lines) and the Hertz' theory curve (dotted
           line) are given. The point B corresponds to
           P=o.

        As proposed by Berry (32), the various energies can be repre-
sented on the (P,δ) plot, Fig. 16.   The elastic energy U_E is the
area under the loading curve OJB, and the surface energy of the
system U_S, the area between the loading curve OJB and the locus of
Griffith's criterion ODCB (see Appendix 2).   For a variation along
the equilibrium curve, eq. (23) leads to

$$-\Delta U_S = \Delta U_E - \int P d\delta$$

that can be checked on the Fig. 16 for a variation Δa displacing the
points J and B.

5.5 Crack Propagation at Fixed Load

        Kinetics of Unloading: Let us consider the equilibrium (G=w)
under a load P, with a radius of contact a and a displacement δ. At
the time t=0, let us apply a load P'<P. Figure 19 shows that one

must observe an instantaneous variation of δ at a constant area of
contact, followed by a simultaneous variation of δ and a along the
curve $(\delta)_{P'}$. The instantaneous displacement at constant a is that
of a punch under a load P-P', and it is

$\Delta\delta = 2(P-P')/3aK.$

Figure 19 shows that evolution along the curve $(\delta)_{P'}$ can lead to a
new equilibrium if $P' > -\frac{3}{2}\pi wR$ or to rupture if $P' < -\frac{3}{2}\pi wR$. In the two
cases the kinetics of the crack propagation can be studied by eq.
(32).

During crack propagation, following the curve $(\delta)_{P'}$ the strain
energy release rate is given by

$$G = \frac{(P_1-P')^2}{6\pi RP_1} \qquad\qquad (77)$$

where P' is the new fixed load, and where P_1 varies with the radius
of contact a according to the eq. (59). The only underlying
assumption is that the displacement δ is purely elastic, and that
viscoelastic losses are only at the crack tip.

Thus, the knowledge of the dissipative function $\phi(a_T v)$ is the
only factor, with w, needed to predict all the features. When the
new load P' is applied, P_1 has the value corresponding to the
initial area of contact, so that G>w and the crack starts, with a
speed corresponding to G given by eq. (77). As the crack pro-
gresses, a decreases; hence P_1 and G-w vary, so does the speed. The
evolution of the "normalized motive" (G-w)/w with the actual radius
of contact is given in Fig. 23. One can see that if $P' > -\frac{3}{2}\pi wR$ (i.e.
x>-0.5 on the figure), G gently decreases towards the value w_3-that
is, a decreases towards a new equilibrium value. If $P' < -\frac{3}{2}\pi wR$ a
minimum in G may be observed at a critical radius of contact given
by

$$\left(\frac{\partial G}{\partial A}\right)_{P'} = \frac{P_1^2-P'^2}{4\pi a^2 RP_1} = 0$$

i.e.

$$\frac{a^3K}{R} = -P'. \qquad\qquad (78)$$

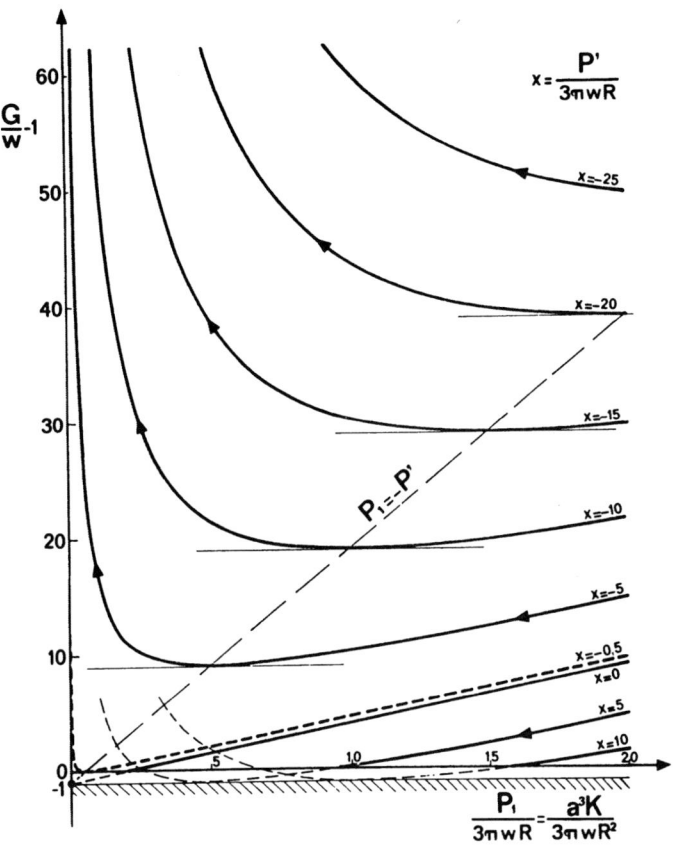

Fig. 23.   "Motive" of crack propagation (G-w), versus instantaneous load P_1 or (radius a)^3 for different unloadings P', in reduced coordinates. If $P' > -\frac{3}{2}\pi wR$, (G-w) gently decreases towards zero value and crack slows down; if

$$-\frac{a_o{}^3K}{R} < P' < -\frac{3}{2}\pi wR$$

(a_o being the radius of the initial area of contact), (G-w) first decreases, then increases and crack slows down then speeds up; if

$P' < -\dfrac{a_o{}^3K}{R}$, (G-w) immediately increases and crack thus continuously accelerates. The straight line $P_1 = -P'$ is the locus of minimum value of (G-w) at fixed load conditions.

The value of this minimum* is

$$G = - \frac{2P'}{3\pi R}.$$

Thus, for a given initial radius of contact a_o, this minimum of G may or may not be observed according to the value of P'. If P'<-$a_o^3 K/R$, G continuously increases with time, and the crack accelerates up to the rupture. If $- \frac{3}{2} wR > P' > -a_o^3 K/R$, G first decreases, then increases, i.e. the crack slows down, then speeds up, so that on a plot a(t) one has an inflexion point at the critical radius of contact given by eq. (78).

<u>Unloading Experiments</u>: Experiments were made with spherical caps of radius R=0.219 and 0.715 cm. The radii of contact were so small (<350 μm) that the polyurethane plates (h=3.175 mm) were considered as viscoelastic half-spaces.

The equilibrium eq. (72) and the adherence force $P = - \frac{3}{2}\pi wR$ have been well verified these last years on smooth solids of low elastic modulus** (Johnson et al.(2), Barquins and Courtel (28), Fuller and Tabor (38), Barquins et al.(29)) or with small radius metallic spheres (Easterling and Tholen (39), Kohno and Hyodo (40), Maugis et al.(41)). As stated above, the work of adhesion w was measured before any set of experiments by measuring equilibrium radii of contact under various loads. The only care to be taken is to be sure that the equilibrium is reached. Figure 24 shows that kinetics of loading are different from kinetics of unloading as first pointed out by Roberts and Thomas (42); by loading the equilibrium is reached in 3 min, and further dwell time has no effect; by unloading this equilibrium is not attained after 20 min. Equilibrium points were obtained by loading with a dwell time of 10 min. The reproducibility of radii of contact during a set of experiments was better than 0.2%.***

---

* The existence of such a minimum was noticed by Roberts and Thomas (42). They rearranged eq. (72) to have w versus a, and differentiated the expression thus obtained. Their w so defined is the apparent work of adhesion w_f, and their minimum of w_f is, in fact, the minimum of G (see eq. (33)).

** The observation of finite areas of contact under negative loads seems to have been first published by Drutowski (44).

*** So, as in Roberts and Thomas (42) experiments, no true hysteresis between unloading and overloading was found, contrary to the results of Kendall (45).

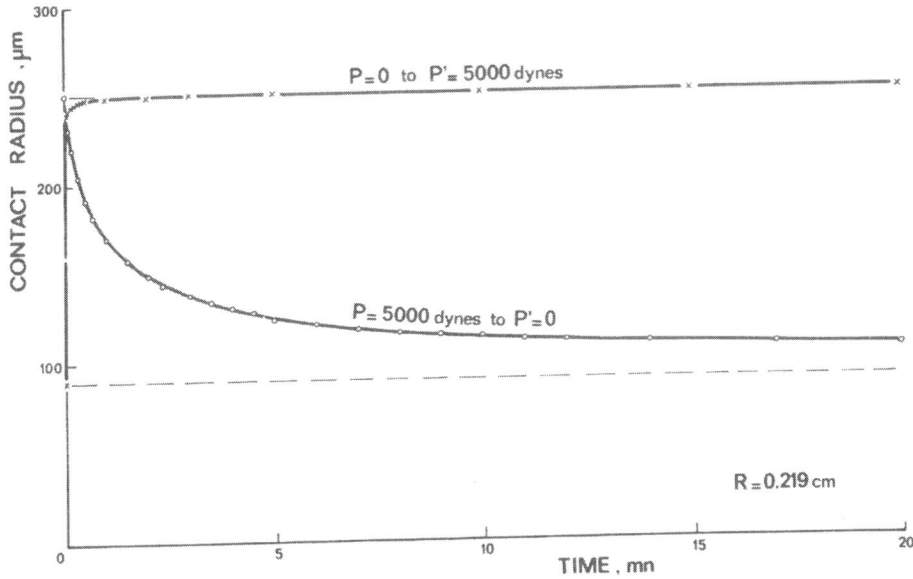

Fig. 24.   Radius of area of contact versus time of a
           glass hemisphere in contact with a polyure-
           thane surface, during an unloading from
           P = 5,000 dynes to P' = 0 and during a
           loading from P = 0 to P' = 5,000 dynes.
           Kinetics of unloading and loading are quite
           different. The equilibrium is more quickly
           reached by loading.

        Figure 25 shows the variation of the radius of contact* with
time for unloading, from P=10000 dyn to various loads P'<P. For
light unloading, a slow return towards new equilibrium is observed,
with continuously decreasing crackspeed (for P'=0 the equilibrium
is not yet reached after 4 h). As expected, a slowing down of the
crack, followed by an acceleration until the rupture is reached, is
observed for strong unloading, i.e. $P' < -\frac{3}{2}\pi wR = -67$ dyn (similar
curves were given by Brunt (43) and Drutowski (44). The critical
radii of contact given by eq. (78) are reported on the figure; the
agreement with the observed inflexion points is quite good. The
dashed curve corresponds to the boundary between the two kinds of

---

* As for punches, a grey annulus due to reflection of light along
  the profile surrounds the area of contact as soon as one unloads.
  Taking the outer radius as the radius of contact would lead to
  large errors.

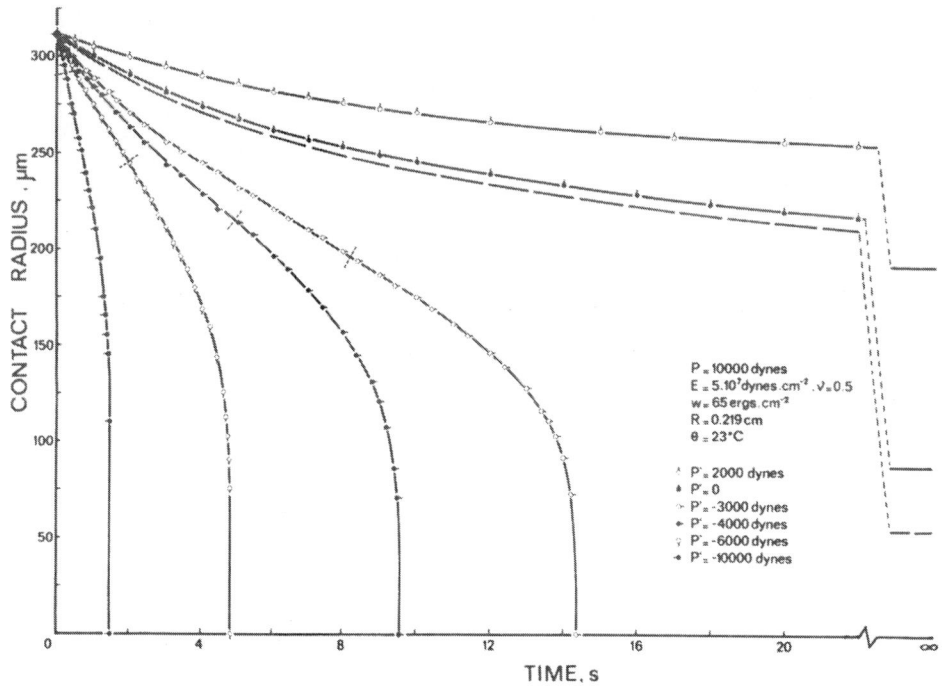

P = 10000 dynes
E = 5.10⁷ dynes.cm⁻² , ν = 0.5
w = 65 ergs.cm⁻²
R = 0.219 cm
θ = 23°C

P' = 2000 dynes
P' = 0
P' = -3000 dynes
P' = -4000 dynes
P' = -6000 dynes
P' = -10000 dynes

CONTACT RADIUS , μm

TIME, s

Fig. 25.   Variation with time of the radius of the
           contact area of a glass hemisphere in con-
           tact with a polyurethane surface, for
           unloadings from P = 10,000 dynes to various
           loads P' < P.   If P' > $-\frac{3}{2}\pi wR$ = -67 dynes,
           a slow return toward a new equilibrium state
           is observed and crack speed continuously
           decreases.   If P' < $-\frac{3}{2}\pi wR$, the crack speed
           decreases, then increases up to the rupture.
           The value of radius a at the inflexion points
           exactly corresponds to the computed minimum
           value of the motive G-w.

unloading (P' = $-\frac{3}{2}\pi wR$); it was computed from data of Fig. 28.   Note
that the time to rupture for an unloading from P to P' = -P is the
time of detachment of a ball of weight P under gravity.

     Figure 26 shows the relation between the strain energy release
rate computed by eq. (77) and the crack speed v (deduced from the
slope of curve a(t) for unloading from P=10000 dyn to P'=-4000 dyn.
The points corresponding to slowing down (($\partial G/\partial A)_{P'}$>0) and these
corresponding to acceleration (($\partial G/\partial A)_{P'}$<0) are indeed on a single

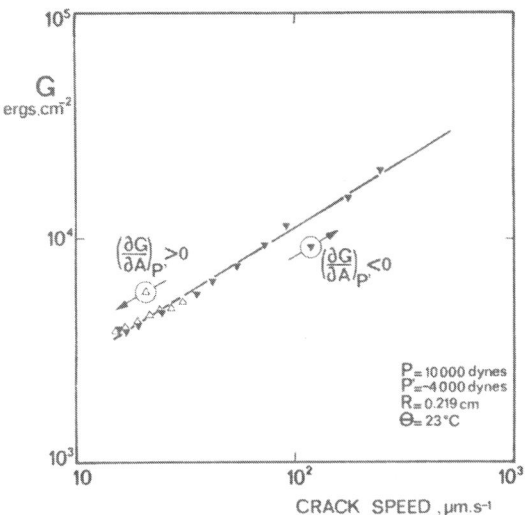

Fig. 26. Relation between strain energy release rate

G and crack speed $v = \dfrac{da}{dt}$, for an unloading

from P = 10,000 dynes to P' = -4,000 dynes
(see Fig. 25). All the points, corresponding
either to decreasing or increasing crack
speed, fall on a single curve.

curve.

The kinetics of unloading at 23 or 29°C are compared in Fig.
27. The crack speed increases with temperature, but the inflexion
points correspond to the same critical radius.

Results concerning various starting loads P, various final
loads P', with two radii R, and two temperatures are collected on
the Fig. 28 (Barquins and Maugis, 46). The results for θ=29°C were
shifted on those corresponding to θ=23°C by the WLF transform (with
T_g=-50°C). This master curve is still the same as for peeling
(heavy curve), and of course for adherence of punches. This figure
thus gives the dissipative function $\phi(a_T v)$ of the polyurethane,
eq. (51).

Elastic displacements have been recorded during crack propagation (Barquins and Maugis, 47). Figure 29 shows unloading from
P=5000 dyn to various loads P'=4000, 2000, 0, -2000 and -4000 dyn.
In two cases, the load was varied during unloading (from P'=0 to -
2000 dyn, and from P'=-2000 to -4000 dyn). The agreement with
theoretical prediction (cf. Fig. 19) is excellent. Thus, as for
punches, the displacements are purely elastic, and viscoelastic

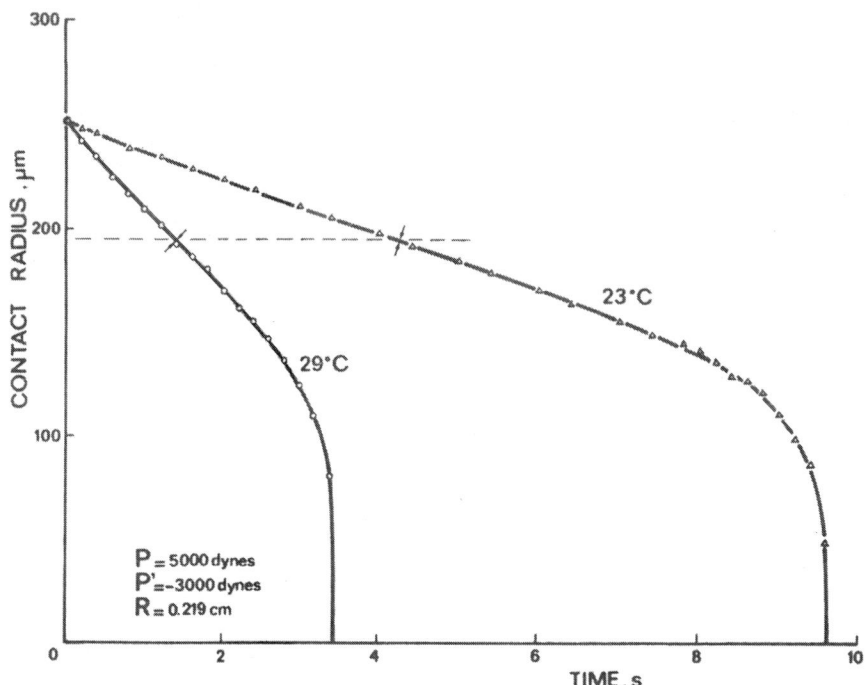

Fig. 27.   Experimental curves a(t) for an unloading
from P = 5,000 dynes to P' = -3,000 dynes
at two different temperatures 23°C and 29°C.
The crack speed increases with temperature
but the inflexion points correspond to the
same critical radius.

losses are limited to crack tips.

Loading Experiments: As shown in Fig. 24 kinetics of loading
are quite different from kinetics of unloading.   This point was
studied (Maugis and Barquins, 49).

On Fig. 30 is shown the path LMN for unloading from equilibrium
under a load P to equilibrium under a load P'<P. Starting now from
equilibrium under the load P' (point N, radius a_N) let us steeply
apply the load P.   A two steps path from N to L can be predicted.
The first step NT begins with an instantaneous displacement NS at
constant radius of contact a_N corresponding to the displacement

$$\Delta\delta = \frac{2\,\Delta P^*}{3a_N K}$$

of a punch under the load

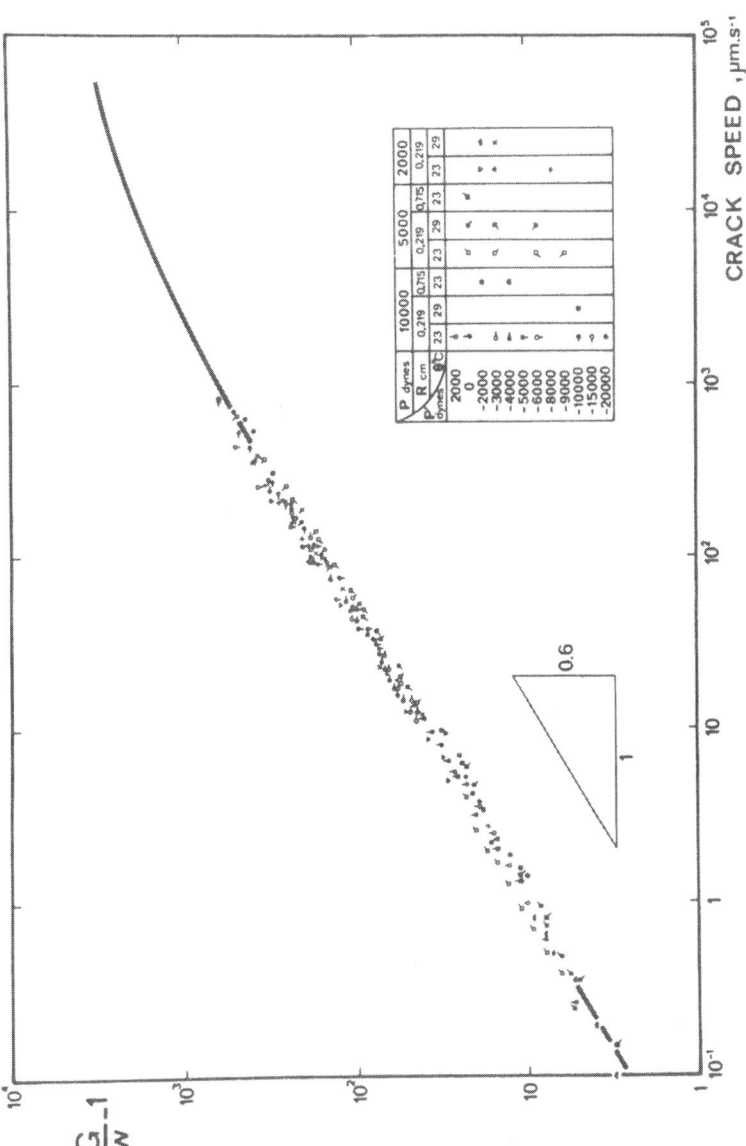

Fig. 28.   "Normalized motive" versus crack speed for adherence of glass hemisphere
on polyurethane, for various unloadings, and two radii and two temperatures.
The same master curve as for peeling (heavy line) and of course for unloading
of flat punch is observed.

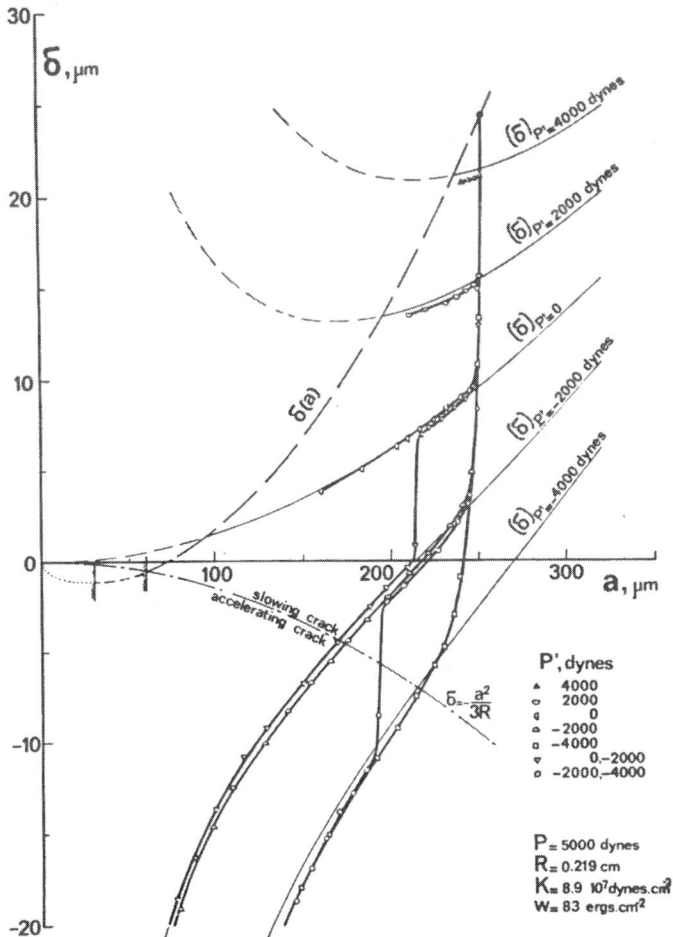

Fig. 29.   Displacement versus radius of area of
           contact for adherence of a glass hemisphere
           on polyurethane, for various unloadings.
           Agreement with theoretical curves is good,
           even when the load is varied during unloading.

$$\Delta P* = \frac{a_N^3 K}{R} - P'$$

$$= 3\pi wR + [6\pi wRP + (3\pi wR)^2]^{\frac{1}{2}}$$

Hertzian conditions are realized at point S, and the path follows
the Hertz curve from the load P'+ΔP* to the load P at point T. The
second step TL corresponds to crack healing at constant load along
the curve (δ)$_p$ during which the crack slows down and stops. For

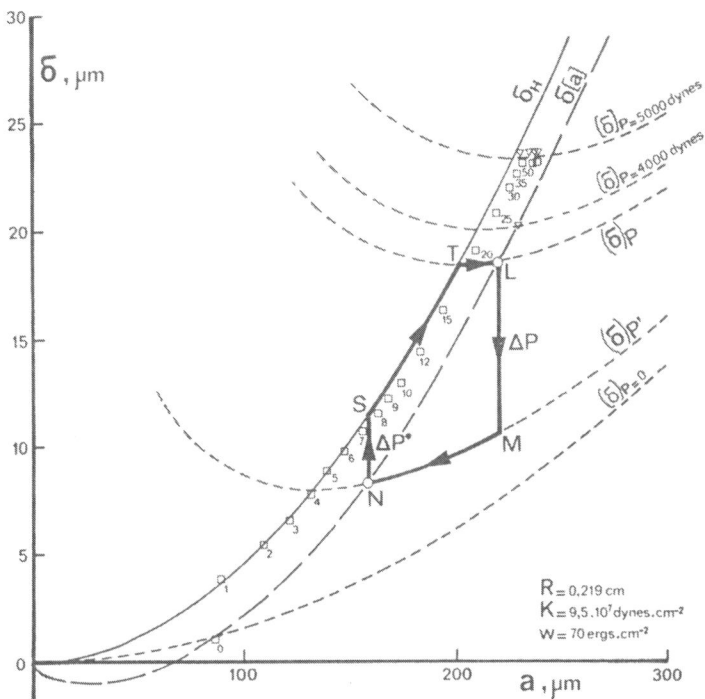

Fig. 30. Relations between elastic displacement δ
          and radius of area of contact a of two
          spheres in an unloading-loading experiment.
          After an unloading from P to P', already seen
          on Figs. 19 and 23 (path LMN), a loading
          from P' to P leads to an instantaneous
          displacement firstly at constant a (branch
          NS) and then along the Hertz' curve (branch
          ST). This path (NST) is followed by an
          evolution along the curve $(\delta)_P$ towards the

          equilibrium point L. Experimental results
          for loading step from P'=0 and P'=4,000
          dynes to P=5,000 dynes are given.

light loading, i.e. $\Delta P = P-P' \leqslant \Delta P^*$, the Hertzian branch ST disappears.

     Thus the first step is the elastic response of a system with
zero surface energy. It is likely that during this fast motion, a
sheet of air is squeezed between the two solids, preventing molecular interactions (analogy with elastohydrodynamics). During the

second step, the sheet of air is extruded and the radius of contact progressively takes its equilibrium value.

Experiments were performed with a glass ball (R=0.219 cm) on the polyurethane. Area of contact was recorded at 50 frames per second, and the elastic displacement was recorded on a memory oscilloscope. To avoid percussion, the loads were applied with a damping system. The experimental points on the Fig. 30 are for loading to P=5000 dyn from P'=0 and P'=4000 dyn. (The numbers are frame numbers.) Experimental results are in perfect agreement with theoretical predictions.

The Hertzian evolution of the contact during loading (branch as ST) is confirmed by the study of the surface profile by Newton rings. For the same radius of contact, Fig. 31a,b,c show the Newton rings and the profile for loading equilibrium at P=0, and unloading. At equilibrium the contact angle is $\theta \# 170°$; during unloading the profile is as shown in Fig. 5c (compare Figs. 31b and 9); but during loading the tangential joining is characteristic of a Hertzian profile.

5.6   Crack Propagation at Fixed Grips

Starting from an equilibrium point one can apply a new displacement δ and record the variation of the force P during crack propagation at fixed δ. The predicted variation LMN towards a new equilibrium is displayed in Fig. 18. For $\delta < -\frac{3}{R} (\frac{\pi w R^2}{6K})^{2/3}$ the final state is the rupture of the joint.

The kinetics of crack propagation can be studied by eqs. (62b) and (64). Figure 32 shows the variation of G with a^3 at fixed δ in normalized coordinates. For any given couple (G,a) the slope at constant δ is larger than at constant P, for $(\frac{\partial G}{\partial A})_\delta > (\frac{\partial G}{\partial A})_P$ (see eq. (47)). As for crack propagation at fixed load, it exists a minimum for G, so that the crack speed can first decrease then increase.

No experimental verification was attempted.

5.7   Crack Propagation at Fixed Cross-head Velocity

Starting from an equilibrium point one can impose a fixed cross-head velocity $\dot\delta = \frac{d\delta}{dt}$ and compute the variation P(t) from eqs. (32), (60), (62) and (51).

The crack velocity is given by

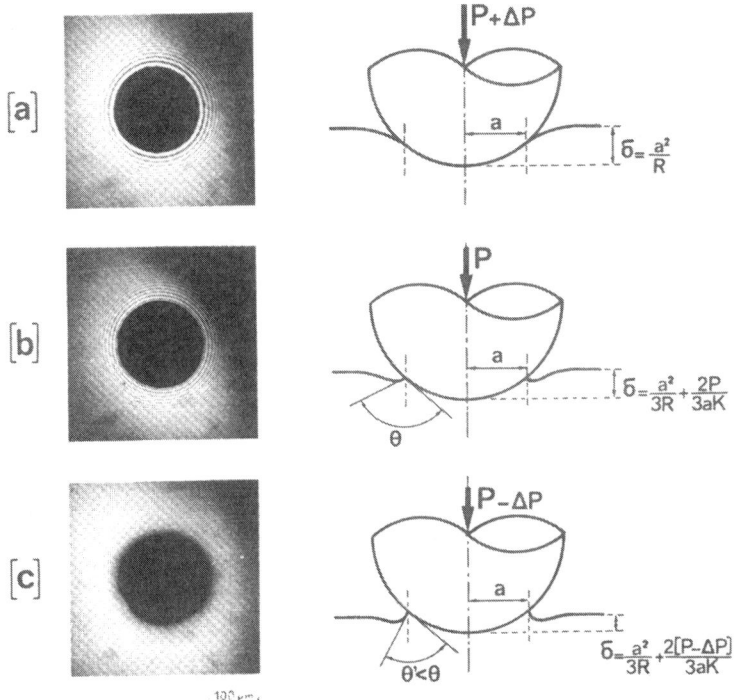

Fig. 31. View of Newton's rings and corresponding
         diagram profile of the surface for loading,
         equilibrium at P=0 and unloading, with the
         same radius of contact. The view (a) confirms
         the Hertzian evolution of the contact during
         loading (branch S T on Fig. 30).

$$\frac{da}{dt} = \left[\frac{3a^3 K}{8\pi R^2 \alpha w} \ (1 - \frac{R\delta}{a^2})^2 - \frac{1}{\alpha} \right]^{\frac{1}{0.6}}$$

and its variation by

$$\frac{d^2 a}{dt^2} = \frac{1}{0.6} \ (\frac{da}{dt})^{0.4} \left(\frac{3a^2 K}{8\pi R^2 \alpha w} \right) (1 - \frac{R\delta}{a^2}) \left[(\frac{R\delta}{a^2} + 3) \ \frac{da}{dt} - \frac{2R}{a} \ \frac{d\delta}{dt} \right]$$

From the same calculation method as for punch (see sec. 4-5) one can
compute the radius a_i, corresponding to $\delta_i = \delta_{i-1} + \dot{\delta}\Delta t$, by

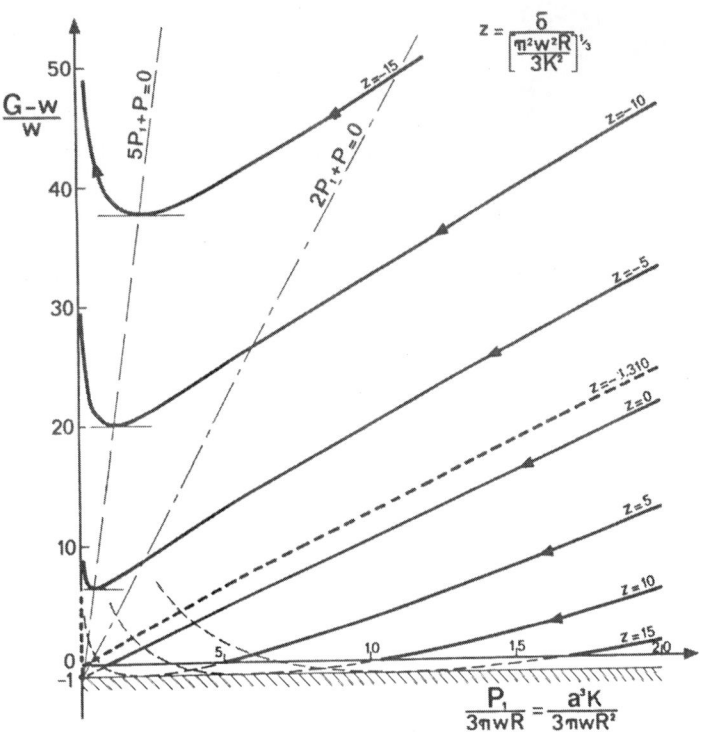

Fig. 32. "Motive" of crack propagation (G-w), versus
instantaneous load P_1 or (radius a)^3 for
different fixed grips conditions, in reduced
coordinates.

If $\delta > -\dfrac{3a^2}{R}$ (i.e. z > -1.310), G-w gently

decreases towards zero value and crack slows

down. If $\delta < -\dfrac{3a^2}{R}$, the "motive" G-w can de-

crease and then increase, or can immediately
increase.

The straight line $5P_1 + P = 0$ is the locus
of minimum values of G-w at fixed grips
conditions, and the straight line $2P_1 + P = 0$
is the locus of inflexion points.

$$a_i = a_{i-1} - \left| \frac{da_i}{dt} \right| \Delta t - \left| \frac{d^2 a_i}{dt^2} \right| \frac{\Delta t^2}{2}$$

hence the force

$$P_i = \frac{a_i^3 K}{2R} \left(\frac{3R \delta_i}{a_i^2} - 1 \right)$$

Figure 33 displays such computed curves for various δ. The maximum P_{max} depends upon δ and the starting equilibrium point. Experimental verifications are in progress.

6.   PEELING

6.1   Theory

Consider (Fig. 34) the peeling at an angle α of a thin elastic strip (thickness h, width b) adherent to a rigid substrate, on to an area A_o. The angle α can vary from $\alpha=0$ (lap shear joint) to $\alpha=\pi$ (π peeling); the film may have been extended to a small strain ε by a tensile force F before application to the substrate so that an elastic energy

$$u = \frac{F^2}{2b^2 Eh} = \tfrac{1}{2} \varepsilon^2 Eh \qquad\qquad (79)$$

is stored by unit area of the strip.   Neglecting the peel bend energy and end effects which remain constants, the potential and elastic energies are

$$U_p = -P\delta = \left(\frac{1-\cos \alpha}{b} + \frac{P-F}{b^2 Eh} \right) P(A-A_o)$$

$$U_E = - \frac{P^2 - F^2}{2b^2 Eh} (A-A_o) + \frac{F^2}{2b^2 Eh} A_o .$$

Hence the strain energy release rate

$$G = \frac{(P-F)^2}{2b^2 Eh} + \frac{P}{b} (1-\cos \alpha) \qquad\qquad (80)$$

and the equilibrium relationship

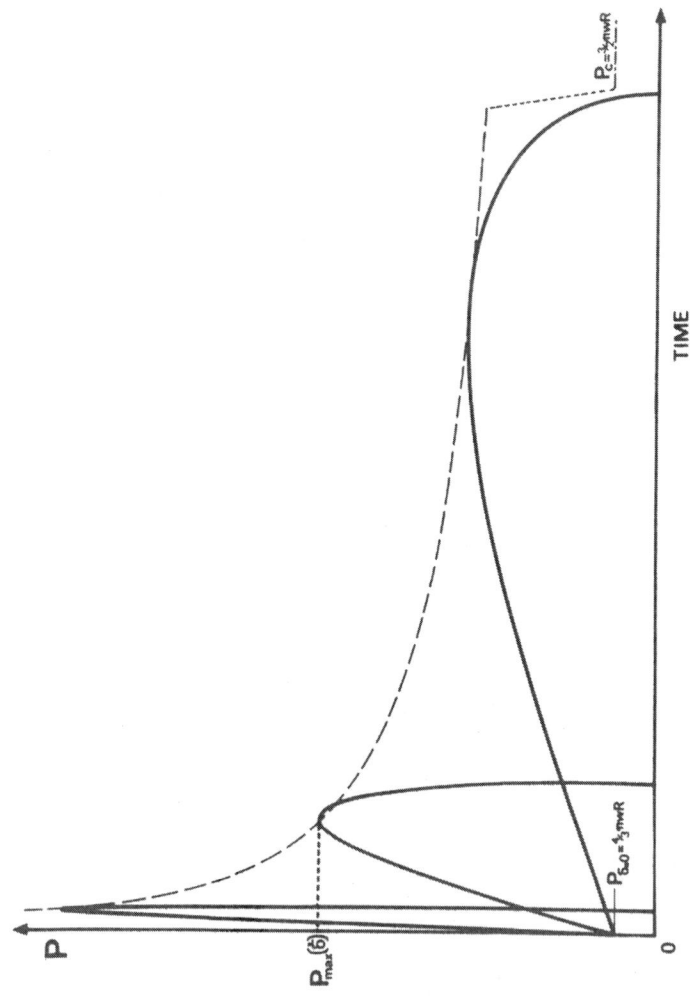

Fig. 33.   Computed curves, for various fixed cross-head velocities,
of the variation with time of the adherence force of a
rigid sphere in contact with a viscoelastic surface.

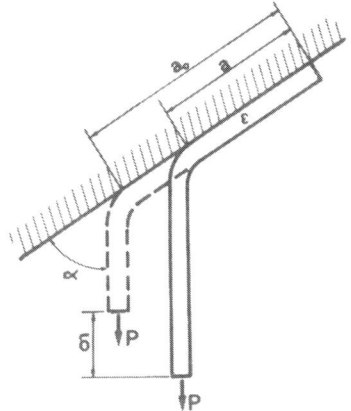

Fig. 34. Thin elastic strip peeling at an angle
          α from a rigid substrate.

$$\frac{(P-F)^2}{2b^2Eh} + \frac{P}{b}(1-\cos \alpha) = w. \tag{81}$$

As $\partial G/\partial A = 0$ (G independent of A), the corresponding equilibrium
load P_c is the quasistatic force of adherence. Under this load the
peeled film is in indifferent equilibrium. For $P>P_c$, the crack
speed is constant during peeling because G is independent of A.

Equation (81), with F=0, was given and experimentally verified
(at constant crack speed, i.e. constant w_f) by Kendall (4). The
approximation (Rivlin (50), Deryagin and Krotova (49), Kendall (1))

$$P = bw/(1-\cos \alpha)$$

is better than 6% when $1-\cos \alpha > 2(2w/Eh)^{\frac{1}{2}}$.

The total energy of the system is

$$U_T = (G-w)A-(G-u)A_0 \tag{82}$$

and under the load P_c, it is independent of A. Figure 35 shows the
variation of the total energy with the area of contact, in reduced
coordinates. It is interesting to compare this figure with Fig. 6:
the normalized curves $(U_T)_P$ have now a constant slope G/w-1. As
crack propagation corresponds to a positive slope, it is easy to
study spontaneous propagation (i.e. under zero load); for P=0 one
has G=u, and Fig. 35 shows that spontaneous propagation occurs when
u/w>1. Thus, an elastic strip cannot tolerate a stored elastic

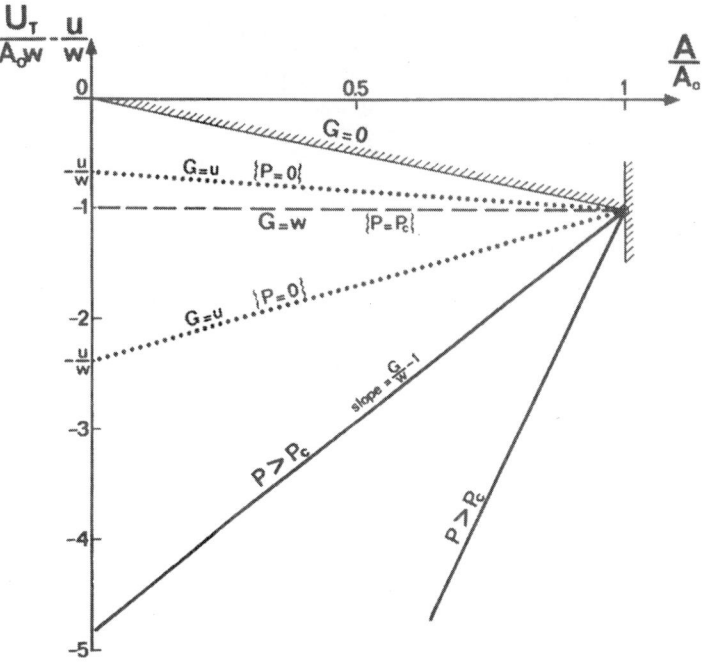

Fig. 35.   Variation of the total energy of peeling
           with the area of contact, in reduced coor-
           dinates. The equilibrium curve (dashed line)
           is a straight line corresponding to a constant
           load. All these curves are thus curves $(U_T)_P$

           with a constant slope $\frac{G}{w}$ - 1. For P = 0,

           spontaneous crack propagation is observed
           when the elastic energy u stored by unit area
           of the strip and due to prestress, is higher
           than w.

energy per unit area higher than w.   But the influence of a
prestress is subtle: a strip has two ends, and if one end is clamped
to avoid spontaneous propagation, the force needed to propagate the
other crack may be increased by a pre-stress.

    The influence of prestress was studied by Kendall for $\alpha=0$
(Kendall, 5,6) and for $\alpha=\pi/2$ (Kendall, 5,3).   In fact for any value
α, there is a critical tensile prestrain ε above which the adherence
force is lowered (Maugis, 53).   Equation (80) may be rewritten as

$$G(F) - G(0) = \frac{F(F-2P)}{2b^2Eh} \qquad (83)$$

which shows that adherence is increased if F<2P. This critical
prestrain is twice the strain in the peeled film, due to the load
P_c. For $\alpha=90°$ its value is about 2w/Eh, i.e. very small.

Note that the displacement δ is still given by the generalized
Castigliano's theorem $\delta = -(\partial U_T/\partial P)A$, and that the compliance of
the system is

$$\lambda = (\frac{\partial \delta}{\partial P})_A = \frac{A_o-A}{b^2Eh}$$

6.2 Experiments

Polyurethane strips (thickness h=3.175 mm, width b=1-3 cm,
length 10-20 cm) were peeled from glass at various peel angles $10°$,
$30°$, $60°$, $90°$. Experimental precautions to have reproducible
results are the following: (i) cleanliness; (ii) constant temperature $(23°C\pm0.5°)$; (iii) absence of prestress* (the strips were
gently laid on the glass, and a dwell time of about 15 min was used,
during which stress relaxations by microslips and squeezing of air
pockets probably occurs); (iv) constant bending (to have no variation in bending energy during peeling). The weight of the peeled
strip was, of course, taken into account in the applied load.

Figure 36 shows the relation between the crack speed v and the
"normalized motive" (G-w)/w with G computed by eq. (80) and w=65 erg
cm^{-2}. From eq. (32) this curve thus displays the dissipation
function $\phi(a_Tv)$ to which the punch and sphere results were compared.
There is a marked dispersion in the results that can be assigned to
temperature variations during experiments, local variation of the
interfacial energy due to dirty areas, local prestresses in the laid
strips, or variation in room humidity. The dissipative function
$\phi(a_Tv)$ varies as $(a_Tv)^{0.6}$, a result often found in peeling.

Birch et al.(54) have shown that humidity increases the peel
rate of rubber on glass, and this result together with peeling
experiments in various liquids by Gent and Schultz (26) or on
various substrate by Andrews and Kinloch (27) can be analyzed by a
reduction of w in eq. (32). (As said in sec. 3.2 the Dupre energy of
adhesion was observed to vary from day to day most probably with the

---

* It was observed that a prestrain $\varepsilon=3\%$ decreases the adherence
  force for $\alpha=\pi/2$ but increases the adherence force for $\alpha=10°$. More
  precisely, G is increased by this prestrain for $\alpha=\pi/2$ and the
  crack speed increases; G is decreased by this prestrain for $\alpha =10°$
  and the crack speed decreases. This is in agreement with eq.
  (80).

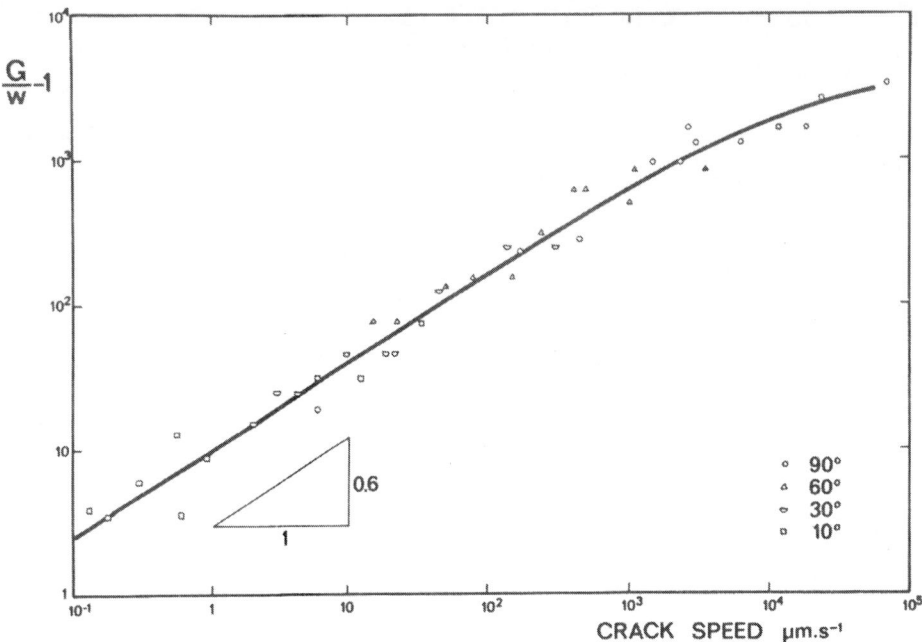

Fig. 36.   Peeling of polyurethane strips from glass
surface: "normalized motive" of crack
versus crack speed at various peel angles
(temperature θ = 23°C + 0.5°C; no prestress;
dwell time 15 mn; w = 65 ergs/cm²).

humidity content of the atmosphere). Roberts (55) has measured the
rolling resistance T of a glass cylinder (length ℓ) rolling on our
polyurethane samples as a function of the humidity content of the
atmosphere (Fig. 37). His data points clearly to a slope 0.6 in
agreement with our results, and as rolling on adhesive substrates
corresponds to peeling at the rear of the contact (Kendall, 45,
Roberts and Thomas, 42), Fig. 35 and Fig. 37 can be compared with
$G = \frac{T}{\ell}$. As $G \gg w$ in Fig. 37, one has $G \simeq w\phi(a_T v)$ and this figure
clearly supports the multiplication effect of w on viscoelastic
losses, as proposed by Andrews and Kinloch (27).

7.   CONCLUSION

The introduction of the concepts of fracture mechanics such as
the strain energy release rate G and its derivative $\partial G/\partial A$ has
allowed us to give a general picture of the adherence of viscoelastic bodies. Two bodies in contact over an area A are in
equilibrium if G=w, where w is the thermodynamic (or Dupre's) work
of adhesion. This equilibrium is stable if $\partial G/\partial A$ is positive,

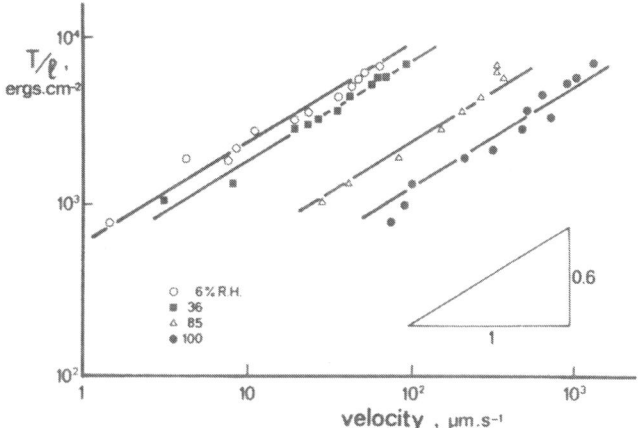

Fig. 37. Effect of relative humidity on the rolling
          resistance of a glass cylinder on a poly-
          urethane surface. $\theta = 21^{\circ}C$. (Roberts, 56)

unstable if negative. The quasistatic force of adherence is the
load corresponding to $\partial G/\partial A = 0$. But equilibrium may be stable at
fixed grips conditions and unstable at fixed load, so that the
quasistatic force of adherence may depend on the stiffness of the
measuring apparatus. When G>w, the separation of the two bodies
starts, and can be seen as the propagation of a crack in mode I. G-w
is the force applied to unit length of crack; under this force, the
crack acquires a limiting speed v which is a function of the
temperature, and one can write

   $G-w = w\phi(a_T v)$.

The second member is the drag due to viscoelastic losses at the
crack tip, and is proportional to w as proposed by Gent and Schultz
(26) and Andrews and Kinloch (27). For a viscoelastic adhering to a
rigid body, the dimensionless function ϕ is a characteristic of the
viscoelastic material, i.e. independent (i) of the nature of the
rigid body, (ii) of the geometry of the contact and (iii) of the
loading system. When ϕ is known, this equation enables us to
predict the kinetics of detachment, provided that the failure is an
adhesive failure, and the viscoelastic losses are limited to the
crack tip. This last condition means that gross displacements must
be elastic for G to be valid in kinetic phenomena. The interest of
the proposed formula is that surface properties and viscoelastic
losses are clearly decoupled from elastic properties and loading
conditions that appear in G.

Three geometries have been investigated: adherence of spheres, adherence of punches, and peeling. The quasistatic force of adherence force, as given respectively by Johnson et al. (2) and Kendall (14) are, of course, found; the variation of energies with the area of contact is given; and the variation of the "motive" G-w with the reduction of the area of contact is studied: G-w is a constant for peeling, it continuously increases for punches, it may decrease, or decrease then increase, or continuously increase for spheres, according to the applied load and the area of contact. Crack propagation at fixed load, fixed grips and at fixed cross-head velocity are studied with implications on tackiness.

ACKNOWLEDGEMENTS

This research was supported by a grant from Directron des Recherches et Moyens d'Essais DRME #77-1037 and a grant from Délégation Générale a'la Recherche Scientifique DGRST #78-7-2667.

APPENDIX 1

The stress distribution for the punch problem is given by eq. (35) with the following asymptotic value near the crack tip:

$$\sigma_z = \frac{P}{2\pi a^2} \sqrt{\frac{a}{2\rho}} = \frac{K_1}{\sqrt{2\pi\rho}} \tag{84}$$

where K_1 is the stress intensity factor:

$$K_1 = \frac{P}{2\pi^{1/2}a^{3/2}} \tag{85}$$

The discontinuity of displacement at the crack tip is given, from eq. (38), by

$$[u_z] = \delta - u_{z1} - u_{z2} = \frac{2\delta}{\pi}\sqrt{\frac{2\rho}{a}} = \frac{4P}{3\pi aK}\sqrt{\frac{2\rho}{a}} \tag{86}$$

where indices 1 and 2 are for solids 1 and 2, and $\rho = r-a$. Eq. (86) can be written

$$[u_z] = \frac{16 K_1}{3K}\sqrt{\frac{\rho}{2\pi}} \tag{87}$$

that reduces to eq. (29) for plane strain in the case of two identical solids.

It can be seen that for two identical solids K_1 and G, eq. (41), are linked by the Irwin relation, eq. (28), for plane strain.

This means that GdA is the work done by punch stresses during crack propagation. Let us verify this point in the general case of contact of two different elastic solids. Let us compute the work dU_E needed to close, at fixed grips, a crack such as in Fig. 5d, from radius a to radius a+da, by progressively applying stresses on the two sides so that the final stresses are given by eq. (84) when the crack is completely closed.

$$dU_E = \int_a^{a+da} \frac{1}{2} \sigma_z [u_z] 2\pi r dr$$

where the factor $\frac{1}{2}$ arises because of the proportionality between traction and displacements. The relevant stresses are those across the annulus, prior to extension, i.e., those corresponding to $\rho = a+da-r$, and the displacements are those across the annulus prior to closure, i.e. those corresponding to $\rho = r-a$:

$$dU_E = \frac{8K_1^2}{3K} \int_a^{a+da} (\frac{r-a}{a+da-r})^{\frac{1}{2}} r dr$$

that gives after a lengthy but simple calculation, and for da→o

$$dU_E = \frac{2K_1^2}{3K} 2\pi a da$$

$$= \frac{P^2 dA}{6\pi a^3 K} = GdA$$

The equation

$$G = \frac{2K_1^2}{3K} \tag{88}$$

is the Irwin relation for plane strain, eq. (28), in the case of two different solids.

The stress distribution for adhesive spheres is given (Johnson 58) by way of the superposition of a Hertzian distribution with a load P_1 and a punch distributiion with a tensile load $P_1 - P$

$$\sigma_z = - \frac{3P_1}{2\pi a^2} (1 - \frac{r^2}{a^2})^{\frac{1}{2}} + \frac{P_1 - P}{2\pi a^2} (1 - \frac{r^2}{a^2})^{-\frac{1}{2}} \tag{89}$$

The asymptotic value near the crack tip is only due to the punch stresses, and is given by

$$\sigma_z = \frac{P_1 - P}{2\pi a^2} \sqrt{\frac{a}{2\rho}} = \frac{K_1}{\sqrt{2\pi\rho}} \qquad\qquad (90)$$

as given by Savkoor and Briggs (56). Relation between G (eq. (62)) and K_1 is still given by eq. (88), that means that only the punch stresses significantly contribute to the work GdA during crack extension.

Thus, the stress intensity approach only takes into account the asymptotic value of the punch stresses near the crack tip. A similar method was followed by Pollock et al.(57) for calculating P_1 in any axisymmetric case by the energetic approach. Suppose that w is at first set equal to zero, while a load P_1 (not necessary given by eq. (59)) is applied to give a contact radius a equal to the observed value. The, elastic and potential energies at this stage are denoted U_E and U_P. Suppose the load to be reduced to the final value P while a remaining constant and the surface energy $U_S = -\pi a^2 w$ added: the elastic and potential energies changes are those due to a punch displacement under the tensile load $P_1 - P$. The total energy is now

$$U_T = U_E' + U_P' - \pi a^2 w - \frac{(P_1 - P)^2}{3aK}$$

As for a punch G-w can be defined as the variation of U_T with a variation da at <u>constant punch tensile load</u> $P_1 - P$ (during which U_E' and U_P' remain constant). Equilibrium is thus given by

$$\left[\frac{\partial}{\partial a} \left(-\pi a^2 w - \frac{(P_1 - P)^2}{3aK}\right)\right]_{(P_1 - P)} = 0$$

i.e.

$$G = \frac{(P_1 - P)^2}{6\pi a^3 K} = w$$

valid for any axisymmetric case.

APPENDIX 2

Calculation of the sector area S bounded by the loading curve OJB and the equilibrium curve BCDO.

To simplify, let us use the reduced coordinates

$$y = \frac{\delta}{\left(\dfrac{\pi^2 w^2 R}{3K^2}\right)^{1/3}} \quad \text{and} \quad x = \frac{P}{3\pi wR} \quad \text{in which:}$$

- the equilibrium curve equation is

$$y_J = x_1^{2/3} + \frac{2x}{x_1^{1/3}}$$

with $x_1 = \dfrac{P_1}{3\pi wR}$ linked to x by $x = x_1 - \sqrt{2x_1}$

- the Hertz curve is defined by

$$y_H = 3x^{2/3}$$

- the loading straight line BJ is given by

$$y_B = \frac{2X}{x_1^{2/3}} + x_1^{2/3}$$

So, the area can be written:

$$S = \int_{OB} y_J dx + \int_{BJ} y_B dX + \int_{JO} y_H dx$$

with $\displaystyle\int_{OB} y_J dx = \int_o^x (x_1^{2/3} + \frac{2x}{x_1^{1/3}}) dx = \int_o^{x_1} (x_1^{2/3} + \frac{2x}{x_1^{1/3}}) \frac{dx}{dx_1} \cdot dx_1$

$$= \frac{9}{5} x_1^{5/3} - 3\sqrt{2} x_1^{7/6} + 3x_1^{2/3}$$

and $\displaystyle\int_{BJ} y_B dX = \int_x^{X_1} (\frac{2X}{x_1^{2/3}} + x_1^{2/3}) dX = 3\sqrt{2} x_1^{7/6} - 2x_1^{2/3}$

and $\displaystyle\int_{JO} y_H dx = \int_{x_1}^o 3x^{2/3} dx = -\frac{9}{5} x_1^{5/3}$

Then $S = +x_1^{2/3}$, which exactly corresponds to the absolute value of the surface energy at the equilibrium point B.

REFERENCES

1. K. Kendall, J. Phys. D: Appl. Phys. 4, 1186-95 (1971).
2. K.L. Johnson, K. Kendall and A.D. Roberts, Proc. R. Soc. Lond.
    A 324, 301-13 (1971).
3. K. Kendall, J. Phys. D: Appl. Phys. 5, 1782-7 (1973b).
4. K. Kendall, J. Phys. D: Appl. Phys. 8, 1449-52 (1975g).
5. K. Kendall, J. Phys. D: Appl. Phys. 8, 1722-32 (1975h).
6. K. Kendall, J. Phys. D: Appl. Phys. 8, 512-22 (1975c).
7. K. Kendall, Proc. R. Soc. Lond. A 344, 287-302 (1975e).
8. K. Kendall, J. Adhesion 7, 137-40 (1975f).
9. K. Kendall, Proc. R. Soc. Lond. A 341, 409-28 (1975a).
10. K. Kendall, J. Mat. Sc. 11, 638-44 (1976a).
11. K. Kendall, J. Mat. Sc. 11, 1263-6 (1976b).
12. K. Kendall, J. Mat. Sc. 11, 1267-9 (1976c).
13. K. Kendall, J. Mat. Sc. 10, 1011-4 (1975d).
14. J.R. Rice, "Fracture: An Advanced Treatise", ed. H. Liebowitz
    (New York: Academic Press), Vol. 2, p. 191-311 (1968).
15. C. Huet, Ind. Minerale 3, 128-41 (1973).
16. D. Maugis and M. Barquins, J. Phys. D: Appl. Phys. 11, 1989-
    2023 (1978b).
17. H.B. Callen, Thermodynamics (New York: John Wiley & Sons)
    (1960).
18. R. Courtel, D. Maugis and M. Barquins, Industrie Minérale 4,
    137-43 (1977).
19. H.D. Bui, Mécanique de la rupture fragile (Paris: Masson)
    (1978).
20. J.D. Ferry, "Viscoelastic Properties of Polymers", 2nd edition
    (New York: John Wiley & Sons) (1970).
21. D.H. Kaelble, J. Colloid Sc. 19, 413-24 (1964).
22. D.H. Kaelble, J. Adhesion 1, 102-23 (1969).
23. A.N. Gent and R.P. Petrich, Proc. R. Soc. Lond. A310, 433-48
    (1969).
24. A.N. Gent, J. Polym. Sc. A-2 9, 283-94 (1971).
25. A.N. Gent and A.J. Kinloch, J. Polym. Sc. A-2 9, 659-68 (1971).
26. A.N. Gent and J. Schultz, J. Adhesion 3, 281-94 (1972).
27. E.H. Andrews and A.J. Kinloch, Proc. R. Soc. Lond. A 332, 385-
    99 (1973).
28. M. Barquins and R. Courtel, Wear 32, 133-50 (1975).
28 bis. A.D. Roberts and A.B. Othman, Wear 42, 119-33 (1977).
29. M. Barquins, R. Courtel and D. Maugis, Eighth World Conference
    on Non-Destructive Testing, Cannes (France), Paper 4 B 11
    (1976).
30. J. Boussinesq, "Application des Potentiels" (nouveau tirage),
    (Paris: Blanchard, 1885 (1969).
31. S. Way, J. Appl. Mech. ASME 7, 147-57 (1940).
31 bis. K. Kendall, J. Adhesion 5, 77-9 (1973a).
32. J.P. Berry, J. Mech. Phys. Solids 8, 194-206 (1960).
33. D. Maugis and M. Barquins, C.R. Acad. Sc. Paris B, 286, 1-4
    (1978a).

34. K.L. Johnson, "The Mechanics of the Contact Between Deformable Bodies", eds. A.D. de Pater and J.J. Kalker (Delft: Delft UP), p. 26-40 (1975).

35. M. Barquins, D. Maugis and R. Courtel, C.R. Acad. Sc. Paris B 279, 565-8 (1974).

36. R.J. Good, J. Am. Chem. Soc. 74, 5041-2 (1952).

37. G.A.D. Briggs and Briscoe, B.J., J. Phys. D: Appl. Phys. 10, 2453-66 (1977).

38. K.N.G. Fuller and D. Tabor, Proc. R. Soc. Lond. A 345, 327-42 (1975).

39. K.E. Easterling and A.R. Thölen, Acta Met. 20, 1001-8 (1972).

40. A. Kohno and S. Hyodo, J. Phys. D: Appl. Phys. 7, 1243-6 (1974).

41. D. Maugis, G. Desalos-Andarelli, A. Heurtel and R. Courtel, ASLE Trans. 21, 1-19 (1978).

42. A.D. Roberts and A.G. Thomas, Wear 33, 45-64 (1975).

43. N. Brunt, Rheologica Acta 1, 242-7 (1958).

44. R.C. Drutowski, Trans. ASME: J. Lub. Techn. 91F, 732-7 (1969).

45. K. Kendall, J. Adhesion 7, 55-72 (1975b).

46. M. Barquins and D. Maugis, C.R. Acad. Sc. Paris B 285, 125-8 (1977).

47. M. Barquins and D. Maugis, C.R. Acad. Sc. Paris B 286, 57-60 (1978).

48. K. Kendall, Wear 33, 351-8 (1975i).

49. D. Maugis and M. Barquins, C.R. Acad. Sc. Paris B 287, 49-52 (1978c).

50. R.S. Rivlin, Paint Technol. 9, 215-6 (1944).

51. B.V. Deryagin and N.A. Krotova, Dokl. Akad. Nank SSSR 61, 849-52 (1948).

52. B.V. Deryagin and N.A. Krotova, Chem. Abstr. 43, 2842 (1949).

53. D. Maugis, Le Vide No. 186, 1-19 (1977).

54. D.A. Birch, J.T. Evans and J.R. White, J. Phys. D: Appl. Phys. 10, 2003-10 (1977).

55. A.D. Roberts, Rub. Chem. Tech. 52, 23-42 (1979).

56. A.R. Savkoor and G.A.D. Briggs, Proc. R. Soc. Lond. A 356, 103-116 (1977).

57. H.M. Pollock, D. Maugis and M. Barquins, Appl. Phys. Lett. 33, 789-9 (1978).

58. K.L. Johnson, Brit. J. Appl. Phys. 9, 199-200 (1958).

NOTES ADDED IN PROOF

1) As K_1 is proportional to the discontinuity of displacement $[u_z]$ (eq. 29), K_1 and G cannot be negative. Hence the minimum value of K_1 and G during crack closure (G < w) are $K_1 = 0$ and G = 0 at which the stress singularities disappear. During crack closure at G = 0 the work done by the stresses at the crack tip and the

viscoelastic losses are negligible.   Crack closure under ΔP is a
very fast phenomenon compared with the reverse crack propagation
under $-\Delta P$.

In loading experiments with spheres (Fig. 30), the branch NS
corresponds to the transition from $G = w$ to $G = 0$, and the Hertz
branch ST to propagation at $G = 0$ (i.e. $P_1 = P$) with no stress
singularity.   No air sheet ($w \neq 0$) is needed to explain this fast
propagation.   The same explanation probably holds for rolling of
cylinders on adhesive substrates: the crack closure at the front is
with $[u_z] = 0$, i.e., $G = 0$, and the rolling resistance corresponds
to $\frac{\pi}{2}$ peeling at the rear with $G \gg w$.

2)   The vertical displacement u_z of the surface of an elastic half-
space indented by a rigid elliptical paraboloid (Hertzian contact)
was given by Boussinesq (1885).   For a sphere it reduces to

$$u_z = \frac{3P}{4a} \frac{1-v^2}{\pi E} \left[(\frac{r^2}{a^2} - 1)^{1/2} - (\frac{r^2}{a^2} - 2) \text{ Arc tg } (\frac{r^2}{a^2} - 1)^{-1/2} \right]$$

that is equivalent to the eq. (15) given in the paper by Derjaguin
et al. (ref. a).   For an adhesive contact ($w \neq 0$) punch displacement
(eq. 36) under tensile load $P_1 - P$ must be added to Hertzian
displacement under load P_1; hence the vertical displacement

$$u_z = \frac{a^2}{\pi R} \left[(\frac{r^2}{a^2} - 1)^{1/2} - (\frac{r^2}{a^2} - 2) \text{ Arc tg } (\frac{r^2}{a^2} - 1)^{-1/2} - \frac{4}{3} \text{ Arc sin } \frac{a}{r} \right]$$

$$+ \frac{1 - v^2}{\pi E} \frac{P}{a} \text{ Arc sin } \frac{a}{r}$$

that describes the equilibrium profiles drawn on Fig. 20, as well as
the non-equilibrium ones on Fig. 31.   The quantity

$$z_o - \delta = \frac{r^2}{2R} - \frac{a^2}{3R} \left(1 + 2 \frac{P_1}{P} \right)$$

would be added to have the profile of an adhesive sphere in contact
with a rigid plane as in ref. b.

REFERENCES

a)   B.V. Derjaguin, V.M. Muller, Yu. P. Toporov, J. Colloid Interface Sci. <u>53</u>, 314 (1975).

b)   D. Tabor, <u>J. Colloid Interface Sci</u> <u>58</u>, 2 (1977); <u>J. Colloid
     Interface Sci.</u> <u>67</u>, 380 (1978).

The Viscoelastic Shear Behavior of a Structural Adhesive

Erol Sancaktar*, Instructor
H. F. Brinson, Professor

Virginia Polytechnic Institute and State University

Blacksburg, Virginia 24061

ABSTRACT

The increased use of adhesives in the automobile, aerospace and other industries is noted. The need to understand the viscoelastic nature of polymeric adhesives is assessed. Previous bulk tensile constant strain-rate, creep, relaxation, and delayed failures studies on an epoxy adhesive are summarized. New constant strain-rate and creep data for symmetric rail shear and symmetric lap shear geometries of the same epoxy adhesive are reported. It is shown that significant viscoelastic shear behavior, including delayed failures, do occur. The linear viscoelastic methods and results from previous tensile studies are extended to the condition of shear. Predictions are shown to be at variance with experimental results. The reasons for deviations are discussed; i.e., the possibility that linear viscoelastic analytical procedures are inappropriate and the likelihood that viscoelastic phenomena are different in bulk tensile, bulk shear and lap shear geometries.

INTRODUCTION

In recent years a great deal of attention has been given to the characterization of light but strong polymeric and polymer based composite materials. The application of these materials in the

---

*   Presently, Assistant Professor, Clarkson College, Potsdam, N.Y.

automotive, aerospace and other industries have made the reliable
design of adhesively bonded joints very desirable. Often adhesive
bonding techniques either augment or eliminate the need of bolted or
riveted joints. A desirable feature of adhesive bonding is that it
is readily adapted to assembly-line manufacturing procedures and
stress concentrations arising from penetration techniques may be
avoided or minimized.

The purpose of the present investigation was to study the
viscoelastic shear properties of a particular epoxy adhesive, Metlbond 1113 and 1113-2**. This was to be done using a symmetric rail
shear test method on bulk panels and symmetric single lap shear test
method for bonded joints. Specifically, constant strain rate and
creep tests were to be performed on these specimens to ascertain
viscoelastic properties including delayed failures. While such
properties had been previously determined using bulk tensile testing (1,2), the primary objective of the current program was to
determine if bulk and lap shear viscoelastic properties could be
predicted from the earlier bulk tensile results. Prior to describing the current effort, it is therefore appropriate to describe the
earlier bulk tensile studies and our characterization procedures
used there (1,2).

Tensile Properties and Their Characterization

First, consider an idealized stress-strain response which we
will subsequently show is a reasonable approximation to the adhesives studied herein as well as other polymeric materials (3,4).
This idealized stress-strain response shown in Fig. 1 contains a
region of linear elastic behavior followed by a region of viscoelastic behavior terminating in a region of plastic flow. It is
important to note that the linear elastic limit stress, θ, is not
identified as a yield stress as is often done. Rather, the maximum
stress, Y, is identified as the yield stress. The values θ and Y
are rate and time dependent. Thus, the idealized material of Fig. 1
exhibits time dependent incipient yielding or failure as opposed to
the time dependent subsequent yielding discussed elsewhere (5).

The idealized response described above is compatible with a
mathematical theory of viscoplasticity by Nagdi and Murch (6) and a
delayed yield response of Crochet (7). The latter is shown schema-
tically in Fig. 1 and is written analytically as,

**   Metlbond 1113 is described by its manufacturer, Narmco Corp.,
       as a 100% solid modified epoxy film with a synthetic carrier
       cloth and is supplied in tape form. 1113-2 is the same as 1113
       without the carrier cloth. Each is rubber toughened.

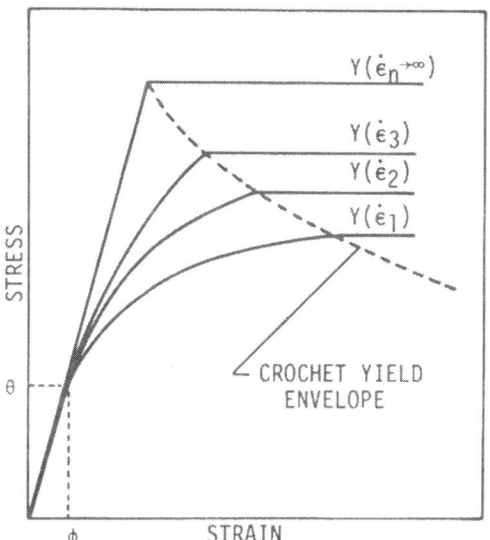

Fig. 1. Idealized Stress-Strain/Strain-Rate
         and Yield Behavior.

$$Y(t) = A + B \exp(-C\chi) \tag{1}$$

where Y is the time dependent yield stress, A, B, C are material
constants and where χ is a time dependent parameter defined by

$$\chi = [(\varepsilon_{ij}{}^V - \varepsilon_{ij}{}^E)(\varepsilon_{ij}{}^V - \varepsilon_{ij}{}^E)]^{\frac{1}{2}} \tag{2}$$

In (2) the terms $\varepsilon_{ij}{}^V$ and $\varepsilon_{ij}{}^E$ represent the viscoelastic and
elastic strains respectively.

In order to use equations (1) and (2) it is necessary to have a
mathematical representation of $\varepsilon_{ij}{}^V$ and $\varepsilon_{ij}{}^E$. A relatively simple
procedure to accomplish this task is through the use of a mechanical
model which includes a sliding friction or a yield element. Several
models which have been proposed for this purpose are shown in Fig.
2. The modified Bingham model used by Brinson (3) demonstrates the
same phenomenological behavior as that shown in Fig. 1, i.e., a
region of linear elasticity, followed by a region of linear viscoelasticity terminating with a region of plastic flow.

The differential one-dimensional constitutive equation for the
modified Bingham model can be written as,

$$\epsilon = \sigma/E \quad , \quad \sigma < \theta$$

$$\dot{\epsilon} = \dot{\sigma}/E + \frac{\sigma - \theta}{\mu} \quad , \quad \theta < \sigma < Y \tag{3}$$

where E is Young's modulus, μ is the viscosity coefficient and θ and Y are as previously defined. It is interesting to note that equations (3) are similar to those used by others such as Malvern (10) and Perzyna (11) which were based on an empirical non-linearization of a simple Maxwell model. Again, the interpretation of yielding herein is quite different than that of the persons just cited.

Solution of equations (3) for constant strain rate, $\dot{\epsilon}$, gives,

$$\sigma(\epsilon) = E\epsilon \quad , \quad \sigma < \theta$$

$$\sigma(\epsilon) = \theta + \mu\dot{\epsilon} \left[1 - \exp \frac{-(\epsilon - \phi)E}{\mu\dot{\epsilon}} \right] \quad , \quad \theta < \sigma < Y \tag{4}$$

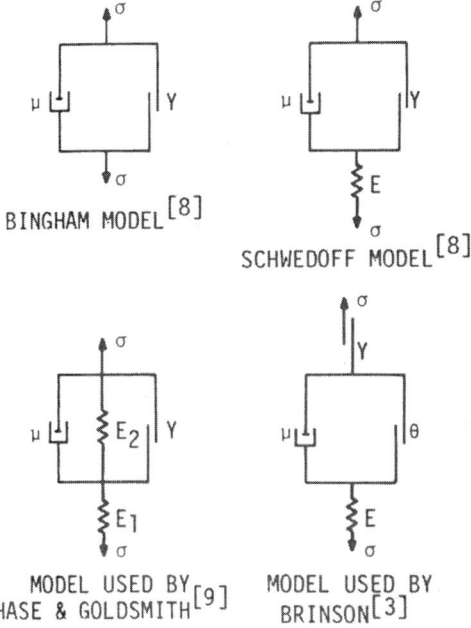

Fig. 2.   Various Viscoelastic-Plastic Models.

Solution of equations (3) for the case of creep and substitution into equation (2) gives the following specific form for the time parameter, χ, needed in equation (1),

$$\chi = \frac{\sigma_o - \theta}{\mu} \, t \, [1 - 2\nu^2]^{\frac{1}{2}} \qquad\qquad (5)$$

where σ_o is the creep stress and ν is Poisson's ratio which has been assumed to be a constant. Crochet's delayed yield equation can now be written as

$$Y(t) = A + B \exp[-K(\sigma_o - \theta)t] \qquad\qquad (6)$$

where $K = C(1 + 2\nu^2)^{\frac{1}{2}}/\mu$ is a constant and a modified Bingham model has been used to describe the stress-strain behavior. Equation (6) represents a creep to yield phenomenon.

It might be noted that Crochet's model is not readily adapted to quantifying the variation of elastic limit stress and yield stress as a function of strain-rate. As a result, in our previous studies we have used the following empirical equations due to Ludwik (1),

$$\theta = \theta' + \theta'' \, \log(\dot{\epsilon}/\dot{\epsilon}')$$

$$Y = Y' + Y'' \, \log(\dot{\epsilon}/\dot{\epsilon}') \qquad\qquad (7)$$

where θ', θ'', Y', Y'' and $\dot{\epsilon}'$ are material constants.

As mentioned earlier, constant strain-rate, creep and relaxation studies were performed on bulk tensile coupons of Metlbond 1113 and 1113-2. Figure 3a shows the constant strain-rate response of 1113. It may be observed that the material displays the same elastic, viscoelastic and plastic response regimes as the idealized material shown in Fig. 1 and the modified Bingham model shown in Fig. 2. Further, it is demonstrated that the modified Bingham model can be used to closely approximate the measured behavior. It should, of course, be mentioned that a different relaxation time must be used for each strain-rate in order to achieve a close fit of the data. The relaxation times needed are shown to vary nearly linearly with strain-rate in Fig. 3b. While the need for a variable relaxation time renders the modified Bingham model to be empirical, it does not negate the utility of the constant strain-rate stress-strain response curve fitting procedure. Obviously, the procedure can be used quite accurately for interpolation and extrapolation purposes (4). That is, given the stress-strain response for a single strain rate, the response for other strain rates can be closely predicted.

Figure 3c shows the variation of yield (maximum) stress with strain-rate compared to equation (7). Apparently, Ludwik's equa-

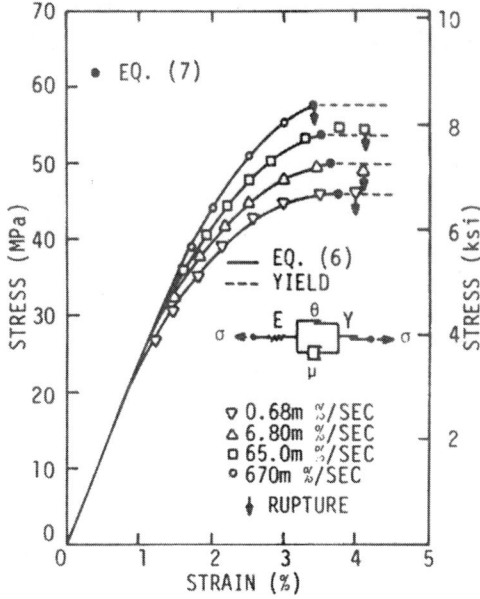

Fig.3a.   Stress-Strain/Strain-
Rate Behavior of Metlbond
1113 and Comparison with the
Modified Bingham Model.

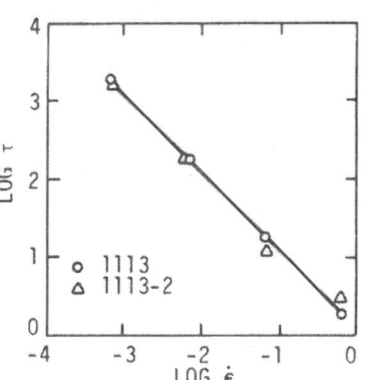

Fig. 3b.   Relaxation Time
Versus Strain Rate for the
Modified Bingham Model.

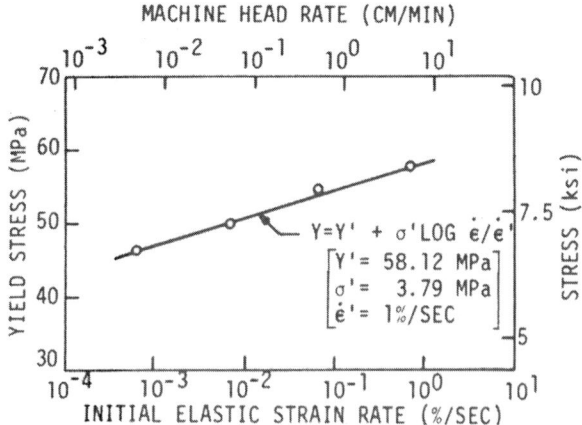

Fig. 3c.   Variation of Yield Stress and Comparison
with Ludwick's Equation for Metlbond 1113.

tion is an excellent representation of the variation in tensile yield stress. Similar results were found for elastic limit stresses also (1).

Figure 3d shows the time for yielding to occur at various creep stress levels. It is to be noted from Fig. 3a that the yield (maximum) and failure (rupture) stresses are the same. As a result, the data points shown in Fig. 3d actually represent the time for creep to failure (rupture) to occur. The results are compared with the predictions of equation (6) which indicates Crochet's model to be a reasonable representation of a bulk tensile delayed failure* phenomenon.

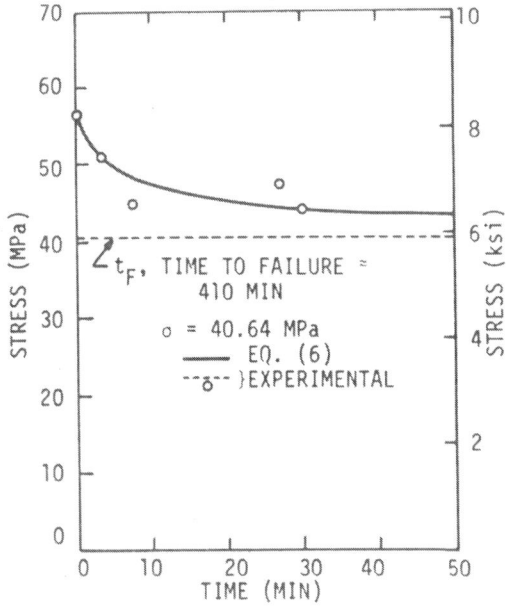

Fig. 3d. Comparison of Creep to Yield Data and Theory for Metlbond 1113.

---

*   It should be recognized that Crochet's model was developed only for a delayed yield process. We have elected to interpret his model to represent a delayed rupture process as well.

Shear Properties and Their Characterization

As stated in the introduction, the objective of the current effort was not only to measure shear viscoelastic-plastic-failure properties but to attempt to predict shear results from the tensile results described in the previous section. The symmetric rail shear and symmetric lap shear specimen geometries shown in Fig. 4 were used to obtain bulk and bonded shear properties respectively. These specimen geometries have been shown to be reasonable approximations of pure shear behavior (12-14).

It is easy to reformulate equations (1) - (7) for the case of pure shear. All that is necessary is essentially to replace tensile stress and strain variables everywhere by their shear counterpart. However, the prediction of shear properties from tensile properties using equations (1) - (7) requires a clearly defined relationship

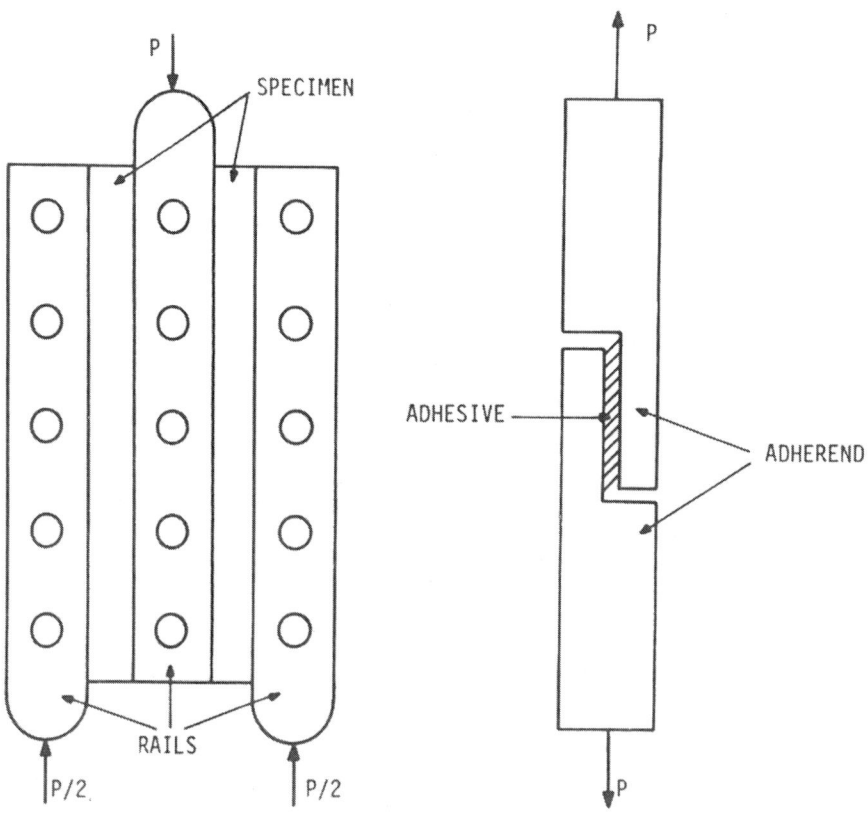

Fig. 4a. Symmetric Rail Shear   Fig. 4b. Symmetric Lap Shear
          Test Specimen.   Specimen.

between shear and tensile elastic limit and yield variables. Equations are needed at the termination of two response regimes--linear elastic and linear viscoelastic.

First consider the linear elastic region in which the following relations are valid for uniaxial tension:

$$\sigma_t = E\epsilon_t \quad , \quad \tau_t)_{max} = \frac{1}{2}\sigma_t)_{max} \qquad (8)$$

where the subscript, t, refers to tensile variables. Hookes law for shear and the relation between moduli are

$$\tau_s = G\gamma_s \quad , \quad G = E/2(1 + \nu) \qquad (9)$$

where the subscript s refers to shear variables. Using the Tresca or maximum shear stress failure theory, it readily follows from (8) and (9) that,

$$\gamma_s = (1 + \nu)\epsilon_t \qquad (10)$$

The use of the Tresca failure criterion and equations (8) - (10) allows the calculation of elastic limit stresses and strains from their tensile counterpart.

By using Laplace transforms, it is easy to show for linear viscoelasticity that equations (8) - (10) are valid where the stresses and/or strains are a function of time providing Poisson's ratio is a time independent constant. Conversion of equations (4) - (6) from tension to shear requires additional information on relaxation times or viscosity coefficients. If Tobolsky's assumption of equivalent relaxation times in shear and tension is invoked, it readily follows that (15),

$$\mu_s = \frac{\mu_t}{2(1 + \nu)} \qquad (11)$$

Using relations (8) - (11) shear predictive equations from tensile counterparts can easily be obtained for constant strain-rate, creep to delayed yield (failure or rupture) and elastic limit or yield stress variation with strain-rate phenomena. These may be written respectively as,

$$\left.\begin{array}{l} \tau_s = G\gamma_s \\[6pt] \tau_s(\gamma_s) = \theta_s + \mu_s\dot{\gamma}_s\left[1 - \exp\dfrac{-(\gamma_s - \phi_s)G}{\mu_s\dot{\gamma}_s}\right] \end{array}\right\} \qquad (12)$$

$$Y_s(t) = A_s + B_s \exp[-K_s(\tau_0 - \theta_s)t] \tag{13}$$

$$\theta_s = \theta_s' + \theta_s'' \log\left(\frac{\dot{\gamma}_s}{\dot{\gamma}_s'}\right) \tag{14}$$

$$Y_s = Y_s' + Y_s'' \log\left(\frac{\dot{\gamma}_s}{\dot{\gamma}_s'}\right)$$

where shear variables τ_s, γ_s, and shear material properties G, θ_s, μ_s, etc., are related to tensile quantities as previously described.

EXPERIMENTAL PROCEDURES

Constant strain rate, creep and relaxation tests were performed on the specimen geometries shown in Fig. 4. The tests were performed utilizing an Instron testing machine in a controlled environment of approximately 25°C (70°F) and 50% R.H. Electrical resistance strain gages were used to measure strains at appropriate locations on the rail shear specimen. Elongations in the lap shear tests were measured using a crack opening displacement (COD) gage. Small holes were drilled into the aluminum adherends on either side of the glue line, steel pins were inserted in the holes and the COD gage was mounted between the steel pins. In this manner a very precise measure of elongation was obtained over a small gage length.

Glued tabs were employed under the steel rails for the rail shear tests. Average shear stresses for both rail shear and lap geometries were calculated by dividing the applied load by the area in shear. Additional test procedures and data reduction details are given in reference (14).

RESULTS AND DISCUSSION

Constant strain-rate shear stress-strain results of the bulk 1113 and 1113-2 adhesives from symmetric rail shear tests are shown in Figs. 5 and 6. As may be observed, the bulk adhesive shows elastic and viscoelastic response regimes much as the bulk tensile data of Fig. 3a. Properties generally increased with increasing strain-rate. The yield (maximum) stresses are not especially meaningful as failure always occurred under a rail at a stress concentration site and not at the location of the strain gage.

Prediction of shear results from tensile data using the equations (12) were successful only in the elastic region. That is, viscoelastic shear stress-strain behavior was not predictable from

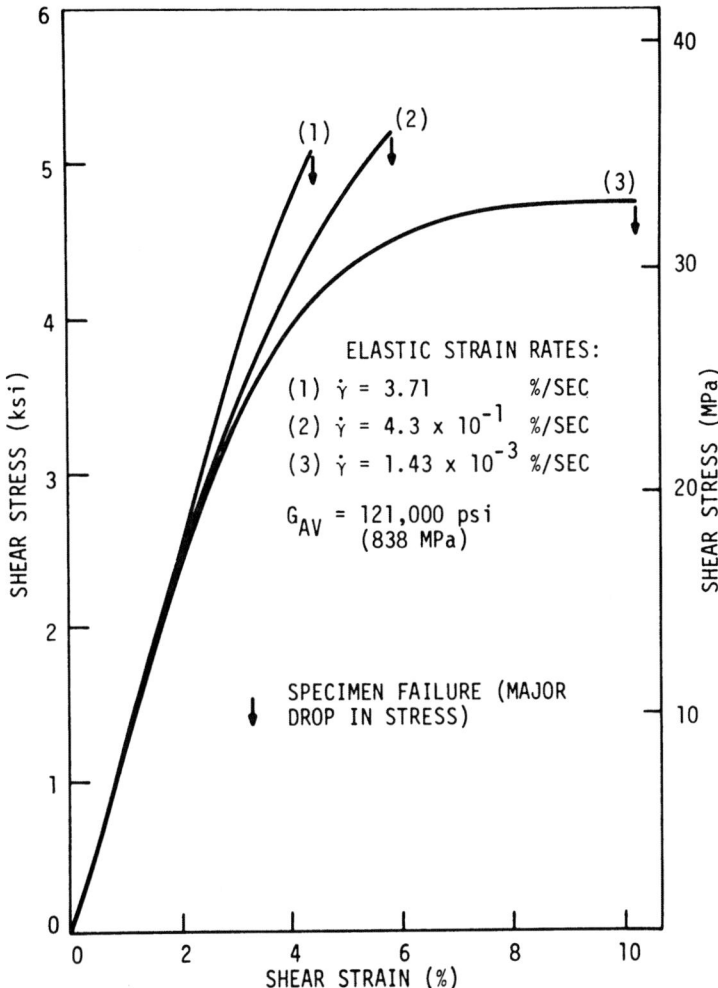

Fig. 5.   Symmetric Rail Shear Stress-Strain/Strain-
          Rate Response of Metlbond 1113.

viscoelastic tensile stress-strain behavior.   It was possible to
fit the data of Figs. 5 and 6 using a modified Bingham model with
new constants and new relaxation times.   Apparently Tobolsky's
assumption of equivalent relaxation times in shear and tension is
not appropriate for the material tested herein.   The linear elastic
limit stresses for different strain-rates from Fig. 6 are shown in
Fig. 7.   Here good agreement was obtained between tensile predictions and measured values.

     The maximum or ultimate (yield as defined herein) stresses are
not shown compared with tensile predictions of equation (14)

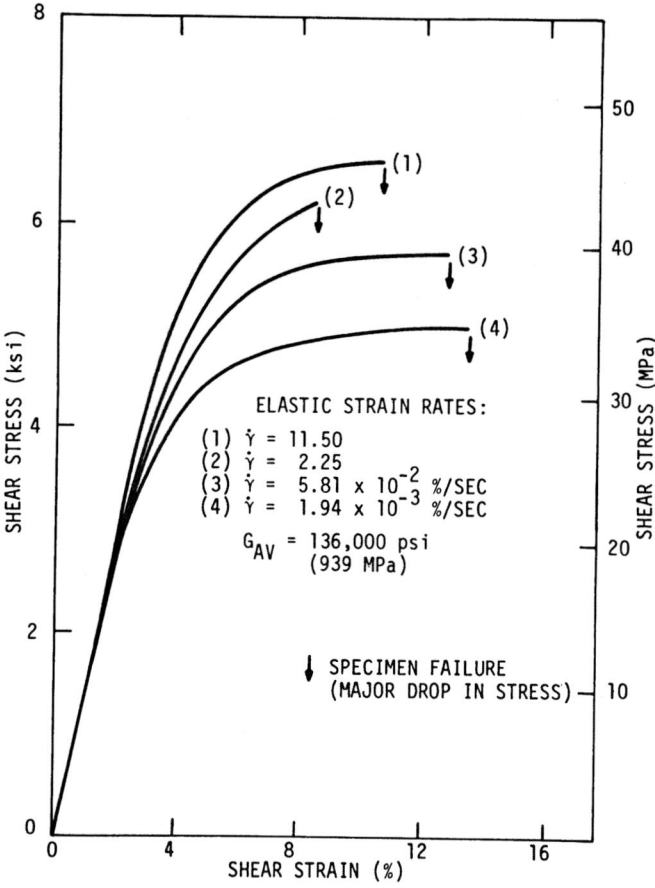

Fig. 6.   Symmetric Rail Shear Stress-Strain/Strain-
Rate Response of Metlbond 1113-2.

because, as will be shown subsequently, large differences were
found.

Symmetric rail shear creep response of 1113-2 is shown in Fig.
8. It is apparent that large creep strains were found in bulk shear
which were not predictable from tensile creep values. Maximum
rupture strains for tension were about 4-6%. Further, rupture did
not occur in most cases and the creep tests were discontinued after
the times indicated. Examination of the creep data clearly indicates the bulk shear response to by non-linear. These observations
would tend to indicate that Metlbond 1113-2 exhibits different

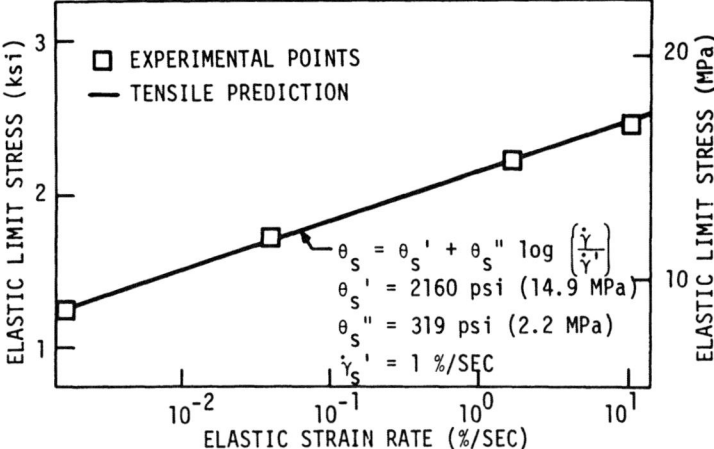

Fig. 7.   Variation of Elastic Limit Stress of
          Metlbond 1113-2 and Comparison with Tensile
          Predictions.

viscoelastic deformation mechanisms in tension and shear.

        Creep to 10% strain levels from Fig. 8 are shown plotted in
Fig. 9. Also shown is the maximum stress reached for the highest
strain rate of Fig. 7. The use of an arbitrary failure strain is
justifiable and clearly Crochet's delayed yield model of equation
(13), if valid at all, should fit such a definition as well as the
actual yield or rupture process. Equation (13) does represent the
data reasonably well but the constants A_s, B_s, C_s are not predictable from tensile values.

        The stress-strain response of Metlbond 1113 for five different
symmetric single lap specimens tested at about the same strain-rate
is shown in Fig. 10. The predicted initial slope from tensile data
is shown and agrees well with the measured results. However, quite
obviously, considerable data scatter was obtained in the viscoelastic region giving rise to substantial variation in failure
stresses and strains. Again, generally large strains were found in
the viscoelastic region not predicted by tensile results.

        Constant elongation (relaxation) tests were conducted on two
lap shear specimens with the results being presented in Fig. 11.
That is, the grips of the Instron machine were held fixed after
reaching a predetermined load and strain level. As is evident,
substantial adhesive strain increase was observed while the machine
load was relaxing. Even though the aluminum adherends were substantially stiffer than the epoxy adhesive, considerable strain transfer occurred between the two materials giving rise to creep of the

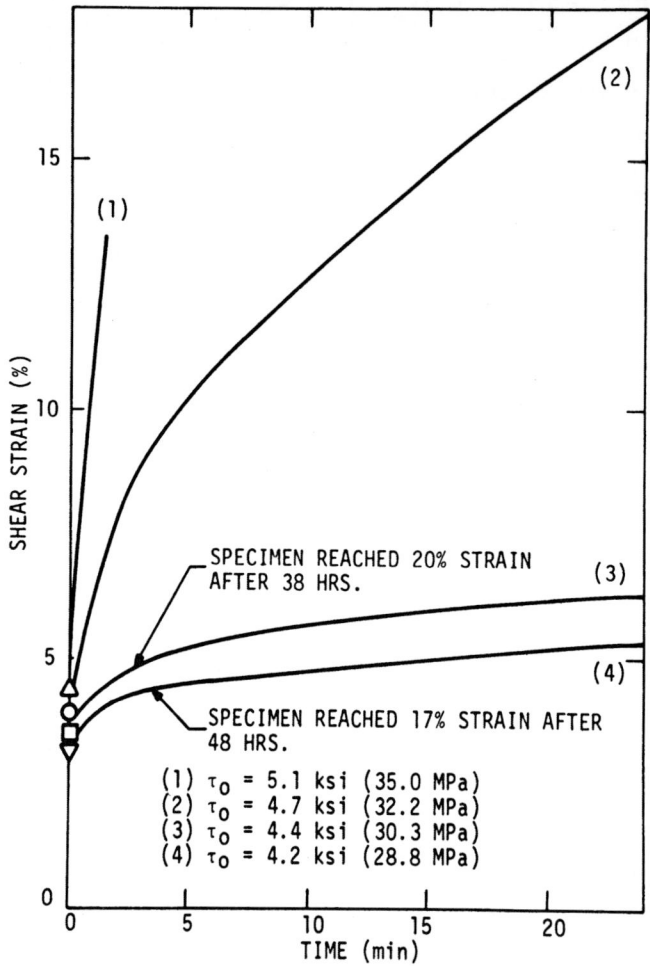

Fig. 8. Symmetric Rail Shear Creep Response of
         Metlbond 1113-2.

adhesive under fixed grip conditions. Obviously a servo-controlled
test procedure not used herein would be the only proper procedure to
determine adhesive relaxation for the symmetric lap shear specimen.

    Creep tests were performed on four 1113 symmetric lap speci-
mens with the results presented in Fig. 12. Again large shear
strains not predictable from tensile results were encountered. At
the highest stress level tested rupture occurred during loading.
For two other specimens, large strain discontinuities were observed
without specimen rupture and were likely the result of an adhesive
fracture and fracture arrest process. Again, the measured creep
strains tend to indicate a non-linear viscoelastic deformation

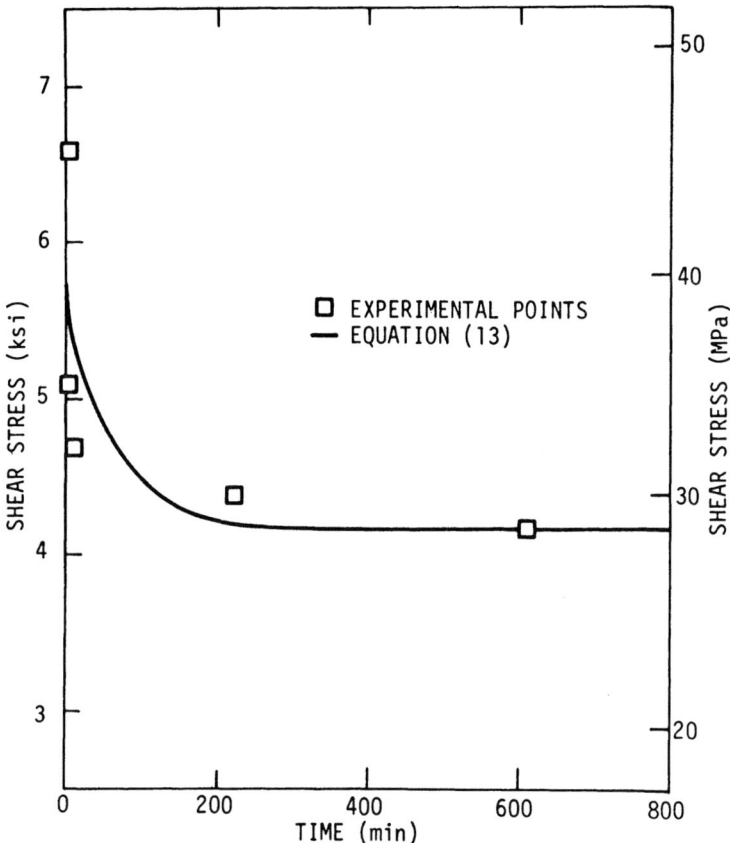

Fig. 9. Creep to 10% Strain Data for Metlbond 1113-2 Symmetrical Rail Shear Specimens.

process.

Creep to failure stresses are shown plotted in Fig. 13 where failure was defined as either rupture, a sudden strain discontinuity or reaching a 10% strain level for creep at the lowest stress level. Obviously, the latter definition has no influence on the character of the delayed failure curve. Again, equation (13) due to Crochet is a reasonable representation of the delayed failure process for the adhesive in the bonded state, but the constants A_s, B_s and C_s are not predictable from tensile results.

Table 1 gives a comparison between the predicted and measured average shear properties of Metlbond 1113. As noted earlier, average elastic shear properties can be predicted reasonably well from tensile properties. However, maximum stresses and strains are not predictable from tensile values and in general larger values are

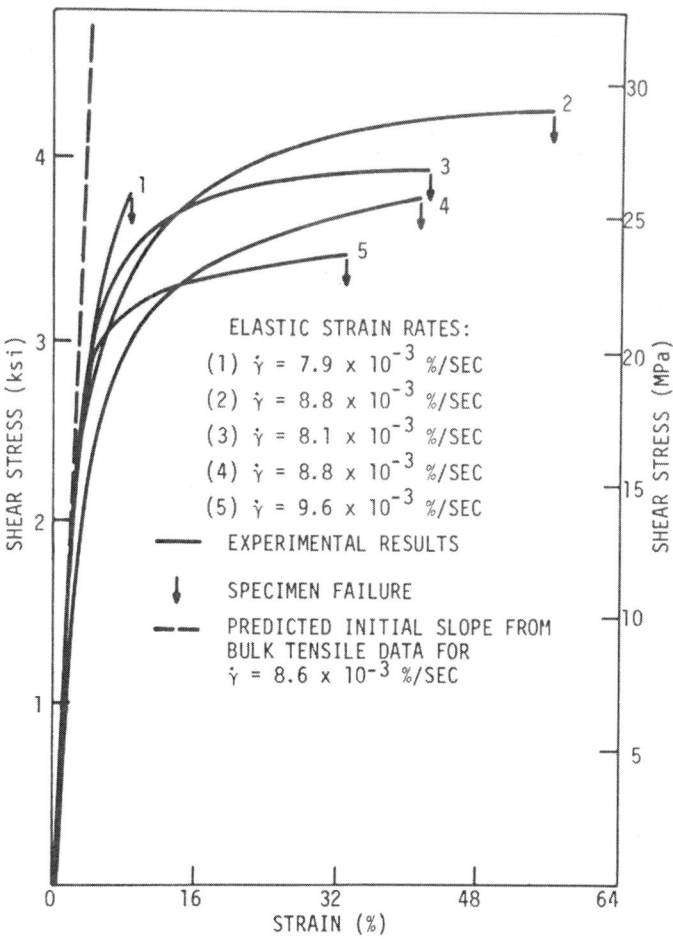

Fig. 10. Symmetric Single Lap Adhesive Stress-Strain
Response of Metlbond 1113. (Specimens
Tested at Nearly the Same Strain-Rate.)

found in shear than would be expected on the basis of tensile
measurements.

SUMMARY AND CONCLUSIONS

Previously reported measurements of viscoelastic adhesive bulk
tensile properties and the analytical procedures to characterize
such properties were presented. Relations were developed by which
adhesive viscoelastic shear properties could be predicted from the
earlier tensile measurements.

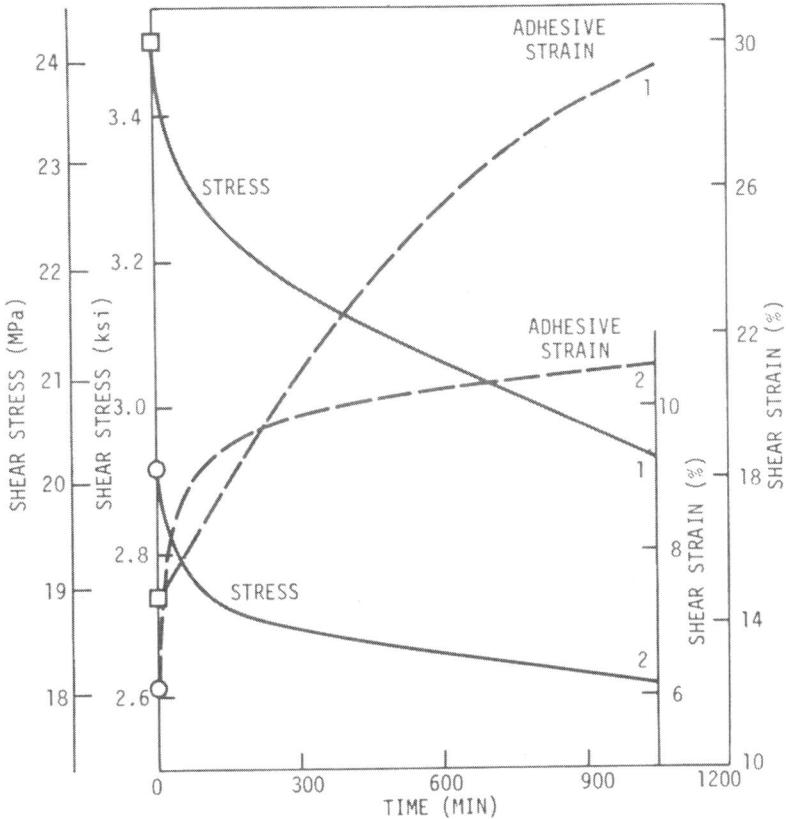

Fig. 11.   Adhesive Relaxation Behavior of Two Metlbond
           1113 Symmetric Single Lap Specimens (Constant
           Deformation was Maintained at the End of
           Adherends.)

Constant strain-rate and creep measurements were presented for
the same epoxy adhesive used in earlier tensile studies. Results
were presented for both bulk and bonded adhesive shear properties
using symmetric rail shear and symmetric lap shear test specimens
respectively.

It was demonstrated that both bulk and bonded shear stress-
strain response include a region of near linear elasticity followed
by a region of viscoelasticity terminating with a yielding region as
was the case with the tensile response. Generally, elastic shear
stresses and strains were shown to be predictable from tensile
results while viscoelastic stresses and strains were not. Large
viscoelastic strains were obtained in shear which were not anticipated on the basis of tensile predictions. Relaxation times in

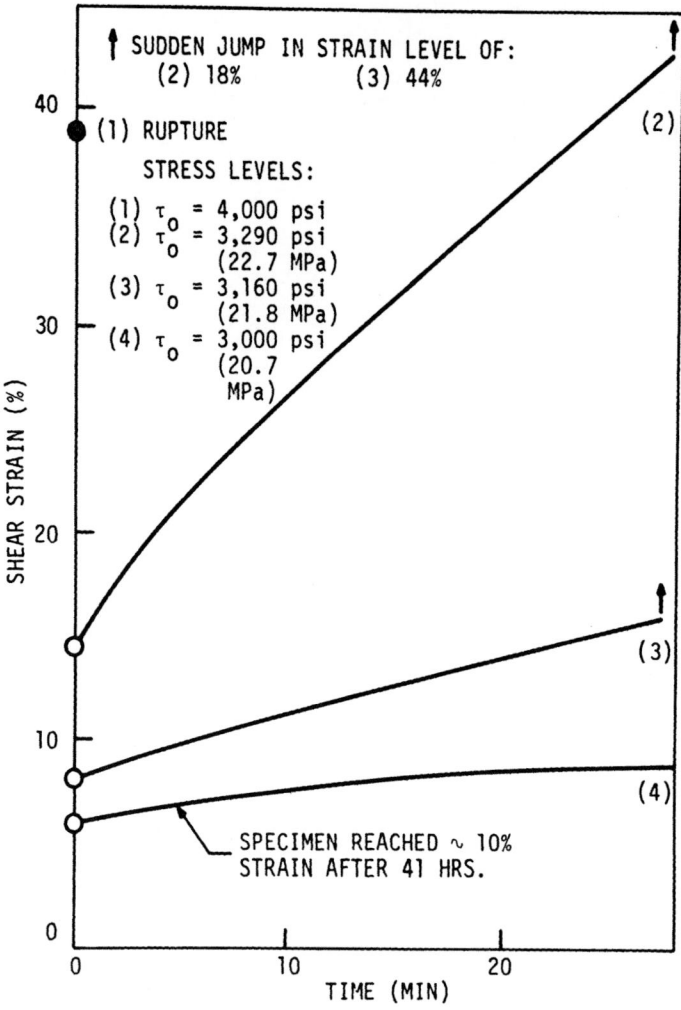

Fig. 12.   Symmetric Single Lap Creep Response of
Metlbond 1113.

tension and shear were not the same in contradiction to Tobolsky's
assumption. For these reasons, it is felt that the materials
investigated tended to exhibit different viscoelastic deformation
mechanisms in shear from tension. Such different deformation
processes may also be evidence that use of a simple Tresca failure
theory for viscoelastic processes is inappropriate and that a more
correct procedure would be to use the general viscoelastic-plastic
theory of Nagdi and Murch (6). Quite obviously, the materials
studied displayed both geometric and stress-strain nonlinearities
in shear. Recently, we have made detailed measurements of tensile

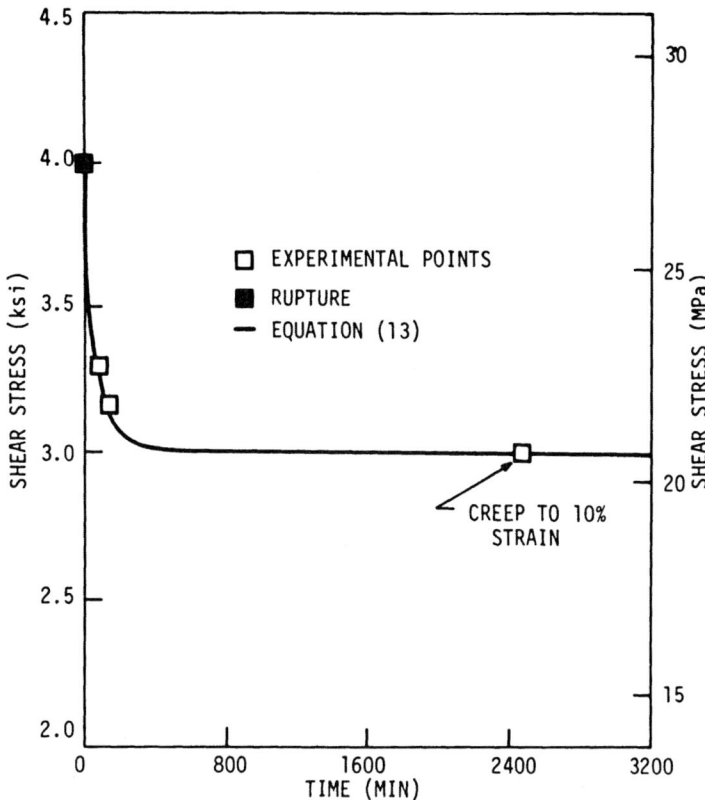

Fig. 13.   Creep to Failure Data for Symmetric Single
            Lap Specimens Bonded with Metlbond 1113.

Table 1.   Comparison of Average Shear Properties
            for Metlbond 1113.

	Elastic Modulus	Yield (Max) Stress	Failure Strains
Tensile Prediction	119 ksi (818 MPa)	3780 psi (26.0 MPa)	5.5%
Symmetric Rail Shear	122 ksi (839 MPa)	5000 psi (34.5 MPa)	6.8%
Symmetric Lap Shear	136 ksi (937 MPa)	3840 psi	36.4%

stress-strain non-linearities of Metlbond 1113-2 and have successfully modeled the results using a non-linear integral equation
approach due to Schapery (16).   These results will be reported
subsequently.

ACKNOWLEDGEMENTS

This effort was supported by NASA's Langley Research Center
under NASA grant NGR 47-004-090. Dr. John G. Davis, Jr., was the
NASA technical monitor and his assistance is gratefully acknowledged. Thanks are also due to Mr. Ted Massa and Mr. W. I. Griffith
for their assistance in preparing portions of the manuscript and to
Ms. Peggy Epperly for her expert typing.

REFERENCES

1. H.F. Brinson, M.P. Renieri, and C.T. Herakovich, "Rate and
      Time Dependent Failure of Structural Adhesives", Fracture
      Mechanics of Composites, ASTM STP 593, 1975, pp. 177-199.
2. M.P. Renieri, C.T. Herakovich, and H.F. Brinson, "Rate and
      Time Dependent Behavior of Structural Adhesives", VPI
      Report, VPI-E-76-7, April 1976.
3. H.F. Brinson, "The Viscoelastic-Plastic Characterization of a
      Ductile Polymer", Deformation and Fracture of High Poly-
      mers, H. Kausch, et al., eds., Plenum Press, NY, 1974.
4. H.F. Brinson and A. DasGupta, "The Strain-rate Behavior of
      Ductile Polymers", Experimental Mechanics, 15, No. 12,
      pp. 458-463, December 1975.
5. W. Prager, Introduction to Mechanics of Continua, Ginn and
      Co., NY, 136, 1961.
6. P.M. Nagdi and S.A. Murch, "On the Mechanical Behavior of
      Viscoelastic-Plastic Solids", J. of Appl. Mech., 30, p.
      321, 1963.
7. M.J. Crochet, "Symmetric Deformations of Viscoelastic-Plastic
      Cylinders", J. of Appl. Mech., 33, p. 326, 1966.
8. M. Reiner, Advanced Rheology, H.K. Lewis and Co. Ltd., London,
      208, 1971.
9. K.W. Chase and W. Goldsmith, "Mechanical and Optical Charac-
      terization of an Anelastic Polymer at Large Strain Rates
      and Large Strains", Experimental Mechanics, 14(1), pp.
      10-18, January 1974.
10. L.E. Malvern, "The Propagation of Longitudinal Waves of Plas-
      tic Deformation in a Bar of Material Exhibiting a Strain-
      Rate Effect", Journal of Applied Mechanics, 18, pp. 203-
      208, 1951.
11. P. Perzyna, "The Constitutive Equations for Rate Sensitive
      Plastic Materials", Quarterly of Applied Mathematics, 20,
      p. 321, 1963.

12. Y.T. Yeow and H.F. Brinson, "A Comparison of Simple Shear Characterization Methods for Composite Laminates", _Composites_, Jan. 1978.
13. W.J. Renton and J.R. Vinson, "Shear Property Measurements of Adhesives in Composite Material Bonded Joints", ASTM-STP 580, 1974.
14. E. Sancaktar, "The Viscoelastic Shear Behavior of Adhesives", Ph.D. Thesis, VPI & SU, June 1979.
15. A.V. Tobolsky, _Properties and Structure of Polymers_, J. Wiley & Sons, Inc., NY, 1960.
16. J.S. Cartner, Jr., "The Nonlinear Viscoelastic Behavior of Adhesives and Chopped Fiber Composites", M.S. Thesis, VPI & SU August 1978.

Surface Energetics and Structural Reliability of Adhesive Bonded Metal Structures

D.H. Kaelble

Science Center, Rockwell International

Thousand Oaks, California   91360

ABSTRACT

All metals, except gold, are chemically reactive with oxygen and moisture at ambient temperature and tend to form oxide or hydrated oxide layers. In addition to oxidative chemical reaction, these high energy surfaces are subject to physical adsorption by water vapor and volatile organic contaminants. Earlier studies have shown for structural aluminum alloy AL-2024-T3 that the kinetics of surface aging, subsequent to acid etching, can be followed by wettability measurements and expressed on surface energy maps of dispersion and polar components of solid-vapor surface tension. Weibull statistics have recently been applied by Wegman and co-workers to investigate the effects of both surface exposure time (SET) and bond exposure time (BET) to high moisture upon the shear bond strengths of metal joints. In this study the surface chemistry and related surface energy of both metal adherends and adhesives are analyzed prior to bonding. The theory of interfacial adhesion and thermodynamics of fracture are combined to predict the effects of both SET and BET on long term bond strength and reliability. Structural adhesive joints of both aluminum and titanium structural alloys bonded with epoxy-phenolic adhesive are shown to degrade according to surface energetics predictions. Aging is shown to lower average strengths without broadening the Weibull strength distributions. Results are discussed in terms of structural reliability of adhesive bonded metal structures.

INTRODUCTION

All metals, except gold, are chemically reactive with oxygen and moisture at room temperature and tend to form oxide or hydrated oxide surface layers (1). In addition to oxidative chemical reaction, these high energy surface films are subject to further physical adsorption by water vapor and volatile organic contaminants. In structural metal to metal joints, the reliability of the bond may be directly related to the chemical stability of the interfacial bond. Weibull statistical analysis has recently been applied to investigate the effects of both surface exposure time (SET) and bond exposure time (BET) on the distributions of single layer shear bonds of aluminum alloys (2). Smith and Kaelble (3) have recently conducted a detailed study of combined SET and BET aging in normal (50% R.H., 23°C), and high moisture (95% R.H., 54°C). The experimental methods of the study by Smith and Kaelble are outlined in Table 1 with full details contained in the published report.

Table 1. Metal Joint Reliability Studies

1. Metal Adherends: Unclad 2024-T3 aluminum alloy surface treated by standard FPL sulfuric chromate etch and T8-6A1-4V titanium alloy treated by standard phosphate fluoride cleaning process. Coupon size 0.063" thick, 1" wide, and 4" long.

2. Adhesive: HT 424 epoxy-phenolic film adhesive (from American Cyanamid) with glass fiber carrier and standard weight 0.0135 ± .005 lb/sq. ft. Unfilled HT 424 primer with parts A and B used with adhesive.

3. Bonding Process: Treated metal coupons spray primed with 0.001" thickness HT 424 primer solution using clean dry argon carrier gas. Primer layers dried 30 min ambient 23°C and 60 min at 66°C. An adhesive film is placed in the 1.000" x 0.500" overlap between two metal adherends. Six such joints are aligned in a bonding jig with the glass carrier acting to provide constant glue line thickness 0.008". Cure cycle with 60 min temperature rise to 171°C and 60 min cure cycle at 171° followed by cooling to room temperature.

4. Tensile Lap Shear Testing: 1.5" x 1.0" x 0.063" aluminum alignment shims bonded to eliminate offset. Tests at 23°C using 0.01"/min Instron crosshead rate and 4.5" jaw separation.

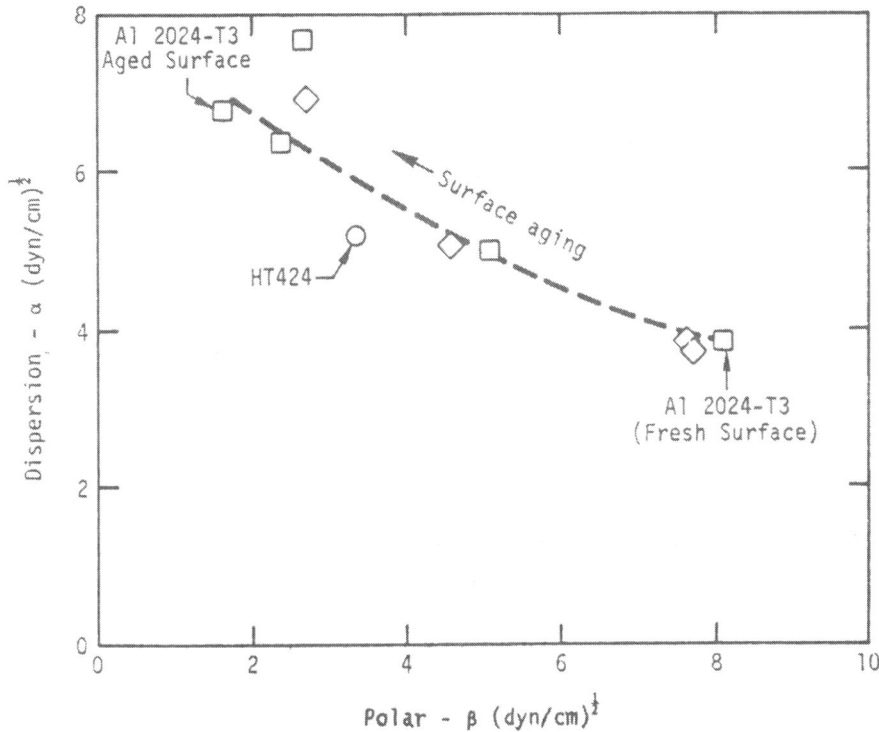

Fig. 1. Dispersion (α) and polar (β) components of the solid-vapor surface tension $\gamma_{SV} = \alpha_S^2 = \beta_S^2$ for HT 424 primer (phase 1) and Al 2024-T3 adherend (phase 3).

DISCUSSION

In this study the surface chemistry and related wettability of both adherends and adhesive were analyzed prior to bonding. The results of this surface energy analysis can be plotted on a surface energy map where the ordinant α and abscissa β as shown in Fig. 1 respectively, refer to dispersion (non-polar) and polar components of surface energy and interfacial bonding mechanisms. The theory of interfacial adhesion experimentally verified in this analysis defines the thermodynamic work of adhesion W_a by the following relation (4,5):

$$W_a = W_{13} = 2(\alpha_1 \, \alpha_3 + \beta_1 \, \beta_3) \tag{1}$$

where α_1, β_1 define the dispersion and polar surface properties of adhesive and α_3, β_3, those of the metal adherend.

As shown in Fig. 1 and further documented in detailed kinetic studies by Kaelble and Dynes (4), the α_3 and β_3 properties of aluminum alloy change dramatically with surface aging time after FPL etch. The α_1, β_1 surface properties of HT 424 adhesive are shown to lie below the curve for Al 2024-T3 at all stages of surface aging which predicts proper bonding between adhesive and adherend in air.

The interfacial work of adhesion W_a as defined by Eq. (1) will decrease with surface aging time (SET) as shown by the upper curve of Fig. 2. As shown in the lower curves of Fig. 2, the lap shear bond strength varies with SET in a fashion that correlates closely with the predictions from work of adhesion calculations. A simple but now widely demonstrated correlation between surface energetics and fracture mechanics is available in the following relations for critical stress σ_c for Griffith type crack initiation under normal stress loadings (5,6):

$$\sigma_c = \left(\frac{2E}{\pi C}\right)^{\frac{1}{2}} \left(R^2 - R_0^{\,2}\right)^{\frac{1}{2}} \geqslant 0 \tag{2}$$

where E and C are a characteristic modulus and crack length which are assumed constant and the surface energy parameters R and R_0 are defined by the following relations (5,6):

$$R_0 = 0.25 \, (\alpha_1 - \alpha_3)^2 + (\beta_1 - \beta_3)^2 \tag{3}$$

$$R^2 = (\alpha_2 - H)^2 + (\beta_2 - K)^2 \tag{4}$$

$$H = 0.5 \, (\alpha_1 + \alpha_3) \tag{5}$$

$$K = 0.5 \, (\beta_1 + \beta_3). \tag{6}$$

In Eq. (4) two new surface energy parameters α_2 and β_2 define the environment (phase 2) at the crack tip. The model for critical stress defined by Eq. (2) can be presented on surface energy coordinates as shown in Fig. 3.

As the adhesive joint changes from dry air immersion with $\alpha_2 = \beta_2 = 0$ to equilibrium response with water immersion with $\alpha_2 = 4.67 (\text{dyn/cm})^{\frac{1}{2}}$ and $\beta_2 = 7.14 (\text{dyn/cm})^{\frac{1}{2}}$, the predicted decrease in critical stress σ_c of the HT 424 to Al 2024-T3 interface is:

$$\frac{\sigma_c(H_2O)}{\sigma_c(\text{air})} = 0.644. \tag{7}$$

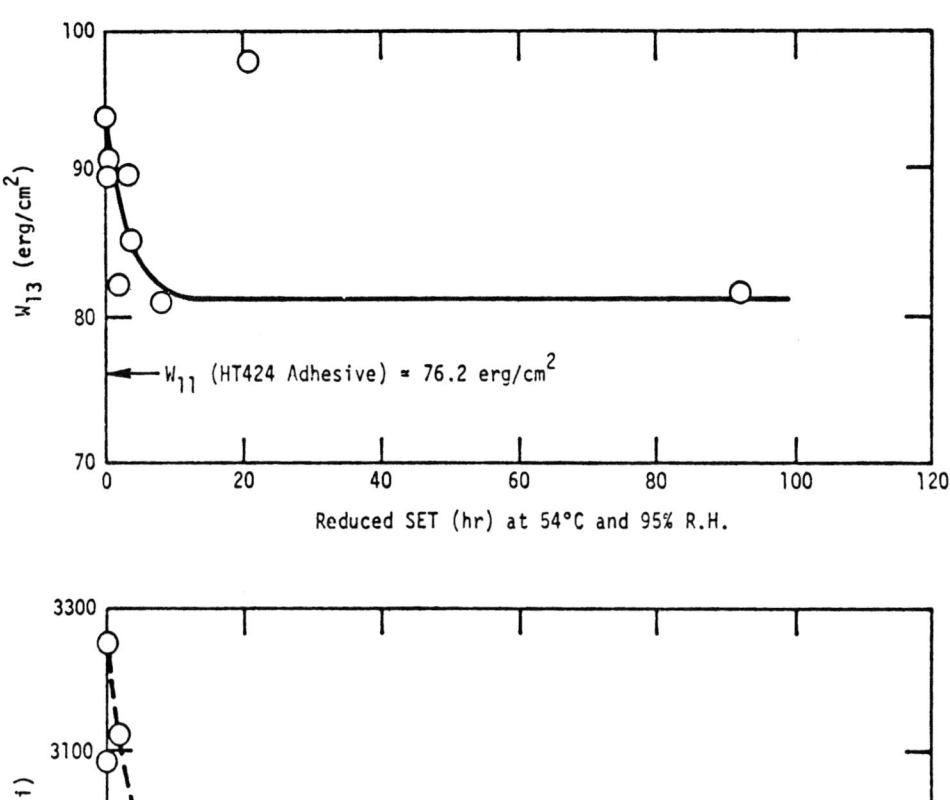

Fig. 2. Dependence of interfacial work of adhesion
         W_{13} (upper curve) and lap shear bond strength
         σ_b (lower curve) at varied SET.

Extensive joint strength testing of this system was completed to
determine the response surface of lap shear strength vs. both SET
and BET under high moisture (95% R.H., 54°C). These results are
summarized on the response surface of Fig. 4 where each point
represents the average of six tests. Comparing joint strengths for
fully aged (20 hr. SET, 1000 hr. BET) and unaged (0 hr. SET, 0 hr.

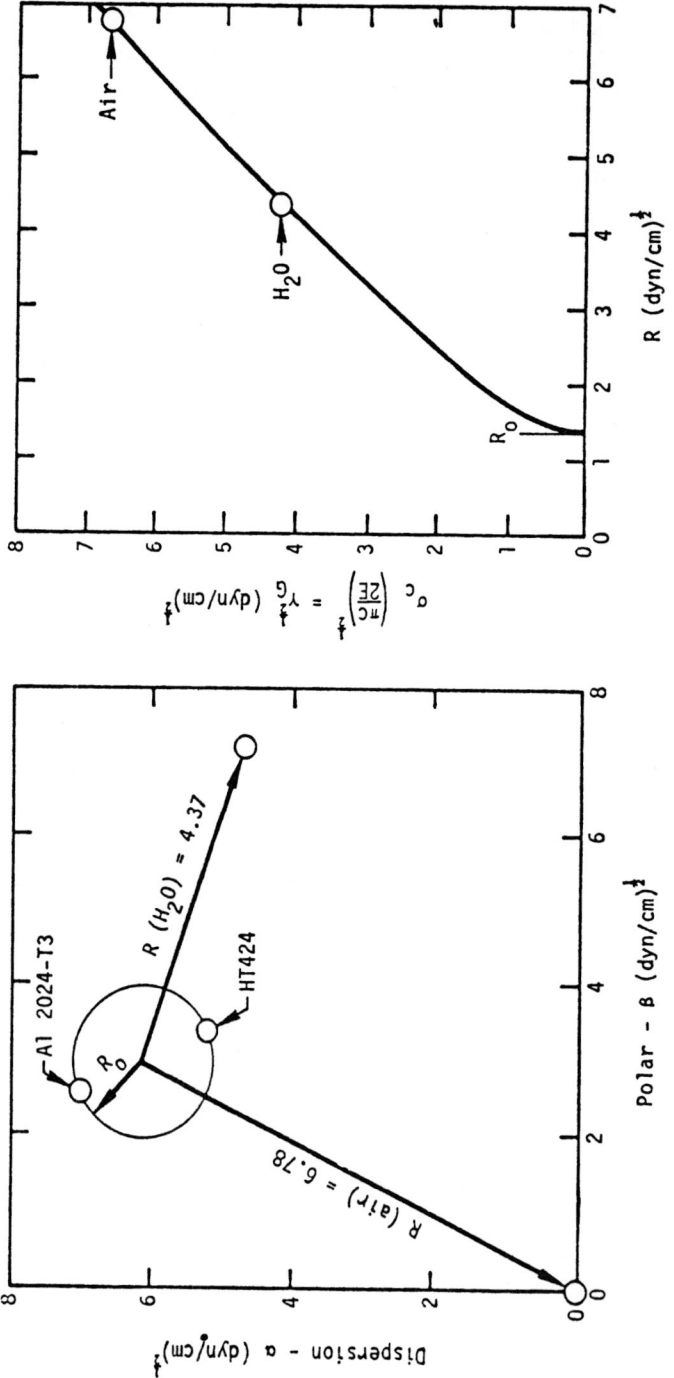

Fig. 3. Modified Griffith analysis of the effect of H_2O immersion in reducing critical failure stress σ_1 for interfacial failure between HT 424 and etched Al 2024-T3.

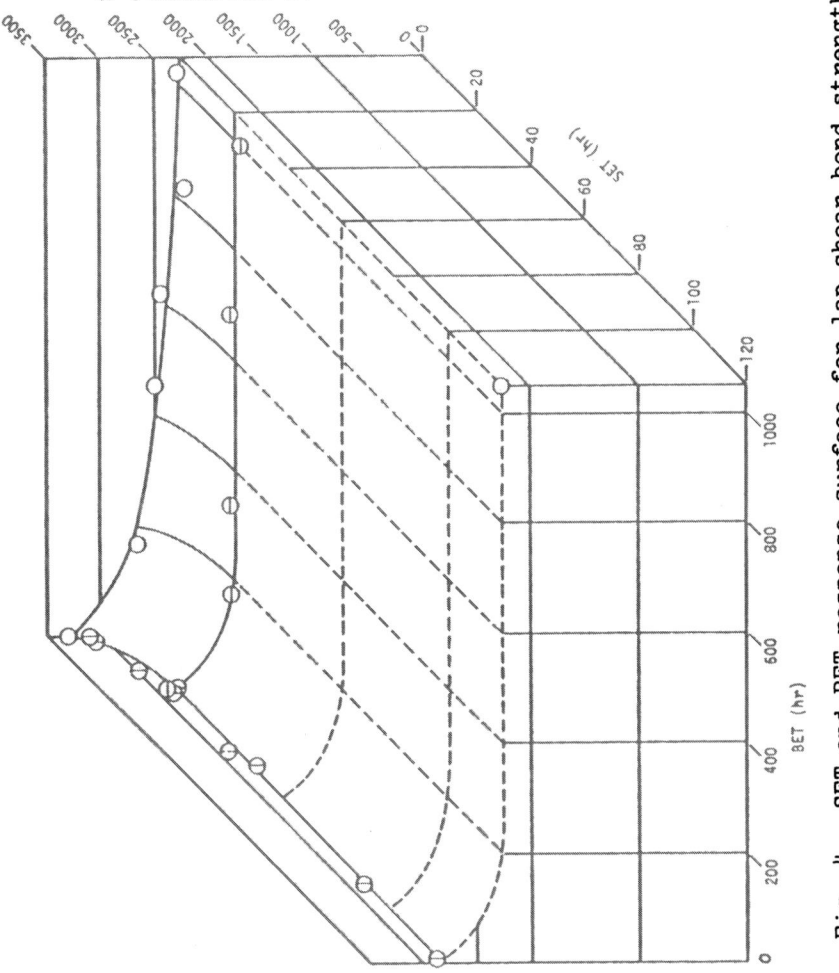

Fig. 4. SET and BET response surface for lap shear bond strength for Al 2024-T3-HT 424.

BET) provides (see Table 2) the following experimental ratio:

$$\frac{\sigma_b \text{ (aged, wet)}}{\sigma_b \text{ (unaged, dry)}} = \frac{2275 \text{ psi}}{3300 \text{ psi}} = 0.69 \qquad (8)$$

which is in close agreement with the prediction of Eq. (7).

An essentially parallel detailed study of surface aging showed a shift in α_3 and β_3 for phosphate-fluoride cleaned titanium alloy similar to that detailed in Fig. 1. Application of the modified Griffith analysis as shown in Fig. 5 provides the following predicted moisture degradation of bond strength at the HT 424 to Ti-6Al-4V interface:

$$\frac{\sigma_c(H_2O)}{\sigma_c(\text{dry air})} = 0.84. \qquad (9)$$

Table 2. Weibull Strength Distributions

Composite Polymer		Test	Strength Distributions $R = \exp -(\sigma_b/\sigma_o)^{m(\sigma)}$	
Metal-Adhesive Joint AL2024T3-HT424 Epoxy SET(hr)	BET(hr)	Single Lap Shear	$\sigma_o(Kg/cm^2)$	$m(\sigma)$
0	0	N=12	232	14.5
0	165.449	12	184	15.4
0	808,1023	12	165	10.0
21	0	12	208	15.0
20	669,983	12	160	18.1
Ti-6Al-4V-HT424 Epoxy SET(hr)	BET(hr)		$\sigma_o(Kg/cm^2)$	$m(\sigma)$
0	0	N=12	270	7.65
0	(670,1016)	12	182	6.22
21	0	12	272	7.65
21	(591,997)	12	202	6.35

SET = surface exposure time, BET = bond exposure time at 54°C and 95% relative humidity.

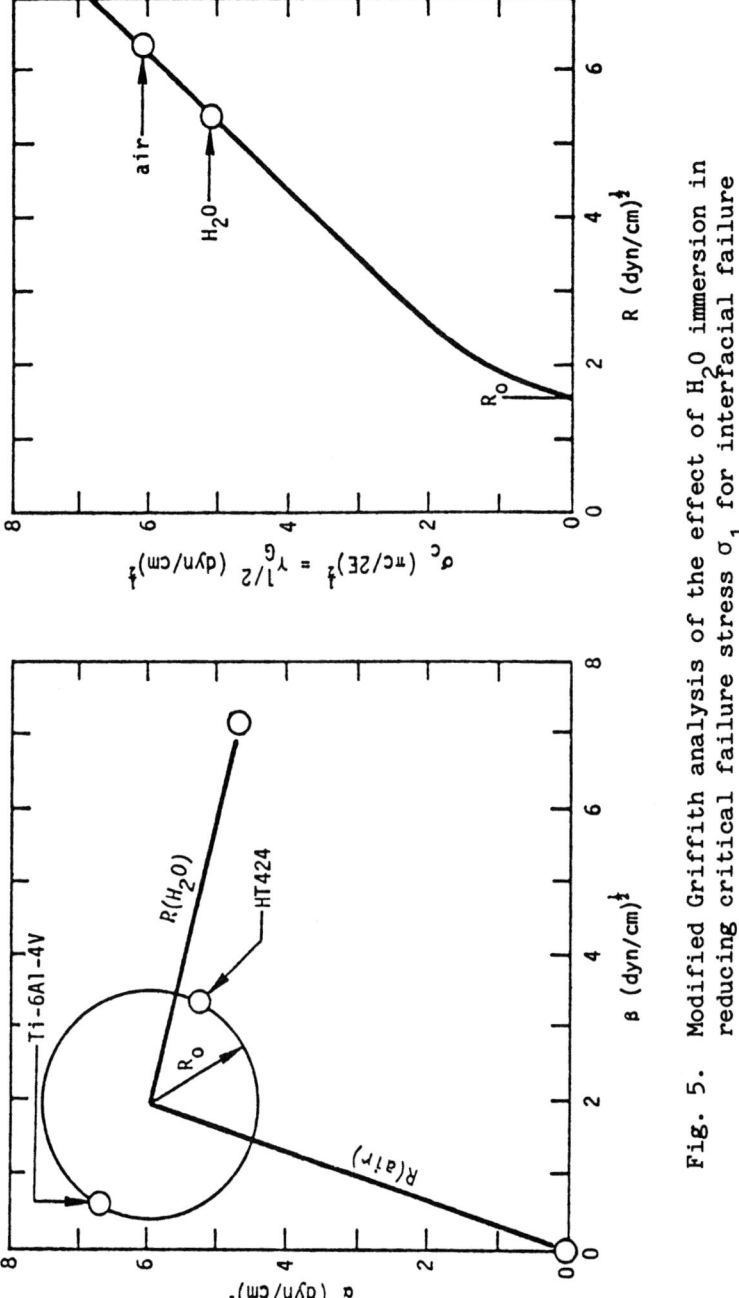

Fig. 5.   Modified Griffith analysis of the effect of H_2O immersion in reducing critical failure stress σ_1 for interfacial failure between HT 424 and phosphate-fluoride treated Ti-6Al-4V.

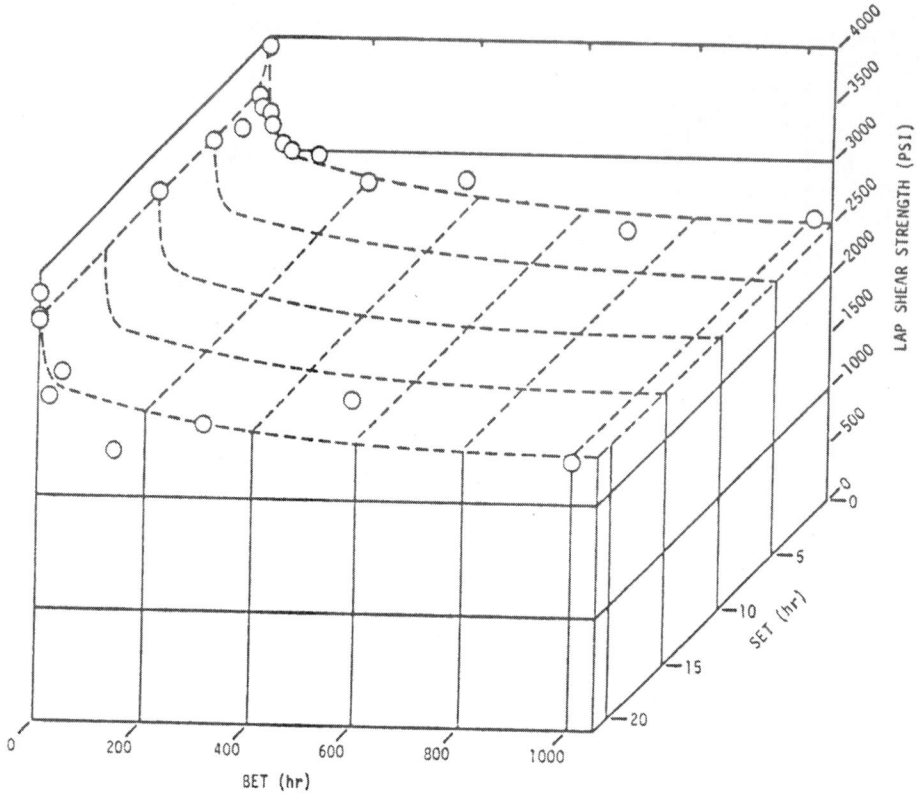

Fig. 6. SET vs. BET response surface for lap
shear bond strength for Ti-6Al-4V - HT424.

An equivalent calculation for moisture degradation of the HT 424
cohesive bond produces the following prediction (3):

$$\frac{\sigma_c(H_2O)}{\sigma_c(\text{dry air})} = 0.63. \tag{10}$$

Joint strength testing of the HT 424 to titanium alloy was
carried out under conditions of separate and combined SET and BET in
high moisture (95% R.H., 54°C). The SET vs. BET response surface of
shear bond strength σ_b is shown in Fig. 6 where each point is an
average of six strength tests. As shown in Fig. 6, the shear bond
strength reaches an equilibrium value under extended moisture
aging. Comparing joint strengths for fully aged (22 hr. SET, 1000
hr. BET) and unaged (10 hr. SET, 0 hr. BET) for Ti-6Al-4V to HT 424
bonds (see Table 2) provides the following experimental ratio:

$$\frac{\sigma_b(\text{aged, wet})}{\sigma_b(\text{unaged, dry})} = \frac{2873 \text{ psi}}{3840 \text{ psi}} = 0.75 \qquad (11)$$

which lies intermediate between the cohesive failure prediction of Eq. (10) and the interface prediction of Eq. (9). Microscopic visual inspection of the fracture surfaces for the HT 424 to titanium lap shear points shows predominant (above 50%) cohesive failure for lap shear bonds described in Fig. 6.

Structural Reliability

The Weibull plots of Fig. 7 show shear bond strength distributions for unaged and fully aged aluminum (upper view) and titanium (lower view) joints. The titanium bonds show lower Weibull m values in both unaged and aged states as evidenced in Fig. 7 and the data summary of Table 2. A design requirement of high reliability shear

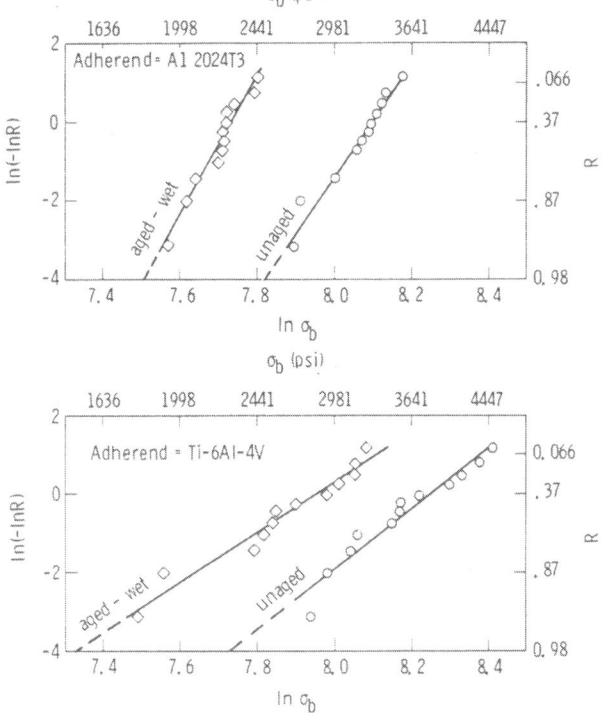

Fig. 7. Comparison of Weibull shear strength
         distributions for aluminum (upper view)
         and titanium (lower view) adherends.

strength where R = 0.98 or ln(-lnR) = -4 is shown by Fig. 7 to predict higher performance for the aluminum alloy joints in both unaged and aged-wet states. Conversely, if mean strength with R = 0.37 and ln(-lnR) = 0 is applied as a design criteria, the curves show titanium alloy joints to display higher unaged and aged strengths.

REFERENCES

1.   J.C. Bolger and A.S. Michaels, in Interface Conversion for
       Polymer Coatings, ed. P. Weiss and G.D. Cheever, American
       Elsevier Publishing Co., NY, Chapter 1 (1969).
2.   D.W. Levi, W.C. Tanner, R.C. Ross, R.F. Wegman and M.J.
       Bodnar, J. Appl. Poly. Sci. 20, 1475 (1976).
3.   T. Smith and D.H. Kaelble, "Mechanisms of Adhesion Failure
       Between Polymers and Metal Substrates", Technical Report
       AFML-TR-74-73, Air Force Materials Laboratory, WPAFB,
       June 1974.
4.   D.H. Kaelble and P.J. Dynes, J. Coll. and Interface Sci. 52,
       562 (1975).
5.   D.H. Kaelble, J. Appl. Poly. Sci. 18, 1869 (1974).
6.   D.H. Kaelble, Polymer Eng. and Sci. 17 (7), 474 (1977).

Composition and Ageing of a Structural, Epoxy Based Film Adhesive

C.E.M. Morris, A.G. Moritz and R.G. Davidson

Materials Research Laboratories

Department of Defence

Ascot Vale, Victoria, 3032, Australia

ABSTRACT

A 180°C curing, structural film adhesive has been analyzed by IR and NMR and found to consist of three different epoxy compounds with dicyandiamide as the curing agent. The ageing of this material at 20°C has been examined by monitoring changes in the molecular weight distribution, epoxy equivalent value and solubility. It was found that the changes which occur are ascribable to a combination of cross-linking and hydrolysis.

INTRODUCTION

The use of structural, film-form adhesives in aircraft construction is now widely adopted. While the one-part nature of these materials has many advantages in terms of ease of use, the accompanying limited shelf life presents significant problems. This paper examines the composition and ageing of one such adhesive.

The material, a 180°C curing, supported film, is extensively used as a honeycomb-skin adhesive in a number of aircraft presently in service. The recommended shelf life of the uncured material is 6 to 12 months at -18°C. As this material must be imported into Australia and only small amounts are needed at any one time for aircraft repairs, the short shelf life has caused considerable problems. A programme has therefore been undertaken to study the composition, ageing and cure reactions of this material. The initial results of this programme are presented here.

313

EXPERIMENTAL

Samples from two batches of the adhesive were used in this work.

The GPC studies were performed on a Clanor instrument fitted with two detectors (differential refractive index and evaporative analyser); the performance of this equipment, which is an Australian instrument, is described elsewhere (1). The column set consisted of four four-foot Styragel columns of nominal porosity 3×10^3, 1×10^3, 500 and 200Å, the solvent (chloroform) flow rate was 1 ml/min and the oven temperature 30°C.

NMR spectra were determined on a Varian HA60-IL spectrometer using deutero chloroform solutions and tetramethylsilane as the internal lock. Infrared spectra were measured on a Jasco IR-C spectrometer as liquid films.

Epoxy equivalent values were determined by a titration method based on those of Jay (2) and Dijkstra and Dahmen (3).

The flow test employed consisted of curing a 3.75 cm square piece of adhesive at 180°C and 0.03 MPa (45 psi), cutting the resultant specimen back to the original dimensions and expressing the flow as the weight of material cut off as a percentage of the original sample weight (4).

RESULTS AND DISCUSSON

Composition

The overall composition of the adhesive was found to be

Epoxy resins	72-74%
Dicyandiamide	7.5%
Asbestos	10%
Glass support	8.5%

There is also a small amount of iron and traces of chromium and nickel compounds. The overall composition varies somewhat from batch to batch and also, as the adhesive is not uniformly spread over the glass support, from place to place within a roll.

The glass support is a plain weave of glass rovings of about 200 ends per roving. The weave is such that the roving is more flattened in one direction than in the other and that the interstitial spaces are rectangular rather than square. (Figure 1)

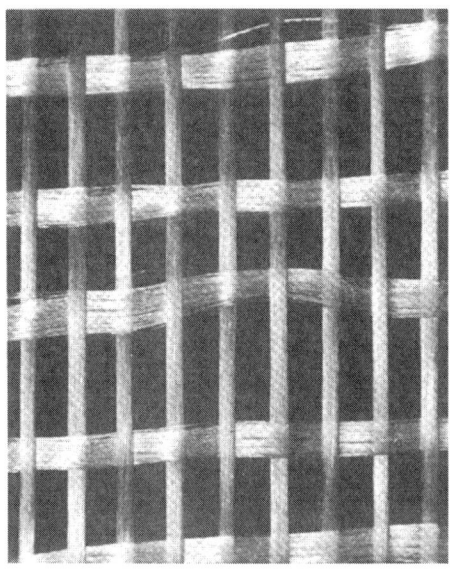

Fig. 1.   Woven glass support (X6 mag.)

        GPC analysis of chloroform solutions was used to examine the
molecular weight distribution and also as a means of fractionating
the material.   The resin dissolves almost instantly in chloroform
but these solutions do not contain the curing agent as dicyandiamide
is only slightly soluble in chloroform.   The molecular weight
distribution of two batches of the adhesive is given in Figure 2:
the molecular weight covers the range approximately from 300 to
25,000.

        The material was collected as four fractions comprising
(1) the lowest molecular weight peak, (2) the next peak, (3) material eluted between about 108 and 134 mls and (4) the highest
molecular weight portion (87 to 108 mls approximately).   NMR and IR
analysis of these fractions showed (5) them to be the following:

    (1) triglycidyl p-amino phenol (I)

<!-- Chemical structure (I) -->

$$\underset{\text{O}}{\overset{/\backslash}{CH_2}}-CH-CH_2$$

(I)

Triglycidyl p-amino phenol

(2)   mainly, an oligomer of I

(3)   chiefly, a cresol-formaldehyde-novolac oligomer (II).
      The NMR spectrum shows little fine structure suggesting a
      mixture of isomers and/or moderately high molecular
      weight material.

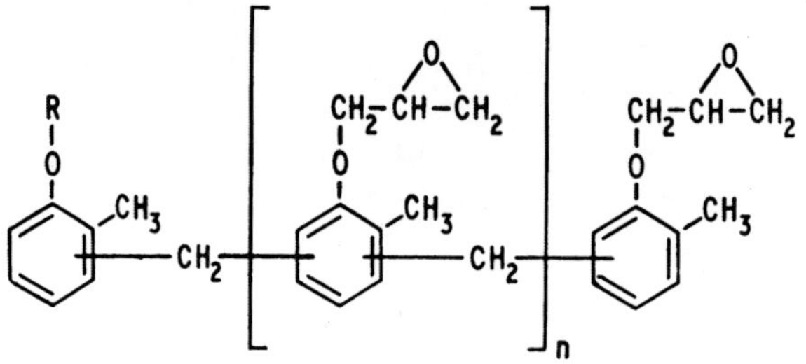

Cresol novolac epoxy

(4)   a high molecular weight oligomer of the diglycidyl ether
      of bis-phenol A (III).

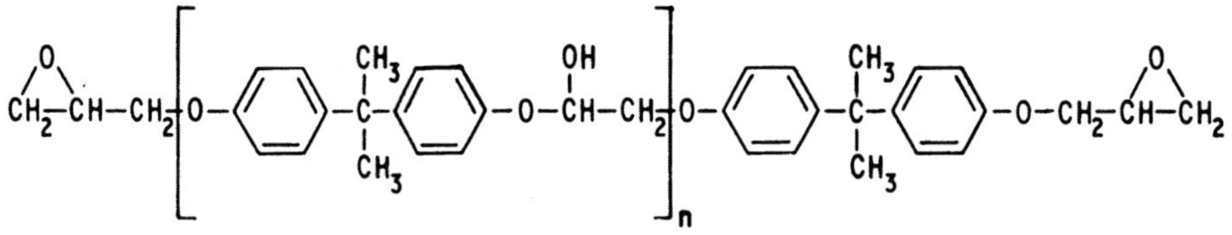

D G E B A

        As might be expected, there were minor amounts of III (of
smaller values of n) in fractions (1), (2) and (3) and significant
amounts of II in fractions (2) and (4).

        It is estimated that the triglycidyl p-amino phenol and its
oligomers constitute about 45% of the epoxy resins and the DGEBA
component 10-15%. The overall epoxy equivalent weight for the resin
fraction of the adhesive is 174.

Ageing

        The room temperature ageing of 2 batches was examined so that
the effect of short times out of a freezer, or loss of dry ice
during shipping, could be assessed. The samples were kept, with

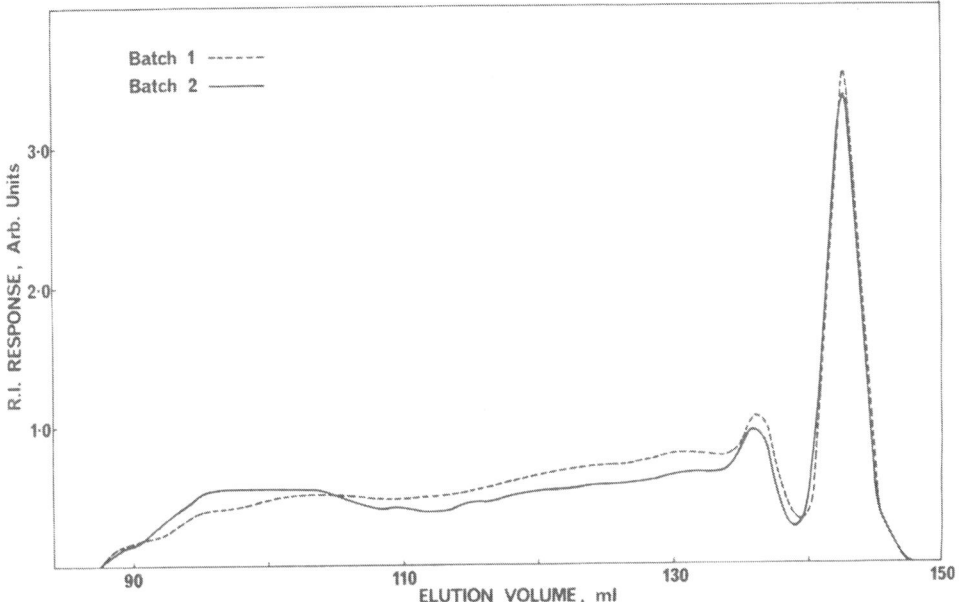

Fig. 2.   Gel permeation chromatogram of two batches
of adhesive.

their backing material in place, inside plastic bags in a laboratory
at 20°C. At the commencement of this experiment batch 1 was some 14
months old and batch 2 about 2 months old.   At intervals of
approximately 14 days, a piece was cut off and a GPC trace run. The
changes which occurred over 4 months are illustrated in Figs. 3 and
4.

Only minor changes were observable in the chromatograms for
the first 30-40 days (less than the initial differences between the
batches). At longer times a tailing at the low molecular weight end
of the chromatogram occurred which eventually became an additional
peak.   IR analysis indicated that this material was hydrolysed
triglycidyl p-amino phenol having a very minor epoxy component and a
greatly increased hydroxyl content. Essentially the same material
was obtained from the aqueous phase after refluxing a fresh sample
of adhesive in water for several hours.   TLC showed it to be a
mixture of five or six similar compounds.

A very small peak, of negative refractive index relative to
chloroform, was discernible in batch 1 at 77 days and in batch 2 at
71 days at an elution volume of about 74 ml, that is, at significantly higher molecular weight than the rest of the material. No
further changes occurred in this region with increasing time. This
compound has not been identified.

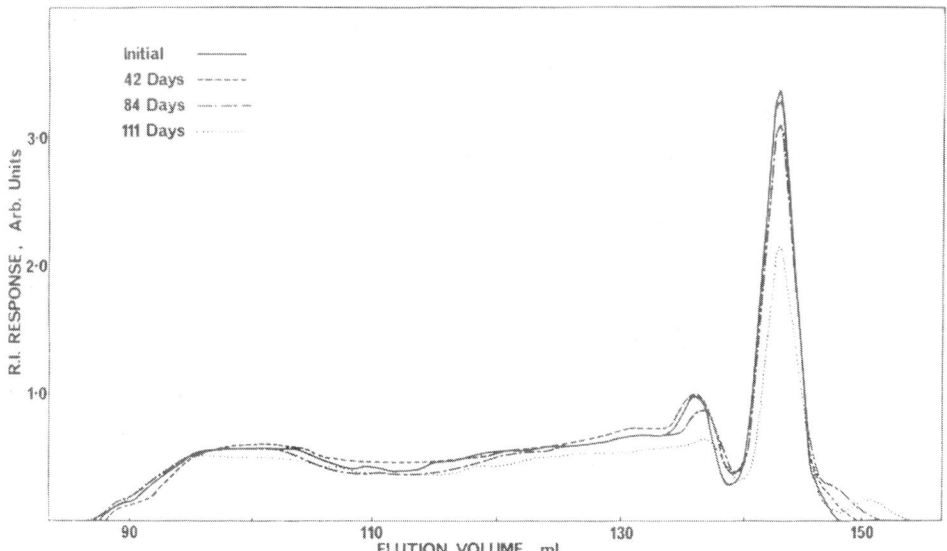

Fig. 3.   Room temperature ageing of batch 1.

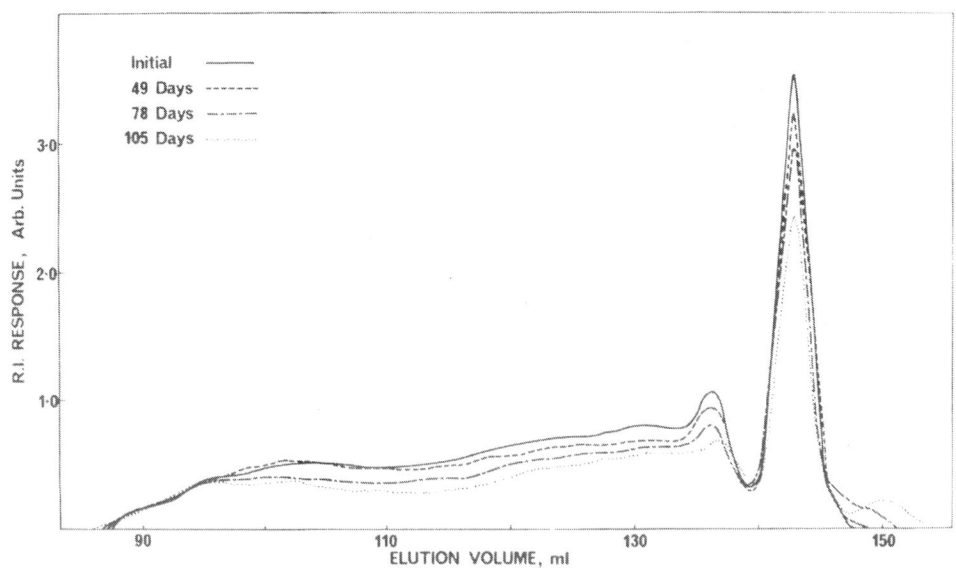

Fig. 4.   Room temperature ageing of batch 2.

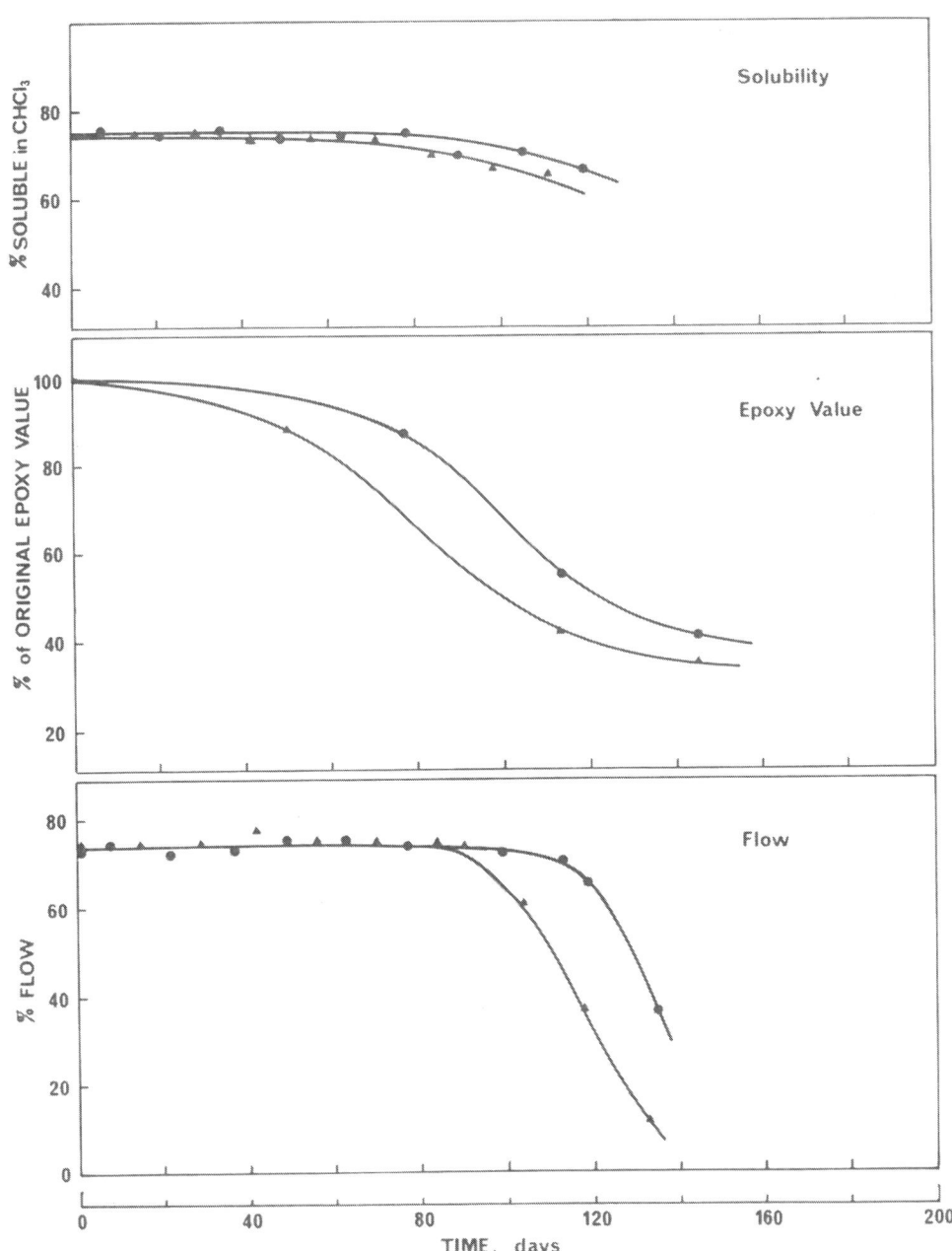

Fig. 5.   Change in resin solubility, epoxy value and
flow during room temperature ageing.
● Batch 1.   Δ Batch 2.

After about 40 days chloroform solutions of batch 2 (concentration about 4.5×10^{-3} g/ml) became more difficult to filter through a 0.45 μ filter and by about 55 days the solutions were extremely difficult to filter. Batch 1 displayed this feature after a somewhat longer interval. After some 60 days the samples were significantly slower to dissolve. However, the solubility in chloroform changed very little until, after about 120 days, the material only swelled. (Figure 5).

The overall epoxy equivalent value (initially 173 for batch 1 and 174 for batch 2) took about 100 days to fall to half its original value (Figure 5). Flow tests on the two batches of adhesive showed essentially no change up to 100 days for batch 1 and about 90 days for batch 2 (Figure 5).

Correlations must be made with a programme of mechanical testing before final conclusions can be drawn on the time which this adhesive can tolerate at ambient temperature and still be used satisfactorily. However, it appears that periods up to several weeks in a temperate climate are accompanied by changes in the adhesive which are minor when considered in relation to batch to batch variations. Batch 2, although actually younger than batch 1, displayed some ageing signs rather sooner than batch 1, but its chromatogram showed less obvious signs of change. It appears that the reactions which occur are a combination of cross-linking and hydrolysis.

ACKNOWLEDGEMENTS

The authors wish to thank P.J. Pearce for determination of the epoxy equivalent values and I. Grabovac for operation of the GPC.

REFERENCES

1. I. Grabovac and C.E.M. Morris, "Performance of a Gel Permeation Chromatograph and its Usefulness in Examining Mixtures of Organic Materials", Materials Research Laboratories, Melbourne, MRL-R-725 (1979).
2. R.R. Jay, Anal. Chem. 36, 667 (1964).
3. R. Dijkstra and E.A.M.F. Dahmen, Anal. Chim. Acta. 31, 38 (1964).
4. C.V. Cagle, Handbook of Adhesive Bonding, McGraw-Hill, p. 32-36 (1973).
5. C.F. Poranski, W.G. Moniz, D.L. Birkle, J.T. Kopfle and S.A. Sojka, "Carbon-13 and Proton NMR Spectra for Characterizing Thermosetting Polymer Systems. I. Epoxy Resins and Curing Agents", Naval Research Laboratory, Washington, NRL Report 8092 (1977).

Viscoelastic Characterization of Structural Adhesive via Force Oscillation Experiments

D.L. Hunston, W.D. Bascom, E.E. Wells (1), J.D. Fahey
and J.L. Bitner

Chemistry Division, Naval Research Laboratory

Washington, D.C.   20375

ABSTRACT

A test device based on the Weissenberg Rheogoniometer was
developed to measure the dynamic viscoelastic properties of solids.
A computerized data acquisition and analysis system with graphics
capability is utilized in this apparatus.   Equations containing
terms for the moment of inertia and damping of the torsion head were
developed for forced oscillation.   The apparatus was employed to
characterize an epoxy with known fracture and fatigue properties.
Despite the epoxies complex morphology, time-temperature superposition could be used to obtain master curves.   The shift factors (and
glassy modulus) for the α transition depended on the thermal history
of the specimen while the shift factors for the β relaxation gave a
linear Arrhenius plot with an activation energy of 15 kcal/mole.

INTRODUCTION

Polymeric systems offer strong, light-weight materials that
are finding important commercial applications and as a result they
are being subjected to a tremendous range of loading conditions.
For example, the use temperature can range from -50°C to $+250^{\circ}$C
while the material's response to stress is often called upon to
exhibit good behavior during the very short time periods corresponding to impact loads (10^{-6} - 10^{-3} sec.), during the somewhat
longer time periods corresponding to fatigue or rising load situation (10^{-3} - 100 sec.), and during the very long intervals charac-

teristic of static loading ($100 - 10^6$ sec.). In order to understand
such complex behavior as fracture, fatigue and wear, it is important
to characterize the basic mechanical properties of the materials
involved. Since polymer properties exhibit a strong dependence on
time and temperature, it is desirable to characterize samples over a
wide range of temperatures and times (or, in the case of cyclic
loading, the equivalent variable-frequency).

At present most characterization experiments involve measuring
the linear viscoelastic properties of the materials over a very wide
range of temperatures at a single frequency (1 Hz) or over a small
range of frequencies (3 to 110 Hz). The low and high temperature
data are then used to qualitatvely interpret the behavior at short
and long times, respectively. Such qualitative predictions, however, are not entirely satisfactory and thus it is desirable to
directly measure the properties over a much wider range of time
scales. Such experiments can identify rheologically simple materials where time-temperature superposition can be used (2). Moreover, commercial polymers are often blends which contain fillers,
additives, toughening agents, etc. As a result they can have time-
temperature regions in which their rheological behavior is complex
and thus measurements over a wide range of conditions is essential
(3).

Although some authors (4,5) have conducted measurements of
this type--generally using creep or stress relaxation experiments--
a much broader application of such testing techniques is needed.
Creep (6) and stress relaxation (7) experiments are particularly
useful for investigating behavior at the longer time scales, while
forced oscillation measurements (8,9) provide a convenient technique for shorter time scale experiments. It is desirable therefore
to combine both methods to obtain data over a wide range of times.
The purpose of this paper is to report the development of a forced
oscillation instrument, based on the Weissenberg Rheogoniometer,
and the use of this apparatus to investigate the viscoelastic
properties of a thermosetting epoxy.

THEORY

There are several potential difficulties that arise in the
analysis of forced oscillation tests. These complications arise
from two particular features of the experimental design. First, the
stress and strain are measured at different ends of the sample and
this complicates the analysis of the results. A recent paper by
Schrag (10) has discussed this topic in some detail. Second, some
motion at the stress measuring device is always required to obtain
data. This can make the sample deformations more complex and
introduce effects due to the inertia and natural damping of the

the stress measurement device. To minimize this problem, efforts are often made to stiffen the stress detection system (11). For some experiments, such as transient measurements, such modifications are extremely important. Some comments on the value of a stiff system for the forced oscillation testing of solids will be given later in this paper. Although the remarks presented here are directed primarily at the use of a Weissenberg Rheogoniometer for testing of solids, the ideas are equally valid for other forced oscillation experiments.

The mathematics of the forced torsional oscillation test were analyzed many years ago by Markovitz et al (12). They consider a cylindrical sample of radius, R, and length, ℓ. The bottom of the sample (axis of the cylinder mounted vertically) is driven in torsional oscillation with an angular displacement, $\theta_o(t)$, and this produces a simple wave motion in the sample. The top of the sample is connected to a torsion head consisting of a torsion bar, sample grip, air bearing, etc. The torsional elastic constant of this device is k, the moment of inertia is I, and the natural damping force association with torsional motion is assumed to be proportional to angular velocity with a proportionality constant, U. Although Markovitz et al. do not consider the natural damping of the torsion head, this term can easily be included simply by replacing $(I\omega^2-k)$ with $(I\omega^2-k-i\omega U)$ where ω is the angular frequency and $i = \sqrt{-1}$. The motion at the top of the sample, $\theta_\ell(t)$, is then found by balancing the stress produced by the wave motion in the sample with the elastic and damping properties of the torsion head.

By modifying the results of Markovitz et al. the motion at the ends of the sample can be written as an infinite series involving powers of the reciprocal of the complex shear modulus, G^*.

$$1 - \frac{\theta_o}{\theta_\ell} = \sum_{j=0}^{\infty} \left[\frac{(\ell\omega)^2 \rho}{(2j+2)} + \ell B \right] \frac{(\ell\omega)^{2j} (-\rho)^j}{(2j+1)!} \frac{1}{(G^*)^{j+1}} \quad (1)$$

where

$$B = (I\omega^2 - k - i\omega U)2/\pi R^4 \quad (2)$$

and ρ is the sample density. The relative motion at the two ends of the sample $\theta_\ell/\theta_o = me^{i\phi}$ where m is the ratio of the amplitude at the top to that at the bottom and ϕ is the phase angle by which the top leads the bottom.

The solution of this problem gives rise to an infinite series because the stress and strain are measured at different ends of the sample. As a result, the only way to increase the rate of convergence for the series is to decrease the sample length. There is a practical limit to this approach, however, since grip effects become more of a factor for small ℓ. Thus, a series solution cannot be avoided. This has consequences not only for measurements in the linear viscoelastic range but also for experiments at larger amplitudes where a non-linear response is achieved. When analyzing the results of non-linear experiments, it is not sufficient to solve a constitutive equation for oscillation at a point. Instead, the boundary value problem for the actual geometry must be addressed. Such an analysis is needed to clarify both the effect of sample length and the nature of the deformation in the sample.

To solve eq. (1) for G^*, it is necessary to terminate the series. For most samples the convergence is very rapid and only the first term is needed; however, certain lower modulus materials may require two terms. Although the equations become rather complicated in this case, the solution is straightforward. Since several quantities in eq. (1) are complex numbers, it is difficult to specify the exact conditions where the second term becomes important. As an example, however, consider $(\ell^2\omega^2\rho/2G^*)$ which is approximately the ratio of the second to the first term. When $\ell = 5$ cm, $\omega = 120$ rad./sec. (19 Hz), $\rho = 1$ g/cm^3, and $|G^*| = 10^6$ dynes/cm^2, the magnitude of this quantity is 0.18 and thus, the second term makes a major contribution. Obviously, this term must be examined when testing low modulus elastomers.

For many materials, however, only the first term is needed and thus

$$G' = \frac{m\left[\left\{\left(\frac{\ell^2\omega^2\rho}{2}\right) + \ell B'\right\}(m-\cos\phi) + \ell B'' \sin\phi\right]}{m^2 - 2m\cos\phi + 1} \qquad (3)$$

$$G'' = \frac{m\left[-\left\{\left(\frac{\ell^2\omega^2\rho}{2}\right) + \ell B'\right\}\sin\phi + \ell B''(m-\cos\phi)\right]}{m^2 - 2m\cos\phi + 1} \qquad (4)$$

where

$$B = B' + i\, B'' \qquad (5)$$

These equations provide a method to calculate viscoelastic properties from suitably treated experimental data.

When the stress detection system is very stiff (small m) and the elastic constant dominates the other terms, the equations can be greatly simplified to give the well known relationships

$$G' = m \frac{2 \ell k}{\pi R^4} \cos \phi \tag{6}$$

$$G'' = m \frac{2 \ell k}{\pi R^4} \sin \phi \tag{7}$$

Although this simplification is convenient, it is not required to measure linear viscoelastic properties. In the linear range the compliance of the stress detection system does not affect the type of deformation present in the sample or the accuracy with which it can be measured (unless the compliance is very high). Although the amplitude of the oscillation will be less in a more compliant system, this is taken into account by the equations. Consequently, linear viscoelastic properties can easily be measured. For non-linear studies, a stiff detection system may be an advantage; however, this can only be determined if a solution with an appropriate non-linear equation is obtained.

Another interesting aspect of eqs. (3) and (4) is the inclusion of terms involving U. Many previous papers have addressed the importance of inertia terms but damping is usually neglected. In most cases this is understandable; however, it may not be justified for very low loss samples. An examination of eqs. (3) and (4) indicates that the second term in the numerator of eq. (3) is unimportant since both U and $\sin \phi$ are small. In eq. (4), however, both terms in the numerator involve small quantities. Consequently, when the sample becomes more elastic (smaller $\sin \phi$) or the inertia contribution becomes larger (smaller $I\omega^2 - k$), the damping term increases in importance. In the present series of tests, when the higher frequencies were used, the damping of the torsion head was often found to be more important than the inertia although the contributions of both quantities were usually small. By using the appropriate equations, however, all effects can be taken into account.

EXPERIMENTAL SECTION

Apparatus

The modified Rheogoniometer is illustrated in Fig. 1. The sample (S) is held between two specially designed grips. The top and bottom sections of each grip are made from aluminum while the center portion is constructed from glass-reinforced nylon to minimize heat flow along the shaft. The lower grip can be driven in constant-speed rotation, in torsional oscillation, or in any combination of the two; separate synchronous motors and 60-speed gear boxes (A) are used for the oscillatory and rotational drives. The

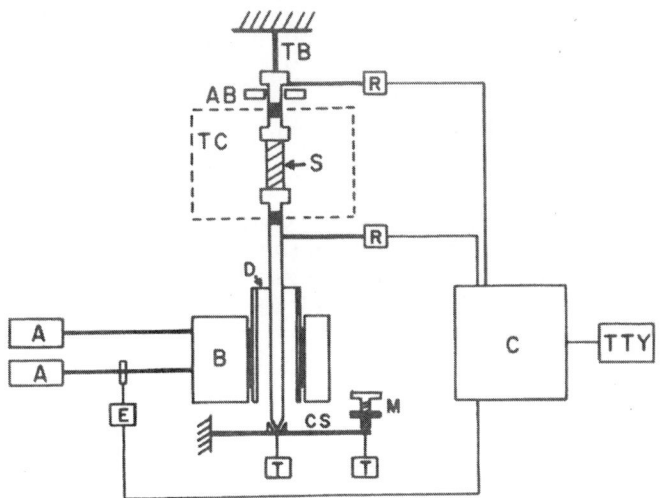

Fig. 1.   Diagram of Modified Rheogoniometer:
          (A) motors and speed adjustment gear boxes,
          (B) main gear box, (C) minicomputer,
          (D) flexible diaphragm, (E) shaft angle
          encoder, (M) micrometer, (R) transducers
          for rotational motion, (S) sample, (T) dis-
          placement transducers, (AB) air bearing,
          (CS) cantilever spring, (TB) changeable
          torsion bar, and (TC) temperature control
          chamber.

rotational speed and the frequency of oscillation can therefore be
independently varied over a range of almost six orders of magnitude
while the amplitude of oscillation is also variable (B). The stress
levels associated with the motion are measured by monitoring the
twist of the top grip which is attached to a changeable torsion bar
(TB). An air bearing (AB) is used to maintain alignment without
introducing significant frictional forces. A compressive or tensile stress can also be applied to the sample along the axis of
rotation by a cantilever spring (CS) attached to the bottom grip
through a rotating connection and adjusted with a micrometer (M) and
displacement transducers (T). Since the torsional drive mechanism
is connected to the bottom grip through a flexible diaphragm (D),
the rotational and axial forces can be applied independently in any
combination. The sample temperature is controlled by a specially
designed chamber (TC) that maintains a constant sample temperature
over the range from $-180^{\circ}C$ to $+275^{\circ}C$.

The most crucial aspect of the instrument, however, is the data
acquisition system, Fig. 2. To fully utilize the versatility of

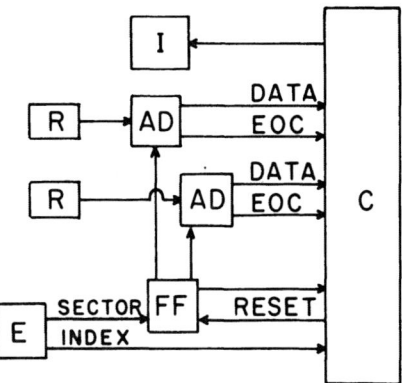

Fig. 2.   Computer interface: (E) is the shaft
          angle encoder (Baldwin Model 3700AZS),
          (I) is indicator lights, (FF) is the
          flip-flop, (R) is the transducer,
          (AD) is the sample-and-hold unit and
          A-to-D converter, and (C) is the com-
          puter (Texas Instrument 960A).

this apparatus, it is necessary to obtain very accurate data over
the entire range of time scales. Furthermore, real-time data
manipulation and evaluation are essential for determining what
experiments must be performed. This problem has been solved by
interfacing the instrument to a minicomputer (C) using an expanded
version of the design developed by Krieger et al (13). The data
acquisition rate and timing accuracy are assured by using high-
speed A-to-D converters equipped with sample and hold units attached to the rotational displacement transducers (R). A shaft-angle
encoder (E) is attached to the oscillatory drive shaft after   the
gear box. Since all of the speed adjustment occurs before the
encoder, each revolution of the encoder corresponds to one cycle of
oscillation regardless of the frequency (13).

    The encoder provides one index pulse and 360 sector pulses per
revolution. The value of 360 was selected because it is the maximum
number of points that can be conveniently handled at the highest
frequencies. The data acquisition cycle proceeds as follows, Fig.
2. When the computer senses the index pulse it clears the interface
by resetting a flip-flop (FF). The first sector pulse after the
interface is cleared triggers the flip-flop which initiates the
acquisition of a data point from each transducer and notifies the
computer that acquisition is in progress. When the computer senses
the end-of-conversion signal from each A-to-D converter, the data is
transferred. Only then is the interface cleared by resetting the
flip-flop. This procedure makes it impossible for the computer to

acquire the same data twice. To be certain that no points are
missed, after 360 data points are collected from each transducer,
the computer checks to see that the index pulse is obtained before
the next sector pulse appears. This verifies that exactly one cycle
has been measured (any number of resolutions other than one can
readily be detected in the data analysis). This interface design
also makes it possible to use the computer in a time-sharing mode so
that other tasks can be performed simultaneously. A set of lights
(I) indicates when the interface is in operation.

In testing a sample, it is essential that the measurements
begin only after all transient effects have decayed away. This is a
major concern at the lower frequencies; while at higher frequencies, it is important to minimize the number of cycles to minimize
heat build-up. In addition, it is desirable, although not necessary, to obtain data for at least 2 cycles so that the base-line
drift can be corrected accurately. For each transducer the computer
stores the sum of data at each sector pulse and the sum of all data
points in a given cycle; thus, for m cycles, 720 + 2m data points
are collected. The total-per-cycle values are used to compensate
for any linear base-line drift and offset from zero. A Fourier
series analysis is then performed on the data points from each
transducer to obtain the amplitude and phase shift (relative to a
sine wave starting at the index pulse) for the fundamental mode and
any harmonics that are present (generally 4 overtones are checked).
The series analysis approach was selected after testing correlation
methods and more general Fourier analysis techniques for two reasons: it can provide a filtering effect that increases the signal-
to-noise ratio and it yields absolute information for the motion at
both transducers and this is useful in the analysis of the results.

When the equipment is operating properly, a pure sinusoidal
wave (less than 1% harmonic content) is obtained for the motion at
the bottom grip. When the deformation is in the range of linear
viscoelastic behavior, the motion at the top grip is also a pure
sinusoidal wave. For the present tests the minimum amplitude of
oscillation was used and a linear viscoelastic response was always
obtained. Under these circumstances the amplitude ratio and difference in phase shifts (i.e. the phase angle, ϕ) can be used in
eqs. (3) and (4) to calculate the dynamic shear modulus. The
computer also determines the contributions made by the damping and
inertia terms in the equations because, although they are included,
they contain quantities that may be difficult to calibrate. The
validity of the series termination necessary to obtain eqs. (3) and
(4) is also checked. The computer system used in this work has
graphic display capability so that the actual data points can be
displayed and examined visually, if desired. The computer not only
plots the data points but also draws the best-fit sinusoidal curves
through the points so this can also be checked (Fig. 3).

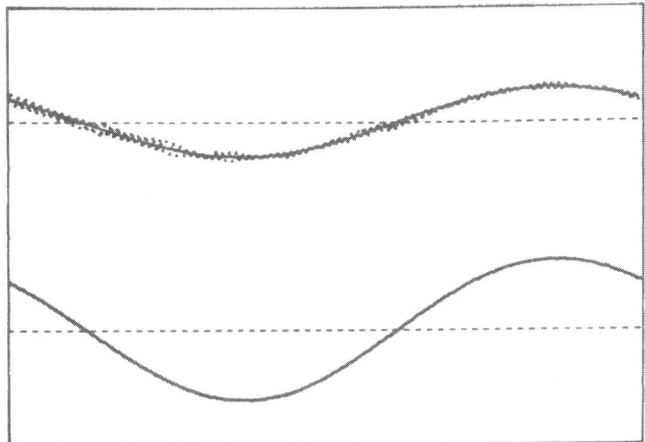

Fig. 3. Data points and best-fit curves for motion
at top grip (upper curve) and bottom grip
(lower curve). Some scatter was introduced
in top grip data so that best-fit curve could
be seen.

Before testing a sample it is essential that the instrument be
calibrated. The torsional elasticity, moment of inertia, and
damping constant of the torsion head were measured as described in
the literature (14). The transducer displacements and relative
phase shifts must also be corrected for effects introduced by the
mechanics and electronics of the system (15). The transducers are
first calibrated at each conditioner setting with a micrometer. The
grips were then clamped together and the amplitudes and relative
phase shifts were determined at each combination of frequency and
setting on the filter unit contained in the electronics package. No
effort was made to independently measure the absolute amplitude of
oscillation since it is the relative motion that is most important
for the calculations. Consequently, the readings for the transducer attached to the bottom grip measured with the highest filter
setting were used as the standard. The corrections determined in
these tests can be quite significant. For example, Fig. 4 shows the
difference in relative phase shifts between the two transducer
signals as a function of frequency for the three possible filter
settings. To obtain phase angle measurements that are reliable to
better than 0.1 degree, it is essential that this correction not be
in the rapidly rising region shown in the figure since the scatter
becomes significant in this area. Thus the filter setting must be
increased at much lower frequencies than those suggested by the
manufacturer. In the apparatus described here, the computer recom-
mends the best filter setting for each set of conditions and

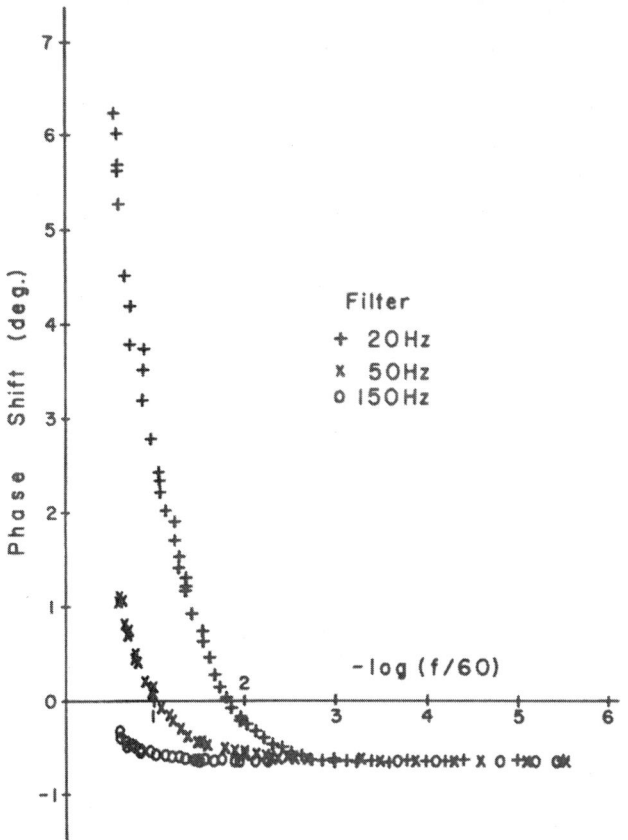

Fig. 4.   Phase-angle correction as a function of
           frequency, f, in Hz for the 3 filter settings.

automatically includes all of the appropriate calibration factors
in the calculations.

The experiments were performed using samples having both rectangular and circular cross-sections. To analyze the results for
samples with rectangular cross-sections, eqs. (3) and (4) were
altered by replacing $\pi R^4/2$ with

$$\frac{\pi R^4}{2} \longrightarrow ba^3 \left[\frac{1}{3} - \frac{64}{\pi^5} \sum_{n=o}^{\infty} \frac{\tanh\,(k_n b/2)}{(2n+1)^5} \right] \qquad (8)$$

where

$$k_n = (2n+1)\pi/a \qquad (9)$$

b is the width of the sample, and a is its thickness. Unfortunately, it is not possible to obtain a complete solution for this
geometry without making some assumptions about the stress or strain
field in the axial direction (16). The validity of these assumptions in an experimental system is difficult to judge. From a
practical point of view, however, the experiments performed here as
well as similar tests by other authors (8) show that, within
experimental error, both rectangular and circular cross-section
samples generally give equivalent results in the linear viscoelastic range. A number of samples with different lengths were also
tested to detect effects resulting from gripping of the sample at
the ends. As with previous studies (8), it was found that these
effects could be ignored when the sample length exceeded 6 cm.

Materials

     To illustrate the use of this apparatus, it was employed to
determine the dynamic mechanical properties of an epoxy that has
been extensively tested for fatigue (17) and fracture (18) behavior. This epoxy is a diglycidyl ether of bisphenol A (Dow Chemical
Co. DER-332) cured for 16 hours at 120°C with 5 phr piperidine.
After the heat treatment the samples were allowed to cool in the
oven in an effort to keep the thermal histories relatively constant.
After each specimen was machined to the proper dimensions, it was
annealed at 75°C for 6 hours. It was felt that even a short
annealing period could affect the sample since significant·thermal
heating was induced during the machining of the specimens.

     All of the temperature tests were performed starting at the
lowest temperature and moving up in steps whose size was determined
by the rate the measured properties were changing. The importance
of thermal history, which will be illustrated later, means that
increasing temperature experiments are the only meaningful way to
obtain data. At each step the frequency dependence was studied
after the sample was given time to reach constant temperature. The
exact number and range of frequencies tested was determined by the
temperature stability and the rate at which the sample properties
changed. In each experiment the final frequency tested was always
the same as the first and the repeatability of this measurement was
used to establish that constant temperature had been reached.

     When all of these experiments were completed, an untested
sample was placed in an oven and annealed for an additional 18 hours
at 75°C. This sample was then characterized in an effort to
evaluate, at least in a qualitative way, effects related to thermal
history.

RESULTS AND DISCUSSION

    The shear mechanical properties of the epoxy sample annealed
for 24 hours were measured as a function of frequency in the
temperature range from -100°C to +110°C. Some typical results are
illustrated in Fig. 5 which shows the thermomechanical data
obtained at 6 Hz. There is an α or glass-to-rubber transition at
about +95°C and a β peak at about -60°C. The α transition was
examined in detail by analyzing data obtained over a 4½ decade range
of frequencies at temperature between +40°C and +110°C. The results
were then considered for the applicability of time-temperature
superposition. The availability of data over a wide range of
frequencies makes it possible not only to accurately determine the
shift factors, a_T, but also to critically examine whether the points
really fall along a single master curve or not. Unless the
frequency range is large enough, a true test of superposition cannot
be obtained. In fact, when the range is too small, the data will
always appear to superimpose.

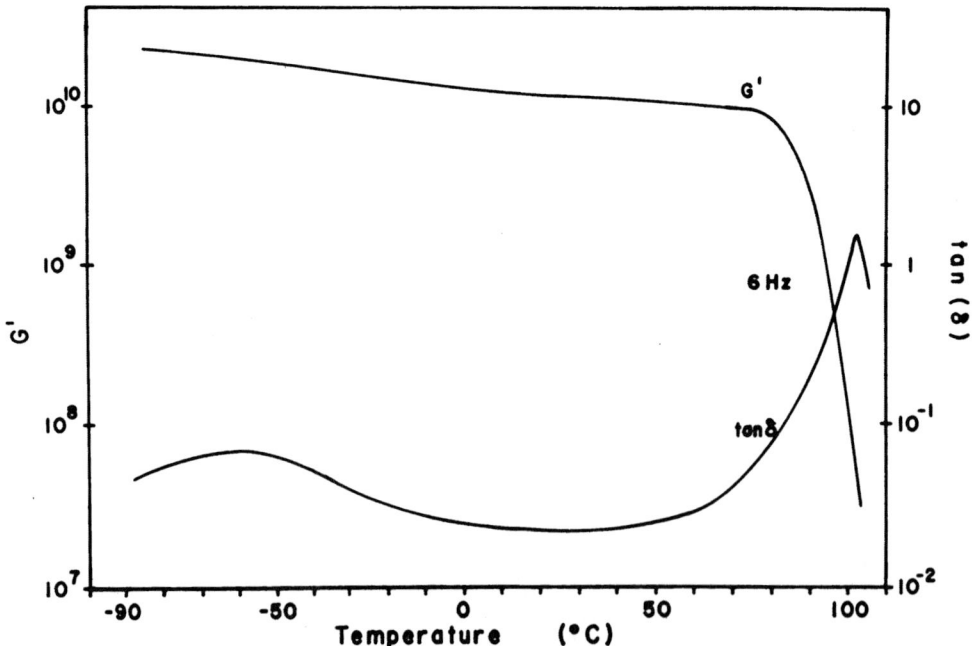

Fig. 5.   Shear storage modulus (dynes/cm^2) and
          loss tangent at 6 Hz as a function of
          temperature.

Recent papers by Koutsky (19) have indicated that the morphology of epoxies is complex in that a nodular structure is involved. Despite this complex morphology, the epoxy samples tested here were found to exhibit rheologically simple behavior in that all of the data could be reduced to single master curves. Figure 6 shows the master curves for the sample (annealed for 24 hours) at a reduced temperature, T_0, of +90°C. A single set of shift factors superimpose all three curves. (The modulus was normalized with the usual temperature-density ratio.)

The shift factors corresponding to the epoxy samples after annealing for 6 and 24 hours are shown in Fig. 7. Unfortunately, it is not possible to obtain data at high enough temperatures to compare the results with the predictions of equations such as the Williams, Landel, and Ferry (WLF) relationships ($T > T_g + 10$ where T_g via DSC is about 80°C) because extended exposure to temperatures near and above the cure temperature (120°C) can change the cure state of the system. There is evidence that WLF behavior might be expected. A series of tests are now being performed on the same epoxy system but with a small amount of an elastomeric toughening

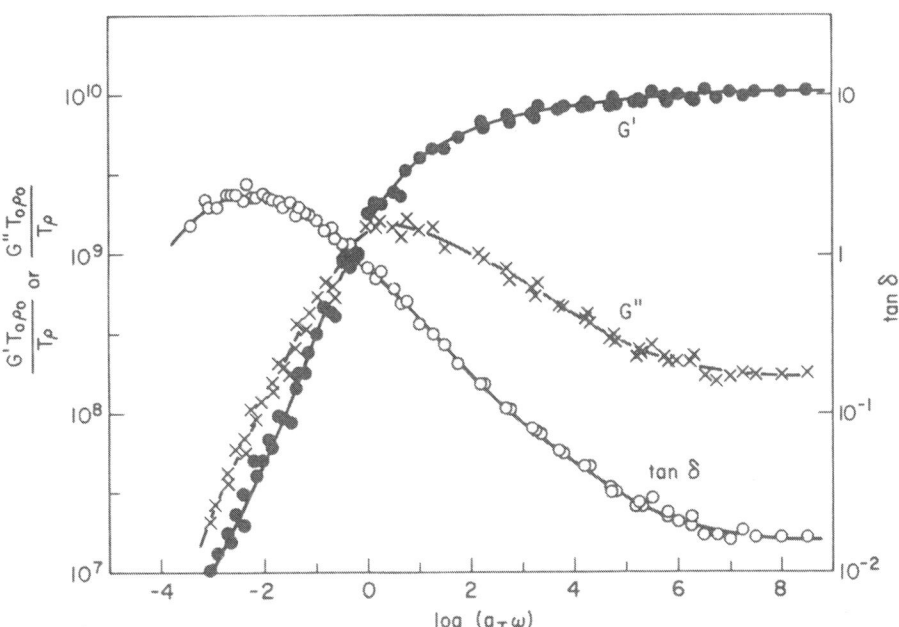

Fig. 6. Curves for dynamic shear modulus (dynes/cm²) and loss tangent as a function of reduced frequency (T_0 = +90°C).

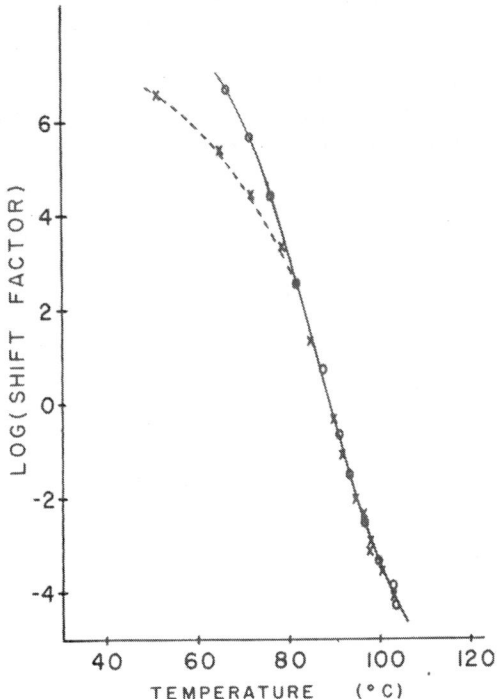

Fig. 7.   Shift factor as a function of temperature
          for the α relaxation of the epoxy after
          annealing for 6 hours (X) and 24 hours (0).

agent added. For this system the cure temperature is still $120^{\circ}C$
but the T_g is lower and so more data above T_g can be obtained. For
the toughened epoxy the preliminary results show that the shift
factors follow the WLF equation. Regardless of the exact nature of
the behavior above T_g, however, the shift factors might be expected
to increase dramatically as the temperature is lowered and approach
infinity at a temperature somewhat below T_g.   Experimentally,
however, a much different behavior is observed.   The shift factors
were not found to increase indefinitely but rather to gradually
level-off.   Moreover, the shape of the curve was found to be a
function of thermal history of the sample; for the specimen annealed
for 24 hours the shift factors were much closer to the expected
behavior.

The explanation for this result lies in the non-equilibrium
nature of the glassy state (20). As a sample is cooled near T_g, the
time required to reach an equilibrium state increases and thus the
cooling rate must be decreased. Eventually, it becomes impossible

to cool the sample slowly enough to maintain equilibrium (21). With systems that are closer to an equilibrium state, for example annealed or aged samples, the temperature dependence of the shift factor is closer to the expected behavior. The difference in shift factors shown in Fig. 7 also has important consequences for the dynamic moduli. For both thermal histories tested here, the master curves were surprisingly similar. Consequently, the moduli for both histories were the same at temperature well above T_g and at very low temperatures. In the intermediate range the sample annealed for 24 hours had a higher storage modulus and lower loss modulus than the sample annealed for 6 hours. This suggests that thermal history (and aging) can be viewed in terms of changes in the temperature dependence of the shift factors. This idea is quite similar to the results of Struik (21). Obviously, a more complete study of these effects is needed. It is clear, however, that no simple theory can adequately predict the viscoelastic properties in this temperature range and thus experimental characterization of the sample is essential if its behavior is to be understood and used to predict more complex properties such as fracture.

In the 6 Hz data (Fig. 5) a β relaxation was observed at about $-60^\circ C$. An effort was made to obtain a master curve in this region by superimposing data obtained over a $3\frac{1}{2}$ decade range of frequencies and the results are shown in Fig. 8. Although the scatter was somewhat greater in this range because the temperature control was not as good, no systematic variations were observed and thus superposition seems to be a valid procedure. Since the data were obtained over a wide frequency range, it is possible to examine the shift factors for the β relaxation in an Arrhenius plot. Figure 9 gives such a graph and the data are found to fall along a straight line whose slope gives an activation energy for the process of 15 kcal/mole. This is typical of the values generally obtained for β relaxations (8).

Figures 7 and 9 illustrate another important point. People often look at isochronal data such as that in Fig. 5 and assert that the motions associated with the α and β processes are "frozen out" at given temperatures, in this case $+80^\circ C$ and $-100^\circ C$. For the glass-to-rubber transition the term "frozen out" has some basis in fact since presumably the motion would stop if equilibrium could be maintained as the temperature is lowered. Figure 7 suggests, however, that it is unlikely that the sample can be cooled slowly enough to totally stop the motion. The β peak, on the other hand, follows Arrhenius behavior over the entire range of temperatures studied here and thus there is no evidence that the motion will cease until absolute zero. Consequently, the expression "frozen out" is not really appropriate. The results shown in Fig. 5 indicate only that the motion associated with the β peak is slow compared to 6 Hz. In reality, the β peak can be shifted to any

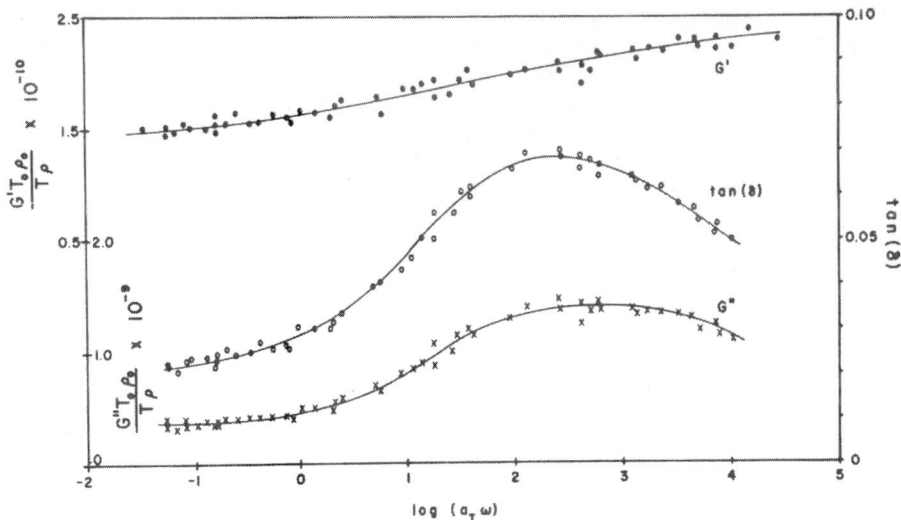

Fig. 8.   Curves for dynamic shear modulus (dynes/cm^2)
          and loss tangent as a function of reduced
          frequency (T_o = -50°C).

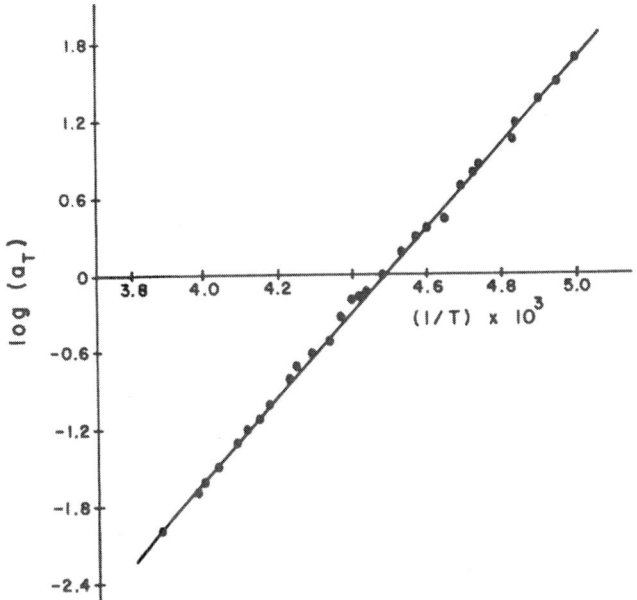

Fig. 9.   Shift factor as a function of inverse
          temperature for the β relaxation of the
          epoxy.

temperature that is desired simply by selecting the appropriate measurement frequency.

Another important question is the mechanism responsible for the β relaxation. Recent studies have combined the results of the present work with those obtained from the Carbon-13 NMR experiments of Dr. A.N. Garroway to address this problem (22,23). This comparison suggests that a rotation in the polymer backbone may be related to the β relaxation (23). Complete details of this study will be published elsewhere.

CONCLUSIONS

A forced oscillation test device, based on a Weissenberg Rheogoniometer, was developed for measuring the viscoelastic properties of solid samples. A computerized data acquisition and analysis system that uses graphic display capability to monitor the acquisition and analysis process was constructed. Appropriate equations containing terms for the moment of inertia and natural damping of the torsion head were developed and computerized.

A bisphenol-A epoxy (cured with piperidine) whose fracture and fatigue properties have been extensively studied was characterized with this equipment. Despite the complex morphology of epoxies, time-temperature superposition could be used to obtain master curves for the dynamic shear modulus. In the range of the transition the shift factors (and the glassy modulus) were dependent on the thermal history of the sample. The shift factors for the β relaxation gave a linear Arrhenius plot corresponding to an activation energy of 15 kcal/mole.

ACKNOWLEDGEMENTS

The authors wish to thank Dr. I.M. Krieger, Dr. H. Markovitz, and Dr. J. Hassell for their helpful comments during the development of this equipment. The assistance of Mr. C.M. Henderson and Mr. T.R. O'Neal in the construction of the apparatus is also gratefully acknowledged.

REFERENCES

1. Present address: Logicon Inc., 8300 Merrifield Ave, Fairfax, Virginia.
2. J.D. Ferry, Viscoelastic Properties of Polymers, Chapter 11, Wiley, New York (1970).

3. D. Kaplan and N.W. Tschoegl, "Time-Temperature Superposition in Two-Phase Polyblend", Polym. Eng. Sci. 14, 43 (1974) and references therein.

4. R.E. Cohen and Y. Sawada, "Extension of the Contour Plotting Method for Representing Linear Viscoelastic Loss Properties of SBS Triblock Copolymers", Trans. Soc. Rheol. 21, 157 (1976).

5. C.B. Bucknall, J.C. Reid, W.W. Stevens, "Fracture Mechanics Studies of High-Impact Polystyrene", Coatings and Plastics Prep. 37 (1), 477 (1977).

6. E. Riande, H. Markovitz, D.J. Plazek, and N. Paghupathi, "Viscoelastic Behavior of Polystyrene-Tricresyl Phosphate Solutions", J. Polym. Sci. Symp. 50, 405 (1975).

7. S.S.Sternstein and T.C. Ho, "Biaxial Stress Relaxation in Glassy Polymers; Polymethylmethacrylate", J. Appl. Phys. 43, 4370 (1972).

8. W.M. Davis and C.W. Macosko, "A Forced Torsional Oscillator for Dynamic Mechanical Measurements", Polym. Sci. Eng. 17, 32 (1977).

9. R.E. Cohen and N.W. Tschoegl, "Comparison of the Dynamic Mechanical Properties of Two Styrene-Butadiene-Styrene Triblock Copolymers with 1,2- and 1,4-Polybutadiene Center Blocks", Trans. Soc. Rheol. 20, 153 (1976).

10. J.L. Schrag, "Deviation of Velocity Gradient Profiles from the "Gap Loading" and "Surface Loading" Limits in Dynamic Simple Shear Experiments", Trans. Soc. Rheol. 21, 399 (1977).

11. R.H. Lee, L.G. Jones, K. Pandalai and R.S. Brodkey, "Modification of an R-16 Weissenberg Rheogoniometer", Trans. Soc. Rheol. 14, 555 (1970).

12. H. Markovitz, P.M. Yavorsky, R.C. Harper, Jr., L.J. Zapas and T.W. DeWitt, "Instrument for Measuring Dynamic Viscosities and Rigidities", Rev. Sci. Instrum. 23, 430 (1952).

13. I.M. Krieger and T.F. Niu, "A Rheometer for Oscillatory Studies of Nonlinear Fluids", Rheol. Acta 12, 567 (1973).

14. Sangamo Controls Limited, "The Weissenberg Rheogoniometer: Instruction Manual, Model R.18", (Sangamo Controls Limited, England).

15. K. Bogie and J. Harris, "An Experimental Analysis of the Weissenberg Rheogoniometer", Rheol. Acta 5, 213 (1966).

16. I.S. Sokolniloff, Mathematical Theory of Elasticity, 2nd ed., pp. 128-134, McGraw-Hill, New York (1956).

17. W.D. Bascom and S. Mostovoy, "Adhesive Fatigue Failure of an Elastomer-Modified Epoxy", Coatings and Plastics Prep. 38, 152 (1978).

18. W.D. Bascom, R.L. Cottington and C.O. Timmons, "Fracture Reliability of Structural Adhesives", J. Appl. Polym. Sci.: Appl. Polym. Sym. 32, 165 (1977).

19. J. Koutsky, "Etching of Polymeric Surfaces: A Review", Polym. Plast. Technol. Eng. 9, 139 (1977).

20. D.J. Plazek, personal communication, 1978.
21. L.C.E. Struik, "Physical Aging in Amorphous Polymers and Other Materials", Elsevier, New York (1978).
22. D.L. Hunston, E.E. Wells and A.N. Garroway, "Linear Viscoelastic Characterization of Thermosetting Epoxy Polymers", présented at the Society of Rheology Meeting, Houston, October 1978.
23. D.L. Hunston, E.E. Wells and A.N. Garroway, "Time-Dependent Mechanical Behavior of Epoxy Polymers", presented at the Adhesion Society Meeting, Savannah, GA, February 1979.

Crack Healing in Semicrystalline Polymers, Block Copolymers and Filled Elastomers

Richard P. Wool

University of Illinois at Urbana-Champaign

Urbana, Illinois 61801

ABSTRACT

Crack healing was found to be a generic property of polymers. Its effect on the mechanical behavior of some semicrystalline polymers, block copolymers and filled elastomers was investigated as a function of time, temperature and strain history. It was found that cracks or microvoids could heal instantaneously or slowly depending on the microstructural damage and molecular rearrangements incurred during the debonding process. Crack healing rates increased with temperature and no healing was observed below the effective T_g of the active molecular healing components in each material. A semiempirical kinetic theory was developed which quantitatively described mechanical recovery due to crack healing in terms of time, temperature and strain and provided a basis for construction of master healing curves by time-temperature superposition. It is apparent that crack formation and healing provides another mechanistic interpretation of viscoelastic behavior in addition to molecular relaxation processes and whose effect on mechanical and fracture properties of polymers is largely underestimated.

INTRODUCTION

Microvoid formation in the interfaces of contiguous microstructural elements is an important aspect of mechanical deformation in multicomponent polymer systems (1). In semicrystalline polymers, filled elastomers, block copolymers and general composites, the interfaces are created by several processes which include impingement growth during solidification, heterogeneous formulation and domain segregation. Cracks or voids may form during deformation, especially in the interfaces and can vary in size from a few angstroms to macroscopic dimensions. The cracks can be permanent, instantaneously reversible (instant healing) or time dependent reversible (slow healing). Figure 1 shows the generic stress-strain behavior resulting from crack healing in multicomponent polymer systems. The microstructural damage is created by the first cycle of the triangular strain input. The stress response for the second cycle depends on the time, temperature and humidity or other environmental factors of the sample in the rest state between cycles. As the cracks heal in the rest state, the mechanical properties of the sample are restored partially or fully to the virgin state depending on the extent of permanent damage.

Crack healing can be studied by a variety of experimental methods. Direct observation of crack closure modes and healing processes can be done in a transmission electron microscope using thin films (2) which have been subjected to mechanical damage. Our recent experiments indicate that solvent cast block copolymers with segregated domain structures, such as styrene-butadiene-styrene (SBS), are useful for this study. Similar work has been done on polyethylene by Miles et al (3). They observed closure of voids in films with a stacked lamellar (hard elastic) morphology. Photo optical methods can also be used to measure the relative amount of crack development and healing (2). The effect of scattered transmitted light due to cracks can be investigated using a photoelectric cell in the ocular of an optical microscope. This method was used to study the contribution of voids during stress relaxation at constant strain in semicrystalline and block copolymers. When the sample is unloaded, the "stress whitened" films heal in a temperature dependent manner as judged by the gradual restoration of the initial clarity and mechanical strength. The degree of void damage which develops at constant strain is found to depend on the magnitude and duration of strain. In addition to direct observation of crack healing by optical and electron microscopy, a number of scattering techniques can be used to monitor crack history. These include small angle X-ray scattering (SAXS), light scattering and neutron scattering. SAXS complements photo-optical studies by monitoring growth or healing of voids before they reach the critical size necessary to scatter visible light. Other spectroscopic techniques can be used in conjunction with the above methods to evaluate mechanisms of healing. Broadline proton NMR has been used to determine the behavior of the amorphous or fluid like chains during deformation and healing of hard elastic polypropylene (4). Infrared spectroscopy can also be used to investigate conformation, orientation and other molecular changes during healing (5,7). Other useful techniques include mass spectroscopy, electron paramagnetic resonance (EPR) and luminescence methods. The success of

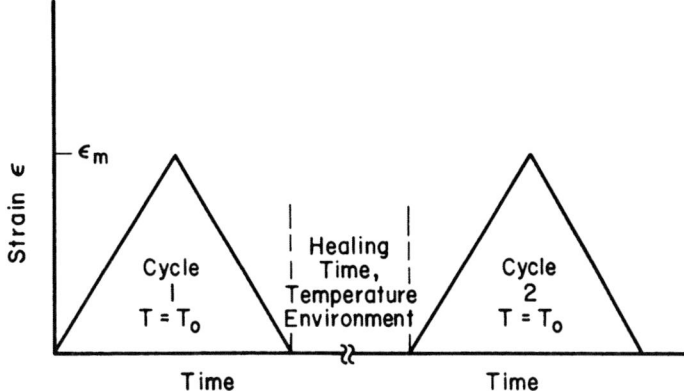

Fig. 1(a). Shows the 2-cycle strain and thermal
            history used to evaluate mechanical
            recovery and crack healing in polymers.

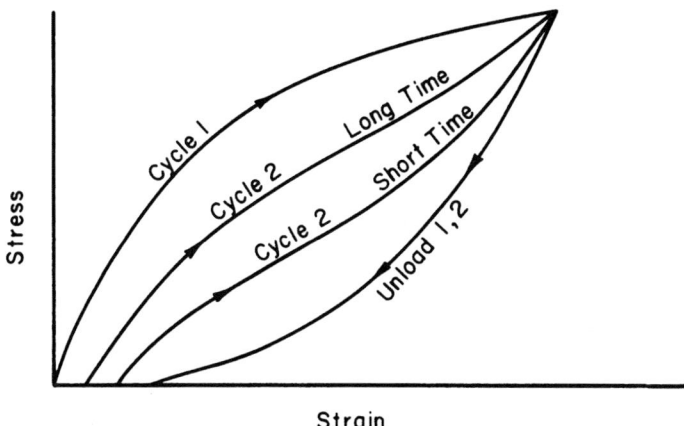

Fig. 1(b). The generic stress-strain behavior
            resulting from crack healing in multi-
            component polymers is shown schematically.

the experimental method depends on the nature of the healing
process.

    Healing rarely involves reformation of primary bonds asso-
ciated with chain scission processes, although radical recombina-
tion can occur in crystal "cages" (8,9) and to a certain extent in
disordered molecular regions. Facile healing occurs by restoration
of secondary bonding between chains or microstructural components
(1). Thus van der Waals or London dispersion forces play a very

important role in healing. Other kinds of bonding which may be important in healing processes include hydrogen bonding and chemisorption of chains on filler particles. The rate of healing depends on the transport of the separated molecular components into each others vicinity and the subsequent reformation of the original or equivalent bonding states. Transport can be achieved by diffusional processes through Brownian motion and bulk translation of microstructural components by release of strain energy. Since molecular mobility is necessary to achieve time dependent healing, it is expected that healing rates will be appreciably affected at temperatures near the glass transition temperature(s) of the healing component(s) in the material.

In this paper the general phenomena of crack healing will be investigated in terms of mechanical properties. Mechanical recovery due to crack healing will be examined as a function of time, temperature and strain for semicrystalline polypropylene, SBS block copolymers and carbon black filled natural rubber. A kinetic theory of healing will be developed which can be applied to all these multicomponent polymer systems. It will become apparent from the physical behavior of damage and healing processes that there is a microstructural analogy to molecular behavior which explains viscoelastic behavior of polymers. The microstructural contribution is very important for understanding nonlinear viscoelastic behavior in addition to bulk or "noncrack" constitutive behavior. While crack healing is advanced as a point of study herein, it is recognized that other "non crack" mechanisms also contribute to the mechanical behavior described in Fig. 1. However, the importance of crack healing will be clearly demonstrated.

MECHANICAL RECOVERY

We can define an adiabatic energy balance for polymer mechanical deformation in terms of three mechanistic quantities as (1)

$$\overset{o}{W} = \overset{o}{U} + \overset{o}{D} + \overset{o}{S}$$ (1)

where $\overset{o}{W}$ = rate of working (mechanical)

$\overset{o}{U}$ = rate of change of strain energy

$\overset{o}{D}$ = rate of dissipation due to viscous and plastic processes

and $\overset{o}{S}$ = rate of change of damage or surface energy due to fracture mechanisms at the molecular or microstructural level.

For the strain history shown in Fig. 1, the mechanical recovery R can be defined in several ways using parameters from Eq. (1). For example, R can be defined in terms of the total work W_1 and W_2 done in cycle 1 and cycle 2, respectively, as

$$R_W = \frac{W_2}{W_1} \qquad (2)$$

The use of Eq. (2) requires few assumptions but it does not provide information regarding specific contributions of the deformation mechanisms during recovery. More detailed information can be obtained by analyzing the stress-strain data in terms of the energetic parameters, U, D and S. In the elastic-fracture case, $D = 0$ and we can define recovery as

$$R = \frac{S_2}{S_1} \qquad (3)$$

where S_1 and S_2 are the total fracture work to create voids, break bonds, etc., in the first and second deformation cycles, respectively, and can be obtained from Eq. (1) as

$$S = W - U \qquad (4)$$

The total stored elastic energy U is obtained by integrating below the unload tensile curve. Thus, during healing, it is assumed that all mechanical recovery is expressed in changes of surface energy as the cracks heal and bonds reunite while no change occurs in the strain energy function of the material. The latter requirement means that the tensile unload curves of both cycles superimpose. Viscous contributions can be partially evaluated by instantaneously reloading (zero healing time) and subtracting the resulting hysteresis energy from the first cycle work. This procedure is reasonable with materials that exhibit a predominant elastic-fracture mechanical response but it may create problems for highly viscous systems. In this study, we have used materials whose viscous contribution is minimal in order to examine the crack healing phenomena. Thus Eq. (3) will be used as a working definition of mechanical recovery, R. If a large viscous contribution is suspected Eq. (2) may be used to represent the data and in this case the kinetics of healing developed later will need to be modified accordingly.

MECHANICAL RECOVERY IN POLYPROPYLENE

Hard elastic polypropylene (HPP) fibers with stacked lamellar morphologies were studied. These materials have been the subject of many recent studies (10-14). Their tensile properties are unusual and are shown in Fig. 2. HPP is highly crystalline (density = 0.92 g/cc) with the molecular chain axes preferentially oriented along the fiber axis. Large deformations are accommodated by void formation between lamellae and aggregates of the microstructure (10-14). The internal crack surface area varies approximately

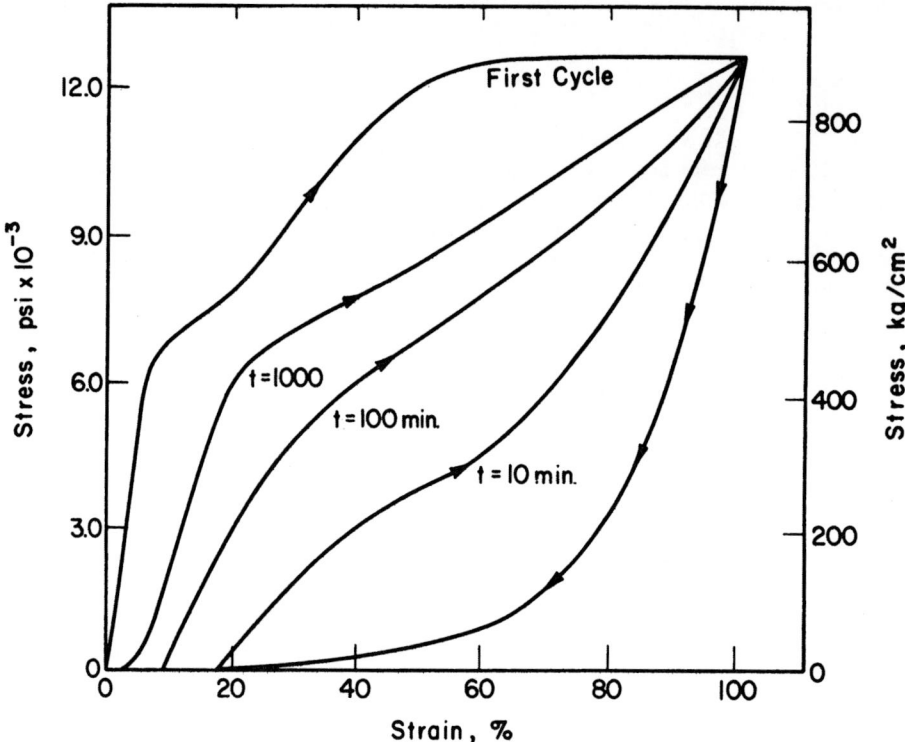

Fig. 2.   The effect of healing time on the stress-
strain behavior of hard elastic polypropylene
fibers is shown at room temperature.

linearly with strain and is about 100 m^2/g at 50% elongation. Thus,
this material is excellent for studying crack healing. As shown in
Fig. 2, the second cycle stress response depends on the healing time
of the fibers between deformation cycles. At short healing times
considerable stress softening is observed but at longer times the
original mechanical properties of the fibers are restored as the
cracks heal (15).

     Recovery in HPP fibers was investigated quantitatively as a
function of time, temperature and strain, using the strain history
shown in Fig. 1. A strain rate of 10/min was achieved on an Instron
tensile tester using a gage length of 2.0 in and a crosshead speed
of 20 in/min. The fibers were tested at ambient room temperature
and humidity for both first and second cycles. The effect of
temperature on healing was studied by placing the fibers in a
constant temperature controlled box, -17°C to 80°C, between deformation cycles for the desired healing time, t. The healing time was
measured as the time between unloading at zero strain in the first
cycle to the beginning of the second cycle. The maximum strain, ε_m

was the same for both cycles.

The effect of healing time and strain maximum on recovery of HPP fibers at room temperature is shown in Fig. 3. About 200 "virgin" samples were tested, each sample being subjected to only the first and second deformation cycles. Using Eq. (3) the recovery R is plotted as the ratio of the fracture energies. As shown in Fig. 3, R is approximately linear with log t and faster recovery occurs at lower strain maxima. This data is qualitatively similar to recovery data obtained by Cannon et al (11) who used the fiber length ratio as a definition of recovery, i.e. the initial "set" recovered with time.

The effect of temperature on recovery is shown in Fig. 4 for strain maxima of 50, 75, 125 and 150%. In each case, the recovery is seen to increase with temperature as might be expected from the enhanced molecular mobility at higher temperatures. At $-17^{\circ}C$ (Tg ≈ $-10^{\circ}C$) no recovery was observed for any of the strain series. In each case the initial or instantaneous recovery occurred at room temperature but further recovery did not occur below Tg. Recovery could be "restarted" by heating the sample above Tg and stopped by dropping below Tg. It was also observed that at 150% strain where considerable damage had been created, almost complete mechanical recovery was obtained at higher temperatures. At temperatures much higher than $80^{\circ}C$, annealing rather than healing processes become

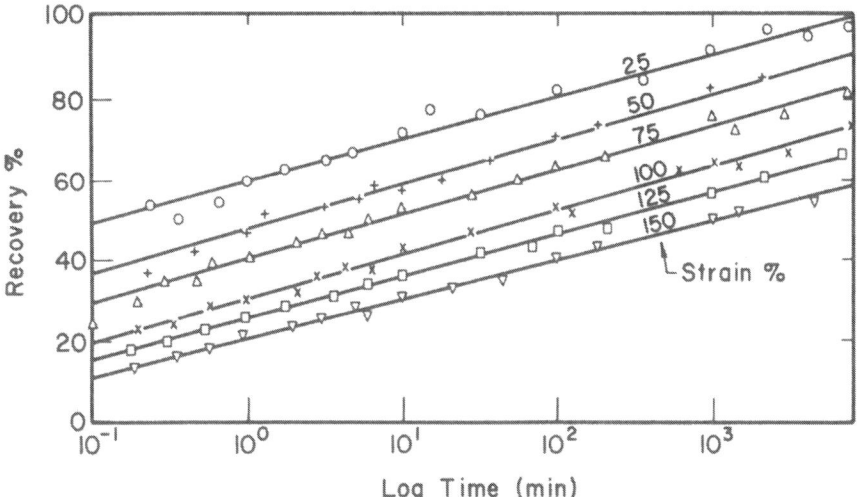

Fig. 3.   The mechanical recovery is plotted as a
          function of log healing time for HPP samples
          subjected to several strain maxima at room
          temperature.

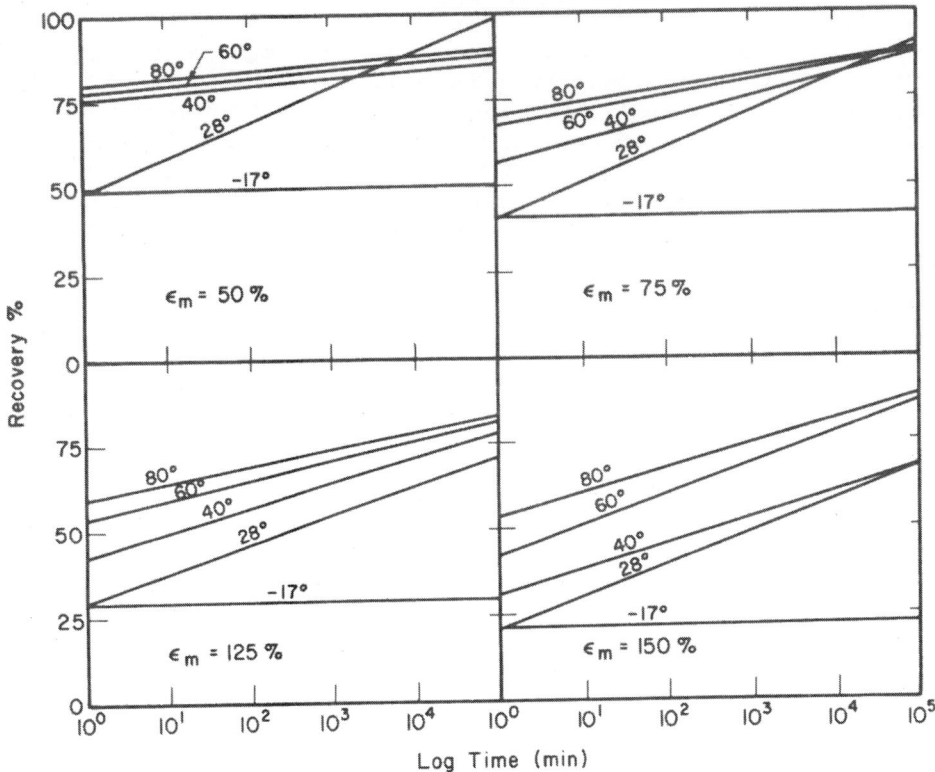

Fig. 4.   The effect of healing temperature on mechanical recovery is shown for HPP samples.

important and the constitutive microstructural character of the
material may change, thus affecting the interpretation of the data
in terms of crack healing arguments.   It should also be noted that
despite the thermal treatments between deformation cycles, superposition of the unload curves was obtained with reasonable accuracy
thus allowing the substraction of the strain energy component from
the total work input.   At strains less than 15%, little hysteresis
was observed even though cracks formed.   As discussed elsewhere (1),
cracks can contribute elastically to the mechanical response without hysteresis effects, i.e. cracks can open and close in a reversible elastic manner especially at low strains.

In HPP fibers crack healing involves molecular motion in the
amorphous regions between the crystalline lamellar surfaces and
healing of internal crystal defects.   Broadline proton NMR studies
(4) of HPP fibers indicate that a large increase in the central
component of the derivative absorption spectrum associated with the
more mobile or fluidlike chains occurs with increasing strain.   In

the rest state between deformation cycles the intensity of this fluidlike central component decreases with healing time in a manner closely paralleling the mechanical recovery. The increased mobility with strain can be attributed both to the increase in internal surface area providing additional configurational freedom to the amorphous chains on the lamellar surface and the generation of new amorphous material. During healing, the debonded lamellar surfaces come together and the intervening amorphous chains intermingle and readsorb on the surfaces assisted by van der Waals interactions. The increasing restriction of mobility of chain segments during healing causes the observed decrease in the fluidlike central component of the NMR broadline spectrum. Density measurements taken on HPP fibers during healing show an increase in density from 0.86 to 0.93 g/cc. If one assumes little density change of the lamellar crystals then very large density changes must occur in the amorphous component. The increased density must decrease the chain mobility which is consistent with the observed NMR behavior. Stress infrared studies (6,7) of HPP films indicate both a decrease in molecular orientation and helical conformation of the PP molecules with deformation. During healing in the rest state, the molecular c-axis orientation and conformational regularity is restored. The increase in orientation is attributed to the deflected lamellae on the void surfaces returning to their initial stacked morphology with a subsequent restoration of the chain stem alignment. This interpretation is supported by X-ray studies of lamellar bending in hard elastic polymers (16,17). The increase in helical conformers is consistent with the NMR results in which the amorphous chain mobility diminishes with void closure and may involve partial ordering of chains in the amorphous interface and elimination of surface defects.

Crack healing in semicrystalline polymers can also be interpreted in terms of fracture mechanics as a process which restores the adhesive fracture energy, Γ, at each point on the interface to its original value. Pursuing this analysis in a recent paper (12) it was observed that in the case of HPP, Γ varied along the crack surface during healing. Thus, healing was observed to occur simultaneously at all points on the crack surface but at different rates. As the crack propagates by lamellar debonding, the molecular damage or defects on the crack surfaces varies with position in a manner such that the damage is least in the vicinity of the crack tip and greatest towards the center of the crack away from the crack tip. This concept of surface damage is useful for interpreting modes of healing in molecular and macroscopic models (18). In the case of PP, the adhesive fracture energy for lamellar debonding was calculated to be about 100 ergs/cm^2 (12,19). As suggested by Kaelble (20), Γ values for interfacial debonding in polymers can be interpreted as the sum of the reversible thermodynamic contribution plus a diffusional demixing component. In PP, the diffusional demixing of amorphous chains represents about 70% of the total

adhesive fracture work.   From this viewpoint healing involves a
rapidly reversible or instantaneous component due to the thermodynamic surface tension effect followed by a slower healing rate
involving diffusional mixing of chains in the amorphous interface.
The latter effect is consistent with the T_g dependence of healing
and is also useful for understanding deformation rate sensitive
adhesive fracture energies.

BLOCK COPOLYMERS

        Similar recovery experiments were conducted on SBS block co-
polymers (Kraton 1101).   Film samples 3 x 0.5 x 0.01 in were cast
from a solution of THF/MEK (90/10).   Under controlled evaporation
conditions, the styrene component segregates into domains embedded
in the polybutadiene matrix.   The resulting material is a thermo-
plastic elastomer in which the hard styrene domains act as cross-
links for the rubbery butadiene matrix (21,22).   During mechanical
deformation, voids form in the interfaces between domains and
domain clusters.   Healing of these voids can be observed by photo
optical techniques or TEM (2).

        Our initial results on healing in SBS films are shown in Fig.
5.   The mechanical recovery, R, is plotted as a function of log time
for healing temperatures ranging from -70 to 43°C.   Again, R was
obtained by separating the fracture energy component from the total
work input using the strain history shown in Fig. 1 and a maximum
strain of 300%.   At the higher temperature of 43°C, complete
recovery is obtained in about 12 hours.   At T = -70°C, time
dependent recovery is not observed.   The effect of T_g on healing is
interesting in this material since we have two discrete T_g values
for the domains and the matrix as well as a range of possible T_g's
in the interfacial regions due to chain mixing. $T_g = 100^{\circ}$C for
polystyrene and $T_g = -85^{\circ}$C for polybutadiene.   Since complete
recovery is observed below 100°C, the polystyrene chains do not
appear to play a major role in the healing process.   However,
healing ceases at temperatures above T_g of polybutadiene and it is
highly probable that the mobile chains which are responsible for
healing are those in the domain-matrix interfaces.   These interfacial regions are mixtures of styrene and butadiene segments and
thus have a resultant effective T_g higher than that of polybutadiene.

        Recovery in SBS is more rapid than HPP.   This is due to the
higher mobility of the butadiene segments compared to polypropylene
and to the crosslinked morphology providing a strong elastic memory
(21).   Also, while SBS was subjected to much higher strains than the
HPP fibers, the degree of microstructural damage was considerably

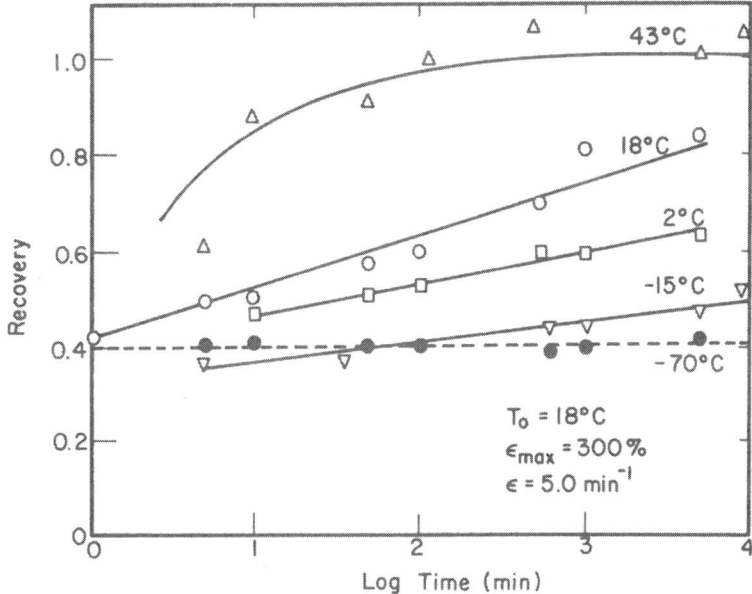

Fig. 5.   Recovery of SBS block copolymers is shown
           as a function of healing time and temperature.

lower in the block copolymer. The mechanical hysteresis and healing
observed in these experiments are in general agreement with results
of other investigators (20-25). A series of papers will be presented elsewhere detailing kinetics and molecular mechanisms of
healing in block copolymers (2).

FILLED ELASTOMERS

    The stress softening effect in filled natural rubber was
extensively investigated by Mullins (26,27), Bueche (28) and others
(29). Many processes contribute to this effect but the dominant
molecular process is considered to be debonding of matrix chains
from the filler particles. Mullins observed that this effect was
reversible, i.e. healing would occur in the rest state after
deformation (27). He determined R values as a function of time and
temperature from stress ratios at the same strain level in the first
and second deformation cycle and found the healing rate to increase
with temperature with an approximate linear logarithmic time dependence.

    We examined mechanical damage and healing in carbon black
filled vulcanized natural rubber using the strain history shown in
Fig. 1. The initial results (30) on the effect of strain level and

temperature on R are shown in Figs. 6 and 7, respectively. Again,
we observe an approximately linear recovery with log time. The
lower strain causes the least damage and thus higher recovery (in
some cases, the recovery can be independent of strain) and as
Mullins observed (27), faster recovery occurs at the higher temperatures. Healing in filled elastomers is considered to involve
readsorption of chains on the filler particle surfaces and closure
of voids in the polymer matrix. Carbon black used in materials such
as auto tires is considered to be a very active filler such that
debonded chains would have little difficulty reattaching once they
diffused to the particle surface. Recovery in these materials is
rapid, similar to recovery rates in SBS and is presumably controlled
by the high mobility of the elastomeric chains.

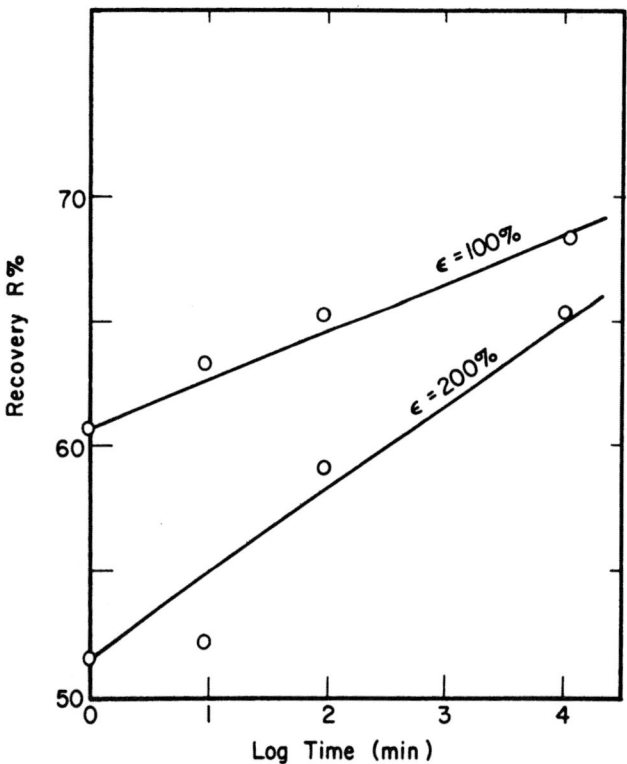

Fig. 6.   Recovery of carbon black filled natural rubber
           (74 phr) is shown as a function of healing
           time at room temperature for strain maxima of
           100 and 200%.

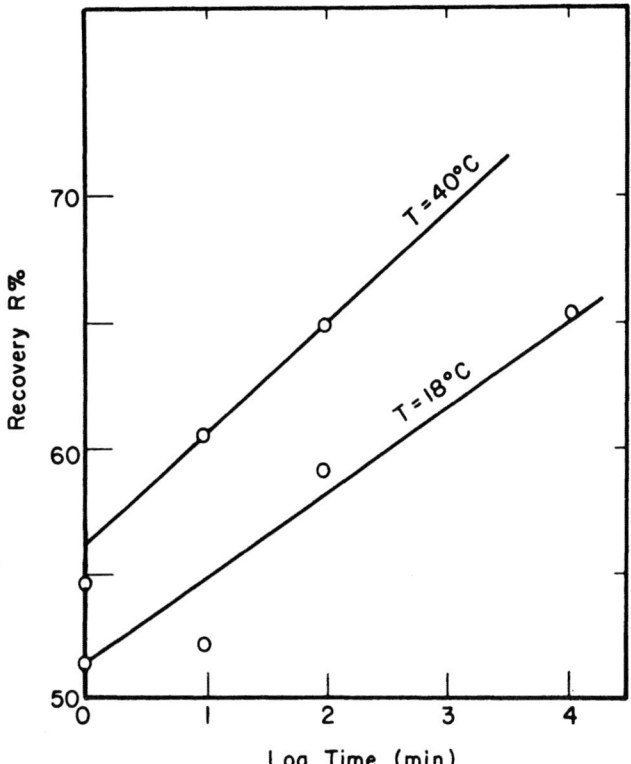

Fig. 7. Recovery of carbon black filled natural
rubber is shown as a function of healing
time at healing temperatures of 40 and
18°C. The maximum strain was 200%.

KINETICS OF HEALING

The generic character of crack healing and its influence on
mechanical recovery has been demonstrated in a number of multicomponent polymers in the past few sections. In this section, a
general kinetic theory of crack healing is presented to describe
mechanical recovery as a function of time, temperature and strain
level.

The mode of crack closure and healing is an important factor in
determining recovery kinetics. Cracks can heal by a "line" mode, a
"point" mode or a combination of such modes. The point mode
involves a "zip-up" mechanism in which healing only occurs at the
crack tip. The line mode involves simultaneous healing at all
points on the crack surfaces. These healing modes are mechanically
and kinetically distinguishable (12,15). Crack healing in brittle

inorganic materials such as NaCl (31) or LiF (32-34) occurs preferentially by the point closure mode but the line mode also contributes especially at higher temperatures. Our observations of crack
healing in polymers indicates a line mode of healing is favored.
The following analysis is based on an averaged line mode of healing
but is also applicable to point mode when consideration is made for
geometric factors at the crack tip.

It is proposed that healing depends on the rate of disappearance of damage or defects in the debonded interface between the
crack surfaces. The defects were created during the debonding
process to make the voids and result from molecular rearrangement
and dislocations on or off the crack surfaces. Healing occurs as
the molecular configuration at the interface is restored to its
original or equivalent bonding state in the virgin material. This
theory assumes that the complex process of healing is controlled by
the slowest step, namely defect or damage disappearance via molecular rearrangement, and that sufficient driving forces exist to allow
the debonded surfaces to approach each other. The driving force for
crack closure prior to healing may arise from a combination of
stored strain energy in the bulk material surrounding the crack, non-
bonded potentials, connecting chains or fibrils between surfaces,
etc.

Let D be the concentration of damage or defects per unit volume
of crack interface. It is assumed that healing occurs at the same
rate at all points on the crack surfaces, i.e. via line mode.
Therefore D is a parameter averaged over the nonuniform damage
distribution on the crack surface. The rate law for damage disappearance is unknown and the following general kinetic expression is
proposed, namely

$$- \frac{dD}{dt} = kD^n \tag{5}$$

where k is a temperature dependent rate constant and n is the order
or power of the healing process. Ideally for crack healing involving multiple cooperative molecular motion, k and n are multi-
ples and sums, respectively, of contributions from individual
kinetic processes but should better be regarded as averaged quantities reflecting a convolution of non-simultaneous molecular mechanisms with varying rate laws.

The mechanical recovery R is related to D by

$$D = S_1 \beta (1 - R) \tag{6}$$

where β is a proportionality constant and S_1 is given in Eq. (3).
Equation (6) reasonably assumes that the time dependent damage is
linearly related to the difference between the fracture energies of

the first and second deformation cycles.

The initial damage existing at $t = 0$ is

$$D_o = S_1 \beta(1 - R_o) \qquad (7)$$

where R_o is the instantaneous recovery component. Integrating Eq. (5) and substituting for D and D_o using Eq. (6) and Eq. (7), respectively, the mechanical recovery is obtained as a function of time, temperature and R_o as

$$R = 1 - \frac{(1-R_o)}{\left\{ 1 + D_o^{\,n-1} (n-1)k_o e^{-E/kT} t \right\}^{\frac{1}{n-1}}} \qquad (8)$$

where $n \neq 1$ and k has been defined as an Arrhenius rate constant in which k_o is the pre-exponential term, E is the activation energy, k is Boltzmanns constant and T is temperature.

R_o is a strain magnitude dependent parameter which can be obtained experimentally and for HPP fibers has the form

$$R_o = J\varepsilon^{-m} \qquad (R_o \leq 1) \qquad (9)$$

where J and m are constants and ε is the maximum strain shown in Fig. 1. The value of ε for which $R_o = 1$ would indicate the strain range for perfectly reversible cracks and/or viscoelastic behavior in each system.

To obtain the kinetic constants from mechanical recovery studies, Eq. (8) is simplified and rearranged as

$$\left(\frac{1 - R_o}{1 - R} \right) = (1 + Kt)^\alpha \qquad (10)$$

where $\quad K = D_o^{n-1} (n-1)k_o \exp{-E/kT}$

and $\quad \alpha = \frac{1}{n-1}$

For $Kt \gg 1$, we obtain from Eq. (10),

$$\log \left(\frac{1-R_o}{1-R} \right) \cong \alpha \log K + \alpha \log t \qquad (11)$$

Thus, a plot of the left hand side of Eq. (11) versus $\log t$, if linear, provides α from the slope, and K from the intercept $\alpha \log K$.

Knowing K at different temperatures, the other kinetic constants can be calculated.

When K and α are known, R can be expressed simply as

$$R = 1 - \frac{(1-R_o)}{(1+Kt)^\alpha} \qquad (12)$$

which is the general expression proposed for healing in multi-component systems.

For a first order healing process not considered above, $n = 1$, we obtain from Eqs. (5), (6) and (7),

$$\ln\left(\frac{1-R_o}{1-R}\right) = kt \qquad (13)$$

for which the kinetic constants can be readily determined experimentally.

The recovery half life, τ, is measured at $R = 0.5$ and can be determined from Eq. (12) as

$$\tau = \frac{1}{K}\left\{\left[2(1-R_o)\right]^{\frac{1}{\alpha}} - 1\right\} \qquad (14)$$

These equations predict that R increases with time and temperature and faster recovery occurs at lower strains (large R_o). If $R_o = 0$, the recovery rate is independent of strain. The permanent damage contribution, D_p can be determined by either modifying the definition of R in terms of energy contributions for permanent damage or experimentally from the relation

$$D_p = 1 - R_\infty \qquad (15)$$

where R_∞ is the observed recovery at very long times.

The recovery behavior predicted by this kinetic approach is shown in Fig. 8. Using Eq. (12), R is plotted as a function of log t. The sigmodial recovery curve is in fact the shape of a master recovery curve which describes healing of cracks and microstructural damage for most polymeric materials subjected to any particular strain history, not necessarily that outlined in Fig. 1. At short times, the lower plateau extrapolates to the instantaneous recovery contribution and at long times the upper plateau describes either full recovery or residual permanent damage. The master curve is associated with a reference testing temperature T_o and strain, and can be generated experimentally by time-temperature superposition of R data at different temperatures (2). Further work is being conducted to investigate the effect of T_o and ε_m on the master curve.

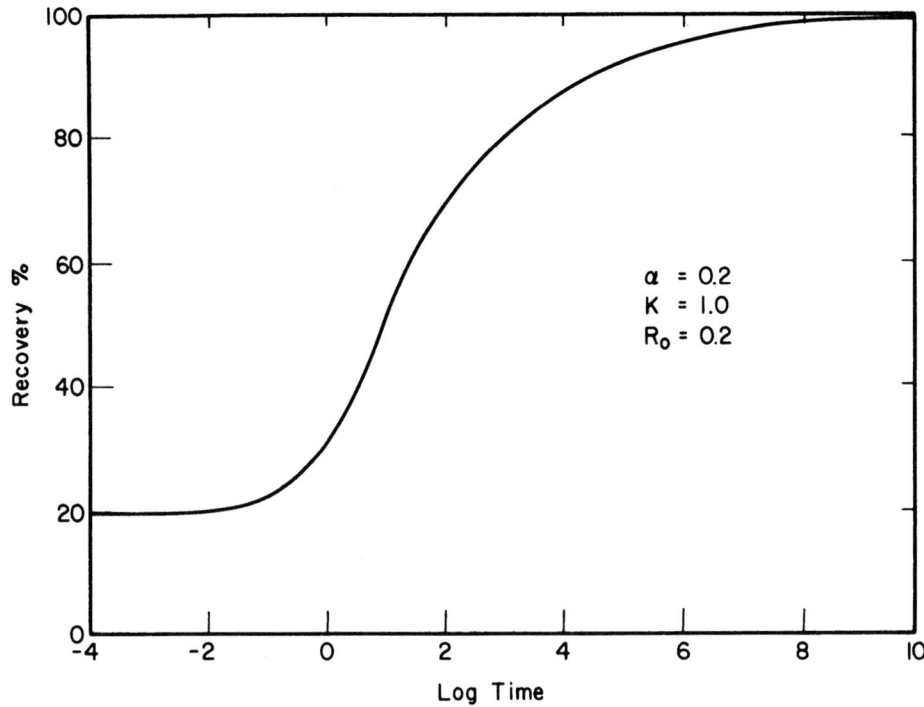

α = 0.2
K = 1.0
R_o = 0.2

Fig. 8.   The theoretical master healing curve is
          shown.

APPLICATIONS

The kinetic theory of crack healing was evaluated by applying
Eq. (11) to the experimental data. Figure 9 shows a plot of log(1-
R_o)/(1-R) versus log t for HPP samples subjected to a maximum strain
of 150% (as in Fig. 1) and several healing temperatures. At long
times, Kt >> 1, the data is linear from which we obtain the slope α =
0.09 ± 0.01 and n ≃ 12. The extrapolated intercept α log K is
temperature dependent and yields an apparent activation energy E ≃
28 kcal/mole. Similar behavior is obtained for other strain levels
over the same temperature range. R_o was determined as a function of
strain at room temperature and can be described by Eq. (9) as

$$R_o = 0.19\,\epsilon^{-0.635}$$

At R_o = 1, ϵ = 7.31% which can be interpreted as the maximum strain
limit for rapid crack reversibility in these hard elastic polypropylene fibers.

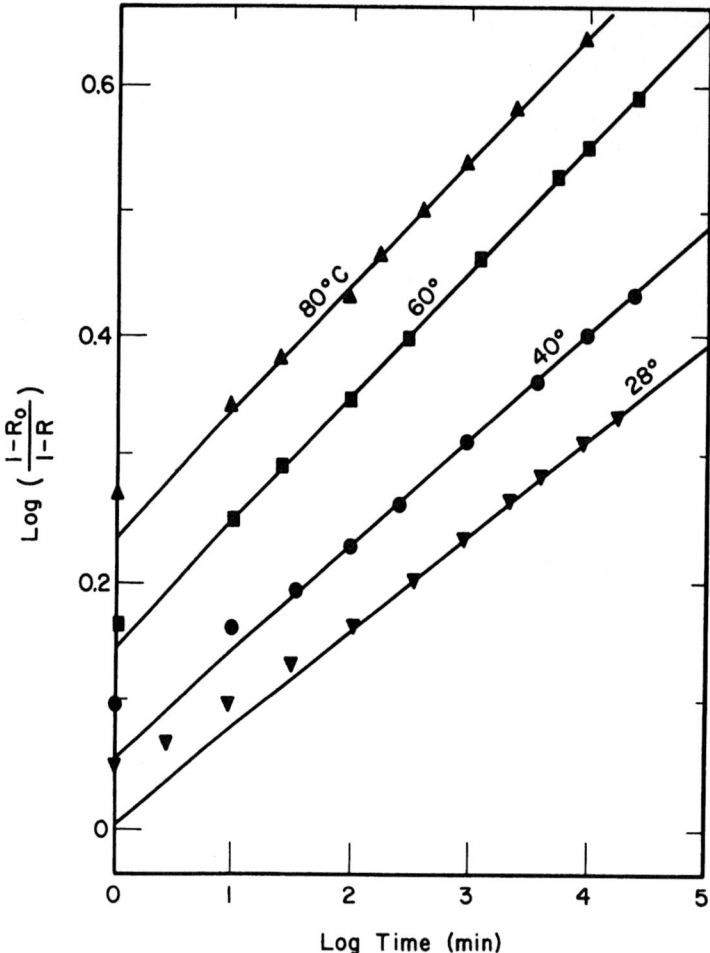

Fig. 9.   Log $(1-R_o)/(1-R)$ is plotted as a function
of log healing time for HPP fibers at
several healing temperatures.   The strain
maximum was 150%.

Using the kinetic constants and the value for R_o at each strain
level, the recovery R was determined as a function of time and
temperature using Eq. (12).   Figure 10 shows a comparison of theory
with experiment for the room temperature recovery at several strain
maxima.   The theoretical plots are slightly curved but fit the data
better than the linear log time plots used in Fig. 3.   Figure 11
shows the comparison of theory with the experimental data for the
125 and 150% strain series at different temperatures.   A reasonably

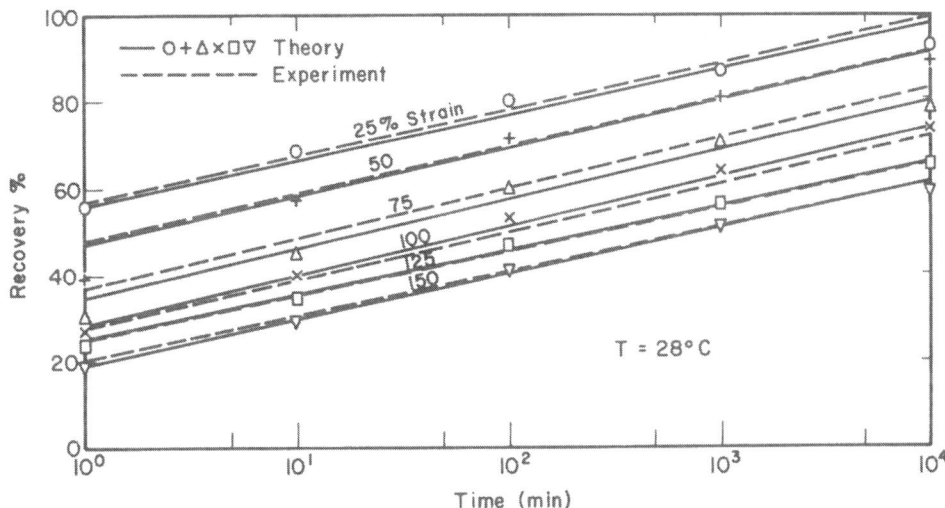

Fig. 10. A comparison of theory with experimental
          recovery data for HPP fibers is shown.
          The strain maxima ranged from 25-150%
          at 28°C.

good correlation is observed. The effect of T_g is not intrinsically
incorporated in the theory although little recovery is predicted
near T_g at -17°C. The temperature dependence can also be introduced
using a WLF approach (2) and provides a rational basis for the
construction of master recovery curves using time-temperature
superposition principles and another experimental method of determining activation energies.

     Despite the simplicity of the kinetic healing model, it is very
useful for quantitatively describing mechanical recovery data and
is in excellent qualitative agreement with the known physical
processes of crack healing in all materials investigated.   The
theory has been applied to block copolymers (2) and filled elastomers (30) with equal success.

SUMMARY

     The concept of crack healing was discussed and shown to
play an important role in determining mechanical properties of
semicrystalline polymers, block copolymers and filled elastomers.
Important aspects and consequences of crack healing are summarized
below:

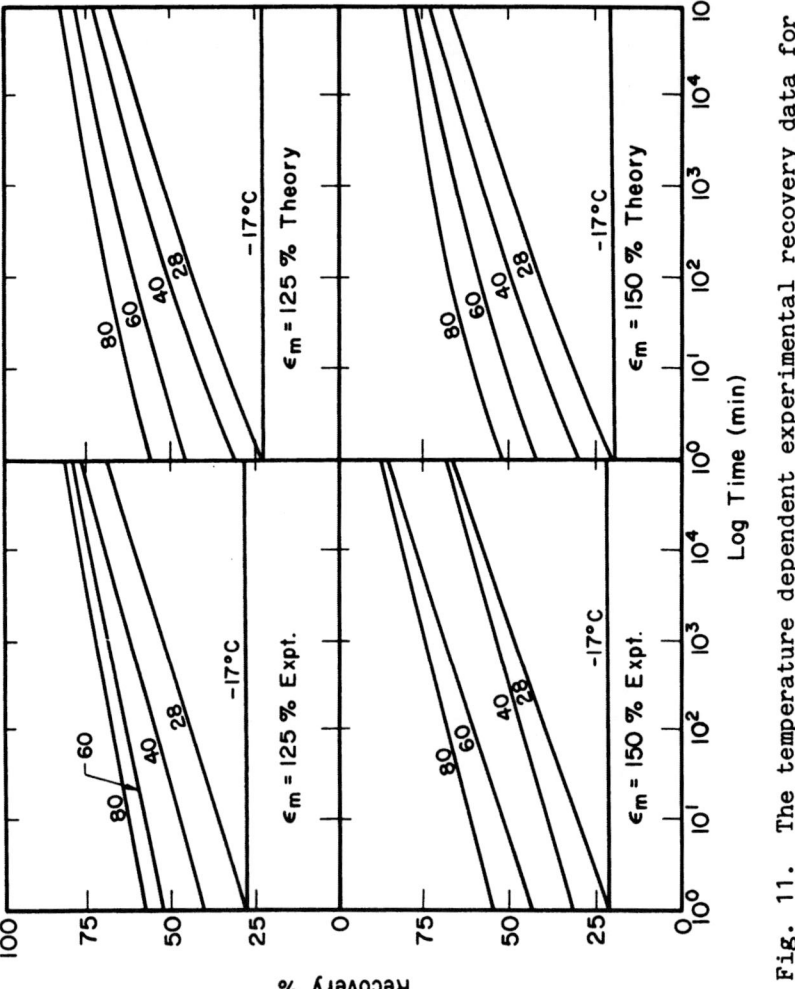

Fig. 11. The temperature dependent experimental recovery data for HPP fibers is compared with the theoretical prediction for strain maxima of 125 and 150%.

1.   Crack healing is a general phenomenon which occurs in many polymer systems.   It affects mechanical behavior and is an important consideration for mechanistically evaluating constitutive relations.

2.   Healing rates are controlled by temperature, strain history, environment and the nature of the microstructural damage. Nonpermanent cracks can heal either instantaneously or in a time dependent manner.

3.   The concept of cracks opening and closing in a time dependent manner provides an alternative microstructural interpretation of viscoelastic behavior in addition to pure molecular relaxation processes.   Rate dependent mechanical properties which are controlled by crack healing can be subjected to time-temperature superposition analyses similar to the WLF approach with viscoelastic polymers.

4.   A semiempirical kinetic theory of crack healing based on the rate of disappearance of damage adequately describes mechanical recovery due to crack healing and provides the basic rationale for constructing recovery master curves.

This work is continuing in an effort to further elucidate the mechanics and molecular mechanisms of crack healing and to evaluate other mechanical recovery processes suggested by several investigators (35-39).

ACKNOWLEDGEMENT

The author is grateful to the National Science Foundation for financial support of this work with Grants ENG-76-09627 and DMR-78-13942 and expresses thanks to Dr. T.J. Rowland and Dr. R.J. Gaylord for helpful suggestions and comments.

REFERENCES

1. R.P. Wool, Poly. Eng. & Sci., 18, 1057 (1978).
2. R.P. Wool and K.M. O'Connor, papers in press; K.M. O'Connor and R.P. Wool, Bull. A.P.S., 24, 289 (1979).
3. M. Miles, J. Petermann and H. Gleiter, Progr. Colloid & Poly. Sci., 62, 478 (1977).
4. R.P. Wool, M.I. Lohse and T.J. Rowland, J. Poly. Sci., Poly. Letters Ed., 17, 385 (1979); M.I. Lohse, T.J. Rowland and R.P. Wool, Bull. A.P.S., 24, 348 (1979).
5. R.P. Wool and W.O. Statton, Chapter 12 in Applications of Polymer Spectroscopy, Ed. E.G. Brame, Jr., Academic Press, New York, 1978.

6. R.P. Wool, _J. Poly. Sci._, 14, 1921 (1976).
7. R.P. Wool, _Poly Eng. & Sci._, in press (1979); C.C. Hsu and R.P. Wool, _Bull. A.P.S._, 24, 349 (1979).
8. K.L. DeVries and D.K. Roylance, _Progress in Solid State Chemistry_, 8, 283 (1973).
9. L.A. Davis, C.A. Pampillo and T.C. Chiang, _J. Poly. Sci._, 11, 841 (1973).
10. I.K. Park and H.D. Noether, _Colloid & Poly. Sci._, 253, 824 (1975).
11. S.L. Cannon, G.B. McKenna and W.O. Statton, _J. Macromol. Sci._, Part D, 11, 209 (1976).
12. R.P. Wool, _J. Poly. Sci._, Physics Ed., in press (1979).
13. B.S. Sprague, _J. Macromol. Sci._, Phys., 8, 157 (1973).
14. E.S. Clark in _Structure and Properties of Polymer Films_, R.W. Lenz and R.S. Stein, Eds., Plenum Press, New York, 1974.
15. R.P. Wool, paper #21S, presented at the 69th Annual Meeting of A.I.Ch.E., Chicago, 1976.
16. H.D. Noether and W. Whitney, _Kolloid Z.Z. Polymere_, 251, 991 (1973).
17. H. Ishikawa, H. Numa and M. Nagura, _Polymer_, 20, 516 (1979).
18. R.P. Wool, _Int. J. Fracture_, 14, 597 (1978).
19. R.P. Wool, _Bull. A.P.S._, 23, 386 (1978).
20. D.H. Kaelble and E.H. Cirlin, _J. Poly. Sci._, C43, 131 (1973).
21. S.L. Aggarwal, _Polymer_, 17, 938 (1976).
22. M. Morton, _J. Poly. Sci._, C60, 1 (1970).
23. J.L. LeBlanc, _J. Appl. Polym. Sci._, 21, 2419 (1979).
24. E. Pedemonte et al., _Polymer_, 16, 531 (1975).
25. Y.D.M. Chen and R.E. Cohen, _J. Appl. Polym. Sci._, 21, 629 (1977).
26. L. Mullins, _Rubber Chem. Technol._, 21, 281 (1948).
27. L. Mullins and N.R. Tobin, _Rubber Chem. Technol._, 30, 555 (1957).
28. A.M. Bueche, _J. Poly. Sci._, 25, 139 (1957).
29. A.F. Blanchard and D. Parkinson, _Ind. Eng. Chem._, 44, 799 (1952).
30. R.P. Wool and S.K. Upadyayula, to be published; S.K. Upadyayula and R.P. Wool, _Bull. A.P.S._, 24, 259 (1979).
31. S.M. Park and D.R. O'Boyle, _J. Mater. Sci._, 12, 840 (1977).
32. R. Raj, W. Pavinich and C.N. Ahlquist, _Acta Met._, 23, 399 (1975).
33. D. Nenow and S. Gueleva, _J. Mater. Sci._, 13, 1195 (1978).
34. D. Nenow and S. Gueleva, _Kristall and Technik_, 13, 429 (1978).
35. H. Brody, _J. Appl. Polym. Sci._, 22, 1631 (1978).
36. J.B. Park and D.R. Uhlmann, _J. Appl. Phys._, 41, 2928 (1970).
37. D.R. Uhlmann and J.B. Park, _J. Appl. Phys._, 42, 3800 (1971).
38. J.B. Park and D.R. Uhlmann, _J. Appl. Phys._, 44, 201 (1973).
39. H.H. Kausch and K. Jud, Macro-79 Proceedings of I.U.P.A.C., Mainz (1979).

Discussion

On the Paper by D. Maugis and M. Barquins

L.H. Lee (Xerox Corporation): Mr. Barquins and Dr. Maugis should be congratulated for this excellent paper presented to this Conference. Both the breadth and the depth of this paper can serve as a good model for a general survey for the subject matter. We are grateful to both authors especially for the excellent, final manuscript to be appeared in the Proceedings. I am glad that Mr. Barquins can come to present this paper.

On the Paper by E. Sancaktar, J.S. Cartner, W.I. Griffiths and H.F. Brinson

A. Silberberg (Weizmann Institute, Israel): It seems to me that failure of the joint is a falure of the material in bulk under the applied deformation field. Failure does not seem to involve the actual seam of attachment to the surface support except that imperfection in the interface zone may be the first points of crack formation. Is this correct and is it that you are attempting to model?

H.F. Brinson (Virginia Polytechnic Institute): I do not fully understand the question. If the discusser is referring to the rail shear tests, then clearly bulk properties are measured. Failure properties may not be very valid, however, since failure only occurred at a stress riser. If the discusser is referring to bonded joints, then I believe our photomicrographic studies (though none are presented in the paper) prove that failures likely initiate at an interface or at least near an interface. They are probably caused by defects in bonding.

R.A. Gregg (Uniroyal): In your three rail test specimens, is the sheet of adhesive chemically bonded as well as mechanically fastened to the metal rails?

H.F. Brinson: No. However, metallic tabs were bonded to the specimen to minimize the effects of stress concentrations. The rails were bolted to the specimen with tabs. Stress concentration was still a problem as all failures first occurred at a bolt hole.

P. Datta (RCA Laboratories): Which shear rates were the shear measurements made? What was the temperature of the samples in the shear measurements? Could this difference be related to the differences in the shear and tensile measurements?

H.F. Brinson: The shear strain rates are given in each figure. Tests were performed under controlled environmental conditions of about $25^\circ C$ $(70^\circ F)$ and 50% R.H. as identified in the text. Thus temperature and humidity were not believed to be an influence.

A.D. Jonath (Lockheed Palo Alto Research Laboratory): (1) In the symmetric rail shear specimen shown for the bulk adhesive, the failure locus appeared to alternate from one adhesive-adherend interface to the other. Can you please comment on this? (2) Are there thus significant tensile stresses introduced during the failure process?

H.F. Brinson: We attach no particular significance to the actual fracture planes shown in the presentation for our rail shear specimen (these photographs are not contained in the final manuscript). The fractures always occurred at a stress concentration site or hole under a rail where a complex stress state different than pure shear prevailed.

The above comment does not negate our conclusion of different deformation process in bulk shear and bulk tension. The tensile stresses and strain measured are lower than would occur if the material at the measurement site were stressed to rupture under pure shear conditions. The low measured values are still higher than those predicted from tension. Obviously, actual rupture values would be higher still.

On the Paper by D.H. Kaelble

D.W. Dwight (Virginia Polytechnic Institute): (1) Would solvent rinse (or other chemical probes) distinguish the SET mechanisms? (2) Fowkes argues that polar forces do not contribute to bonding, so there appears to be disagreement. (3) Are your results independent of locus of failure?

D.H. Kaelble (Rockwell International): (1) Ion sputter profiling, Auger electron spectroscopy (AES), and ellipsometric analysis show definite correlations of oxide film morphology and SET mechanisms (see ref. 3 for details). (2) The special relations utilized in this discussion show that polar forces as analyzed by contact angle measurements of unjoined surfaces directly influence the strength of the bonded joint. A large number of studies (see Ref. 7) now confirm the general validity of these surface energy relations, and their ability to encompass the surface properties of water and many hydrated surfaces. It is important to point out, however, that dispersion-polar bonding is a subset of a much larger class of covalent-ionic mechanisms in which hydrogen bonding and other interfacial mechanisms can be described (see D.H. Kaelble, Physical Chemistry of Adhesion, Wiley-Interscience, New York, 1971). (3) As pointed out by eq. 10, the fracture energy relations can also be applied to predict cohesive failure criteria so as to provide relative assessments of interfacial and cohesive degradation.

J. Perkins (AMMRC): In $CrO_4^=$ etching of Al, have you investigated the possible interaction of adherend on the aging of the resin and/or the adhesive bond? By plasma etching (air or other), there might be a change in aging characteristics.

D.H. Kaelble: The standard FPL sulphuric chromate etch process was not varied in this study. Auger electron spectroscopy of the surface treated aluminum alloy showed no evidence of residual chromium at the surface or in the metal oxide film (see Ref. 3 for details).

P. Dryfuss (University of Akron): You have given data on the effect of aging of the substrate in various environments. Have you studied the corresponding effect of aging of the adhesive?

D.H. Kaelble: The bulk properties of the structural adhesive have been studied by calorimetric and thermomechanical analysis in dry N_2 with no notable aging effects shown (see D.H. Kaelble and E.H. Cirlin, J. Poly. Sci., Part C, 35 (1971), pp. 79-100).

P. Dryfuss: My colleagues at the University of Akron have studied effect of aging of polybutadiene layers. We agree there is minimal effect of aging in nitrogen, but we do find a very large enhancement in autoadhesion after aging in air.

On the Paper by C.E.M. Morris, A.G. Moritz and R.G. Davidson

J. Perkins (AMMRC): Of the two batches tested, one older, were the initial values observed close or diverse? (The graphs were in terms of % of initial values.)

C.E.M. Morris (Materials Research Laboratories, Australia): The initial epoxy values were very similar, 183 for Batch 1 and 174 for Batch 2.

M.A. Grayson (McDonnell Douglas Research Labs): We have observed epichlorohydrin precursors in cuts from LC analysis of the tetraglycidyldiaminomethane epoxy system. Have you found any of these compounds in your analyses?

C.E.M. Morris: No, but we have not specifically looked for such compounds.

L. Masters (National Bureau of Standards): Will the mechanical tests you mentioned be performed on bonded components or on non-bonded specimens of the epoxide?

C.E.M. Morris: Tensile shear tests will be conducted on aluminum single overlap joints. It is also planned to look at samples of the cured adhesive on a Rheovibron.

PART FOUR

FRACTURE STRENGTHS OF
POLYMERIC SYSTEMS

Introductory Remarks

K. Nakao

3-12-1, Makayama-Sakuradai

Takarazuka-shi

Hyogo-ken, Japan 665

The analyses of tensile bond strength, shear strength and peel strength are important problems which should be clarified in the field of the adhesion science and technology. In the case of interfacial fracture of cross lap joint, Mr. Masuoka suggests that the failure of the tensile adhesive joint is governed by an energy criterion. The stress distribution and the maximum stress in shear at the circular area of lap joint carrying the bending moment have been shown by Dr. Yamaguchi. On the peeling of flexible materials, Dr. Igarashi shows: (1) The steady peeling strength for the elastic adhesives is determined by the surface energy of the adhesives. (2) The energy dissipation of the adhesives causes the increase in the steady peeling strength. (3) On the contrary, the initial peeling force is derived from the maximum stress in the adhesive layer.

The molecular weight of adhesives is also one of the most important factors governing bond strength. Dr. Tsuji shows the superposition of peel strength with molecular weight of adhesive, temperature and time.

The composition and reaction of adhesives can also govern bond strengths. Recently for making denture, a new technique, that is, the fluid technique using self-curing acrylic resin has been developed. By Mr. Harashima, the effect of the crosslinking of PMMA has been shown. These two papers are published elsewhere.

To my great regret, Dr. Miyairi could not attend our Conference for personal reasons. However, his paper on the deformation and shear strength of FRP adhesive joint is included in this part.

I was pleased to accept Dr. Lee's invitation to preside at this session concerning the "Fracture Strength of Polymeric Systems."

We, Japanese scientists, are grateful to the Chairman, Dr. L.H. Lee and Co-Chairman Dr. W.H. Grant for organizing and conducting this Conference.

Criterion of Interfacial Fracture on Tensile Adhesive Joint

Mineo Masuoka and Kazumune Nakao

Osaka Prefectural Industrial Research Institute

2-1-53, Enokojima, Nishi-ku, Osaka, Japan 550

ABSTRACT

The present investigation was undertaken to verify the criterion of the interfacial bond failure. The time-temperature superposition, the failure envelope, the dependence of the tensile bond strength on the degree of crystallinity and the fractography of the interfacial region of the adhesive were observed by the use of steel-nylon 12-steel cross lap joint. The ultimate tensile properties of the joint were found to be associated with the viscoelastic nature of the adhesive rather than with a thermodynamic criterion. The experimental results suggest that the Mises hypothesis can be applied to the tensile adhesive joint as the fracture criterion.

INTRODUCTION

The performance of an adhesive joint is usually evaluated in terms of a breaking load of the joint under different test conditions. Therefore, the fracture criterion of the adhesive joint plays an important role in predicting when the bond failure will occur. In order to predict the bond strength quantitatively, the fracture criterion must be combined with the stress distribution in the adhesive layer. It has been reported by Gent (1) that the observed bond strength of the joint subjected to tensile, pure shear and peeling separations is consistent with an energy criterion for adhesive failure and that a fracture mechanism based on Griffith's energy criterion is very significant. In order to explain the mechanical behaviors of the adhesive joint, a criterion for the interfacial failure has been introduced by Hata (2): namely, the bond failure will occur at the interface when the elastic work of deformation of the adhesive reaches a critical value.

In the case of the internal fracture, we assumed the triaxial stress state in the adhesive layer and derived a theoretical formula for the tensile bond strength of a butt joint (3). The formula was given by combining the stress distribution in the adhesive layer under large plastic deformation with the fracture criterion of Mises or Tresca, which is the hypothesis of the maximum shearing strain energy or the maximum shearing stress. This formula includes only measurable parameters and is expressed by the following form,

$$(\sigma_b/\sigma_t) = 1 + \frac{1}{2} (a_o/h_b)^2 \, \epsilon_{zb} \qquad\qquad (1)$$

$$h_b = h_o (1 + \epsilon_{xb})$$

where σ_b and σ_t are the tensile bond strength and the yield strength of the adhesive material, which was obtained from uniaxial tension of dumbbell, ϵ_{xb} and ϵ_{zb} show the longitudinal tensile and the lateral contractive strains at breaking the joint, a_o/h_o is the aspect ratio of the adhesive layer, and h_o means the adhesive thickness. In the case of aluminum-ethylene vinylacetate co-polymers(EVAs)-aluminum butt joint, the following empirical relation was obtained (Fig. 1).

$$(\sigma_b/\sigma_t) = 0.35 + 1.8 (a_o/h_b)^2 \, \epsilon_{zb} \qquad\qquad (2)$$

It is found from the agreement between both Eqs.(1) and (2) that the hypothesis of Mises or Tresca is applicable to the criterion of the internal fracture of the butt joint. The Mises and Tresca conditions describe quite well the mechanical behavior of a ductile material and are adopted frequently as the criterion of a ductile fracture.

The present investigation has been undertaken to verify the criterion of the interfacial fracture and the relationship between the tensile bond strength and the mechanical properties of the adhesive polymer by the use of steel-nylon 12-steel cross lap joint of the interfacial bond failure. It is our opinion for metal - polymer adhering system that the bond strength is determined by the mechanical and viscoelastic properties of the adhesive polymer, by the stress state developed in the adhesive layer and by the local stress which depends on the geometry of the adhesive layer. In connection with them, one of the essential features on the mechanical behaviors of the adhesive joint is the location of the site where the bond failure is most likely to be initiated. Since Bikerman (4) has pointed out that a true adhesional (or interfacial) failure is improbable, the problems on the locus of failure and on the weak boundary layer of the adhesive joint have been argued from various viewpoints by many investigators (2,5,6,7). In the case of the tensile adhesive joint such as butt or cross lap joint, a stress concentration will be developed in the adhesive because the defor-

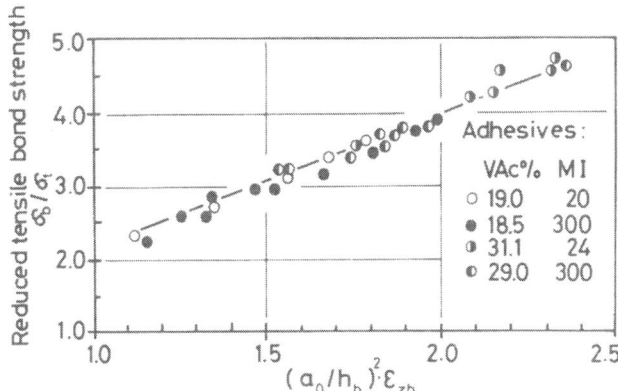

Adhesive: ethylene vinylacetate copolymers (EVAs)
VAc: vinylacetate content in wt %
MI : melt index number

Fig. 1. The plot of (σ_b/σ_t) against $(a_o/h_b)^2$.
ε_{zb} for the internal fracture of the
butt joint, according to Eq. (1).

mation of the thinner adhesive layer in lateral (shear) direction is
restrained by the high modulus or rigidity of metal adherend. That
is, this restraint exerted by the high modulus material on the lower
one is expected to be highly localized at the interface and the
failure always occurs within the weaker phase. Therefore, we
consider that the existence of the interfacial fracture of the
adhesive joint may be valid and even in such case, the principle of
analytical mechanics can be applied. It is more practical, in this
sense, to define the interfacial fracture in terms of the state in
which no significant amount of residual adhesive remains on an
adherend surface after breaking the joint by the use of a simple
method. We confirm such a state by the use of microscopic methods
such as microscopes and X-ray microanalyzer.

However, it is very difficult to set up and solve the equation
describing the cross lap joint by the use of analytical mechanics,
because of a bend of the adherend in response to the applied tensile
loading. In this study, therefore, the criterion of the interfacial
fracture will be evaluated from experimental observations of the
ultimate tensile properties of the cross lap joint.

A crystalline polymer is essentially a composite material made
of two phases, that is, one crystalline and the other amorphous, and
its mechanical properties depend on the degree of crystallinity,
which changes with its thermal history. When a crystalline polymer
is used as an adhesive, there is a large merit that the mechanical

properties of the adhesive can be varied by thermal treatments without any change in wetting behavior at making up the adhesive joint by hot-melt bonding.

EXPERIMENTAL

Nylon 12 powder of XF 5000 supplied from Toray Co. LTD. was used as the adhesive. Before making up the adhesive bond, the powder was dried well under high vacuum at room temperature. The adherend of mild steel with specification JIS SPCC was machined into 25.2mm x 50.4mm x 5.0mm (bonded area = 1 in^2) or 10.0mm x 50.0mm x 5.0mm (bonded area = 1 cm^2). The adherend plates were polished by Emery paper, degreased in distilled toluene by the use of ultrasonic cleaner, and finally dried under vacuum at 100°C for 4 hrs.

The cross lap joint was formed at 200°C for 10 min. in hot press and allowed to cool or quenched by immersing it into dry ice - acetone mixture. The average cooling rates were about 20°C/min and 730°C/min, respectively. In order to change the degree of crystallinity of nylon 12 adhesive, the bonded specimen was annealed at 150°C in nitrogen atmosphere. The annealed specimen was aged at room temperature for a week in vacuum desiccator to relax the residual stress developed in the adhesive layer during cooling process (8). The tensile bond strength decreased to about one-half during 4 days and then held constant for over two weeks.

The tensile bond strength was measured by the use of Instron tension tester in accordance with ASTM D-1344-54T. A universal joint and some special attachments were used to make straight pull of the adhesive joint. The observed tensile bond strength was averaged over five specimens. The deviations from the average were within ±7%.

The dependence of the mechanical properties of bulk nylon 12, such as the yield strength, the yield strain, the Young's modulus and the tensile modulus (ASTM D-638-61T), on the degree of crystallinity was determined from uniaxial tension of nylon 12 dumbbells annealed. These quantities were also averaged.

The density of the adhesive or the dumbbell was measured at 25°C by the use of a density gradient column. The density was also averaged over five pieces for each specimen. The degree of crystallinity was calculated from d_a=0.99 g/cm^3 and d_c=1.10 g/cm^3 by the use of the following relation (9).

$$1/d = x/d_c + (1-x)/d_a$$

where d_a and d_c are the density of amorphous and crystalline regions, respectively, and x is the degree of crystallinity of the crystalline region having the density d.

Reflecting microscope and X-ray microanalyzer (CKα) were used to observe the locus of failure and the fracture surface of the boundary region of the adhesive.

RESULTS AND DISCUSSION

The surface states of the steel adherend were investigated by the use of reflecting microscope and X-ray microanalyzer to confirm the locus of failure. The results are shown in Fig. 2. We did not observe any change in surface appearance under the scope of reflecting microscope (top photograph). White spots, which were detected by CK α-ray and shown in photographs (B), (C) and (D) of Fig. 2, mean a region where carbon atoms are concentrated on the adherend surface. Photographs (B) and (C) show the surfaces before bonding and after breaking, respectively. The density of white spots is the same in both photographs (B) and (C). On the other hand, photograph (D) shows the feature of the adherend surface coated with very thin layer of organic compound, which would be identical with the state having a residual adhesive or a weak boundary layer on the surface. This surface was prepared by immersing a purified adherend into the 1 % toluene solution of polyethylene oxide oligomer having the molecular weight of 700 for 1 min. and then by drying as to keep it standing vertically in a vacuum desiccator. Evidently, the surface of (D) shows the higher density of CKα spots than others. These results indicate that there is no significant amount of a residual adhesive or a weak boundary layer on the adherend surface after breaking the joint. These photographs in Fig. 2 were chosen at random from a large number of similar montages on the scope of both analyses and certain features shown in chosen photographs are considered to represent whole features. In the case of steel-nylon 12-steel cross lap joint adopted in this study, we can conclude that the interfacial fracture occurred under all test conditions.

The temperature-time superposition and the failure envelope are important means for characterizing the ultimate properties of viscoelastic materials and will provide the criterion for predicting when fracture will occur in a specimen subjected to different test conditions. From these viewpoints, the dependence of tensile bond strength on both temperature and strain rate is examined by applying these methods to the interfacial fracture of the cross lap joint. In this study, the apparent strain rate, which was defined as a crosshead speed divided by the adhesive thickness, was employed to represent an average deformation rate of the adhesive because the true rate could not be measured due to the bend of the steel adherend under applied loading.

(A) The adherend surface after breaking
     observed by reflecting microscope.
(B) The adherend surface before bonding
     detected by CKα.
(C) The adherend surface after breaking
     detected by CKα.
(D) The adherend surface having thin layer
     of oligomer detected by CKα.

Fig. 2. The features of adherend surfaces charac-
         terized by the use of reflecting microscope
         (top) and X-ray microanalyzer (bottom).

     The tensile bond strength increases continuously with the
apparent strain rate up to a critical rate and then decreases
abruptly, as shown in Fig. 3.   An abrupt decrease in the bond
strength with the peel rate was reported for the peel separations by
Gent (10), Kaelble (7) and Fukuzawa (11).   In their results, the
change in the fracture mode from "cohesive" to "adhesive" at the
transition point was observed.   The critical strain rate is in the
range between 3×10^{4} and 1×10^{5} %/min.   The transition occurs at
higher rate at higher test temperature, as shown in Fig. 3.   In
accordance with our calculation on the basis of the rate process,
when the time to break is equal to the relaxation time of the
adhesive polymer, such transition will occur (12).   This critical

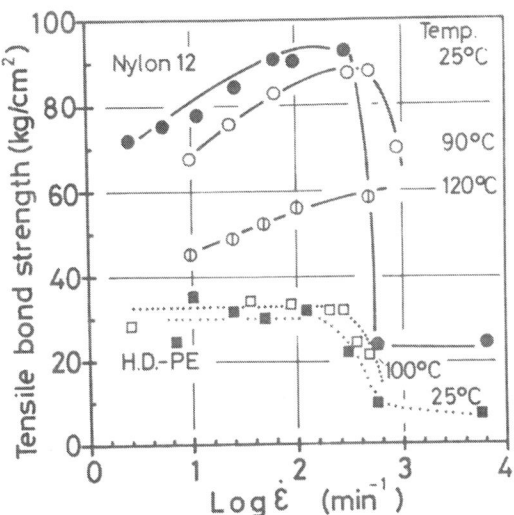

Adhesives: Nylon 12, 0.2 mm in thickness
            and high density polyethylene
            (H.D.-PE), 0.15 mm in thickness.

Fig. 3. The change in the tensile bond strength with
         the strain rate ($\dot{\varepsilon}$) and the test temperature
         in the case of the interfacial fracture of
         the cross lap joint. The bonded area is
         of 1 cm^2.

rate is identical with the "critical deformation velocity" specified in the region of impact test for viscoelastic materials ($10^4 \sim 10^5$ %/min).

The fracture surfaces of the boundary region of the adhesives are given in Fig. 4. It shows the change in the fracture pattern with the strain rate. The fractography obtained from the interfacial fracture indicates that the failure mode changes abruptly from the ductile to the brittle one at the critical rate because of the glassy manner of the adhesive deformation under higher rate than the critical velocity. The interfacial bond failure will be initiated by ductile or brittle rupture of the boundary region. We

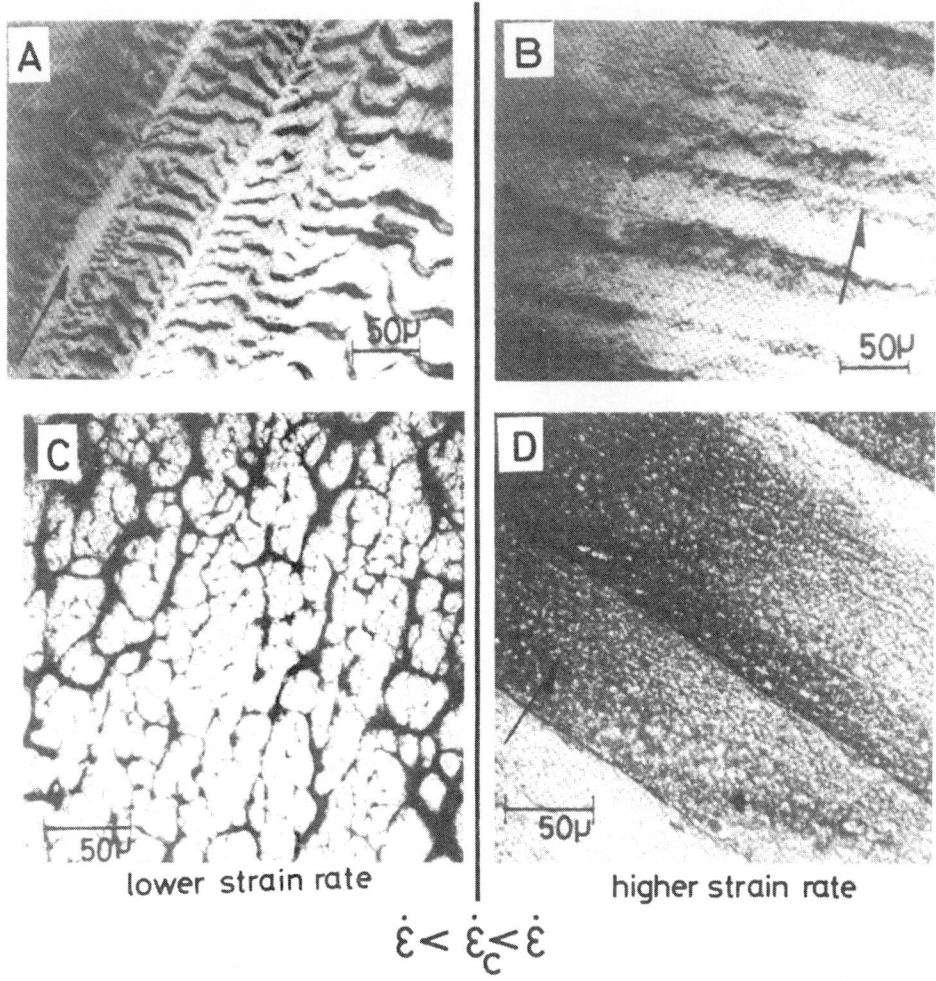

$$\dot{\varepsilon} < \dot{\varepsilon}_c < \dot{\varepsilon}$$

Adhesives: (A) and (B) are high density
            polyethylene.
          (C) and (D) are nylon 12.

Fig. 4. The change in the fracture surfaces with the
        strain rate. ε_c is the critical deformation
        velocity.

wish to emphasize on the basis of the fractographic observation for
the ductile range that the shear stress or strain acting at the
boundary region plays a very important role in the initiation of the
interfacial fracture and consequently, in the determination of the
tensile bond strength. Needless to say, such shear components at
the boundary region depend on the adhesive geometry and on the

mechanical properties of the adhesive polymer.

The reduced tensile bond strength, $\sigma_b \times T_o/T$, is plotted in Fig. 5 against the apparent breaking strain, ϵ_b, which was estimated by the conventional method reported previously (12), where σ_b is the tensile bond strength observed at each test temperature (T) and T_o means the reference temperature ($T_o = T_g + 50\ ^oK$). The failure envelopes obtained for the interfacial fracture indicate a common pattern which would be expected for a viscoelastic material (13).

The tensile bond strength, reduced to 363^oK, is plotted in Fig. 6 against the apparent strain rate ($\dot{\epsilon}$). The results obtained at different test conditions are superimposed satisfactorily to get a single master curve. The master relation for the interfacial bond failure exhibits the Arrhenius's type. The activation energy of 51 kcal/mol is calculated from the dependence of the shift factor, a_T, on 1/T, as shown in Fig. 6. This activation energy agrees well with that obtained from the rate-temperature dependence of ductile fracture for viscoelastic materials. However, the activation energy of 51 kcal/mol is much larger than 9 to 13 kcal/mol reported by Illers for a rheological rate process of nylon 12 (14).

The Ferry's shift factor, a_T, is given approximately by $a_T = \eta_T/\eta_{T_o}$. The shift factor is equal to the ratio of the viscosity (η) at temperatures of T and T_o. Even though the time-temperature dependence of the tensile bond strength would result from that of the viscosity, the activation energy will be expected to be 10 to 20 kcal/mol. The rate process on the fracture of the adhesive joint consists of many elemental steps and the rate-determining step is still uncertain. Consequently, it is very difficult to explain the activation energy of 51 kcal/mol in terms of the rheological models proposed.

The master relation and the failure envelope obtained for the interfacial fracture of the cross lap joint are quite similar to those obtained for a viscoelastic material. This similarity in the ultimate tensile properties indicates that these tensile properties of the adhesive joint are associated with the dynamic effect of a viscoelastic nature of the adhesive polymer rather than with a thermodynamic surface criterion, such as the work of adhesion, wettability and so on. A similar conclusion has been obtained for the internal bond failure, that is, Gent (15) has applied the WLF - superposition and a fracture mechanism in terms of Griffith's energy criterion to the cohesive strength. In accordance with Gent's findings, the observed bond strength reflects the visco-elastic responses of an adhesive polymer and the bond failure will occur when the energy stored elastically in the adhesive in the vicinity of the initial flaw and released by growth of the initial

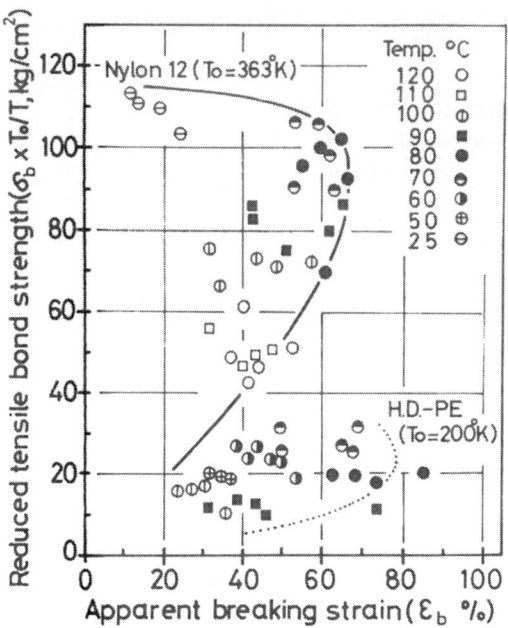

Fig. 5.   The failure envelopes of the interfacial
          fracture of the cross lap joint in the
          lower strain rate region.   The adhesive
          thickness is 0.2 mm for nylon 12 and 0.15
          mm for high density polyethylene (H.D.-PE),
          respectively.   The bonded area is of 1 cm^2.

flaw is sufficient to meet the energy requirement for growth. It is
suggested from the fractographs shown in Fig. 4 that the fracture
mechanism proceeded in the boundary region will be analogous to that
pointed out by Gent (15).   Phenomenologically, the problem of
characterizing the ultimate tensile properties for the "interfacial" fracture would be essentially similar to that for the
"internal" bond failure.

In the case of the interfacial fracture of steel-nylon 12-
steel cross lap joint, the dependence of the tensile bond strength
on the degree of crystallinity of the adhesive is examined in order
to predict the fracture criterion from the relationship between the
tensile bond strength and the mechanical properties of the adhe-

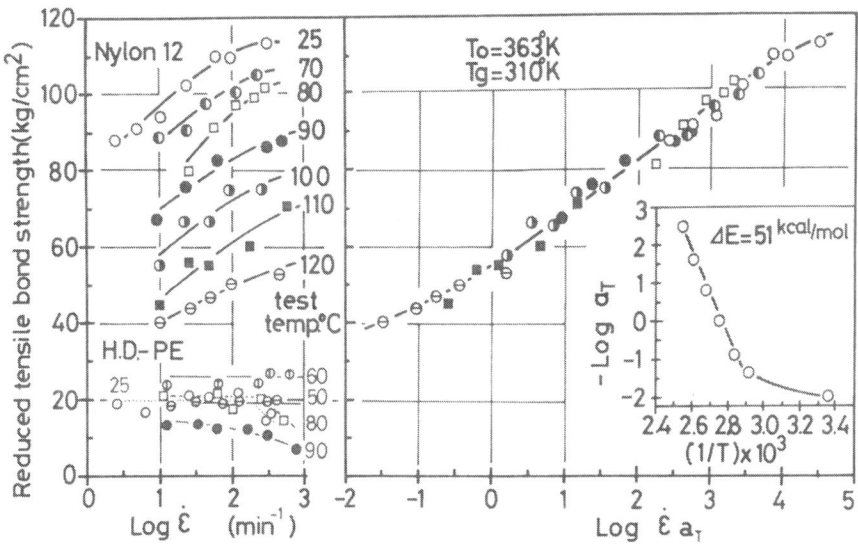

Fig. 6. The time-temperature superposition and
the dependence of the shift factor on the
temperature for the interfacial fracture
of the steel-nylon 12-steel cross lap joint.
The adhesive thickness is about 0.20 mm.
The bonded area is of 1 cm².

sive. An annealing of the bonded specimen resulted in the increase
in the density or the degree of crystallinity of the adhesive. The
results obtained are shown in Fig. 7. The density on the abscissa
in Fig. 7 shows the average value of the bulk adhesive. The degree
of crystallinity at the interfacial region is higher about 5 % or
more than that of the bulk nylon 12 adhesive, regardless of the
crystal morphology (16). The density of the interface at which the
bond failure occurred is expected to be higher than that shown in
Fig. 7. The cross section of the adhesive is given in Fig. 8. The
crystallites in the adhesive allowed to cool are larger than those
in the quenched adhesive. The transcrystalline region is not grown
at the interface in this case.

The observed tensile bond strength increases markedly with the
density but is independent of the crystallite size. For example,
the increase in the tensile bond strength from 50 to 200 kg/cm² can
be attributed to the increase in the degree of crystallinity from 25

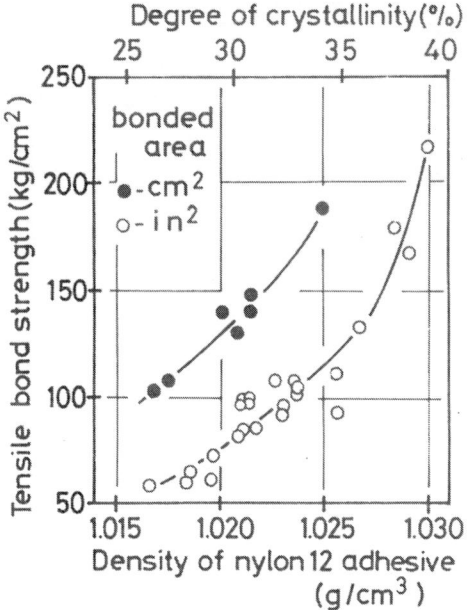

Fig. 7. The dependence of the tensile bond strength
         on the density or the degree of crystallinity
         of the adhesive for the interfacial fracture
         of the cross lap joint. The adhesive thickness
         is about 0.10 mm.

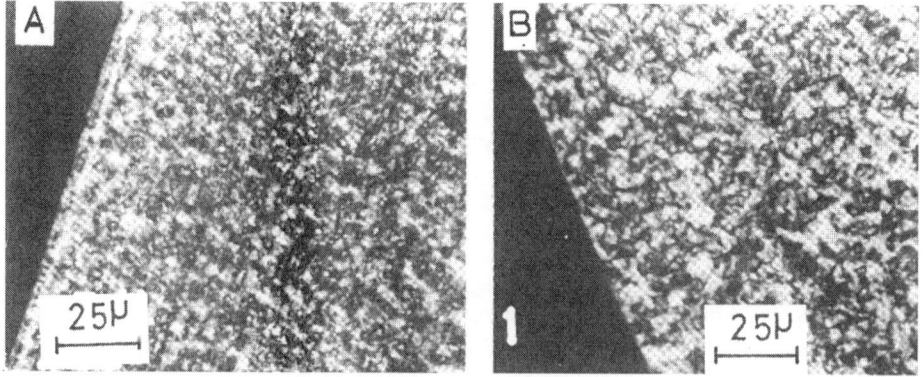

(A) The quenched adhesive.
(B) The adhesive allowed to cool.

Fig. 8. The cross sectional microphotographs of the
         nylon 12 adhesive molded against the steel
         adherend. The adhesive thickness is about
         0.2 mm.

to 40 %. Fig. 7 shows also the well-known fact that the larger is the bonded area, the lower is the bond strength per unit bonded area.   The dependence of the tensile bond strength (σ_b) on the density (d) is expressed by the following equations.

For the bonded area of 1 in^2;

$$\text{Log } \sigma_b = 45.82d - 44.86 \text{ (kg/cm}^2) \tag{1}$$

and for the bonded area of 1 cm^2;

$$\text{Log } \sigma_b = 32.02d - 30.54 \text{ (kg/cm}^2) \tag{4}$$

The higher degree of crystallinity gives the higher modulus of the adhesive. From the analysis of Gent (15) and Hata (2), the tensile bond strength will be expected to increase with the Young's modulus of an adhesive. On the other hand, a thermodynamic or surface chemical interaction force operative at the interface would be expected to be almost constant since an effective contact area at the interface will be altered even slightly by the increase in the density. It is very difficult to consider that the bond failure occurred at the weak boundary layer, which will grow by migrating the low molecular weight compound such as antioxidant to the interface during annealing. A strength of such weak boundary layer is independent of the degree of crystallinity of the adhesive. Therefore, the increase in the tensile bond strength with the density, as shown in Fig. 7, could not be observed, even if the bond failure would occur at the weak boundary layer.

When a crystalline polymer is crystallized in contact with metal surface, transcrystal sometimes grows at the interfacial region. The preferable effect of the transcrystalline region on the adhesive joint in relation to the weak boundary layer has been reported by Sharpe (5) and Schonhorn (17). On the other hand, it has been pointed out by the authors (16) that the crystal morphology of transcrystal itself can not contribute to the increase in the tensile bond strength. The transcrystalline region has a higher degree of crystallinity and higher modulus (18). The fact that the transcrystalline region gives the higher bond strength is due to its high degree of crystallinity rather than its crystal morphology.

Mechanical properties of nylon 12, such as the yield strength (σ_t), the yield strain (ε_t), the Young's modulus (E) and the tensile modulus (E_t), were measured.   The results are given as a linear function of the density or the degree of crystallinity, as shown in Figs. 9 and 10.   These quantities were obtained from uniaxial tension of nylon 12 dumbbells annealed under nitrogen atmosphere. An example of the stress-strain curve of the dumbbell and the definition of these quantities adopted in this study are shown in Fig. 11.

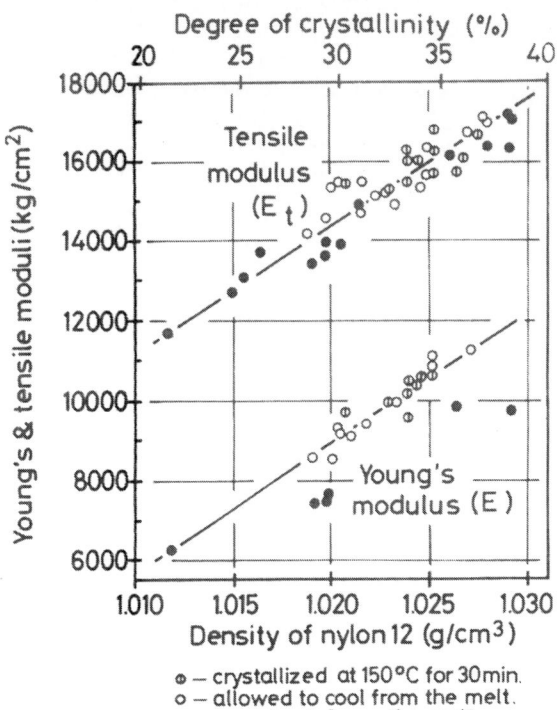

Fig. 9.   The dependence of the Young's and tensile
          moduli on the density or the degree of
          crystallinity. These quantities were
          obtained from uniaxial tension of nylon
          12 dumbbells annealed.

The Young's modulus was measured within elastic limit ($\varepsilon <$ 0.05%). The tensile modulus adopted is identical with that of ASTM D-638-61T. The similar relations for nylons have been reported by Starkweather (19).

The dependence of the tensile bond strength on the degree of crystallinity of the adhesive, as shown in Fig. 7, can not be explained by any single mechanical quantity of the bulk polymer. Therefore, two tensile properties were calculated by the use of the data shown in Figs. 9 and 10. One is the product of the tensile modulus and the yield strain, $E_t \cdot \varepsilon_t$, which means the tentative yield strength of bulk nylon 12 when the linear stress-strain relation is assumed. The other is the half of the product of the yield strength and the yield strain, $\frac{1}{2} \cdot \sigma_t \cdot \varepsilon_t$, which shows the work of deformation during yielding. The dependence of the values, both $E_t \cdot \varepsilon_t$ and

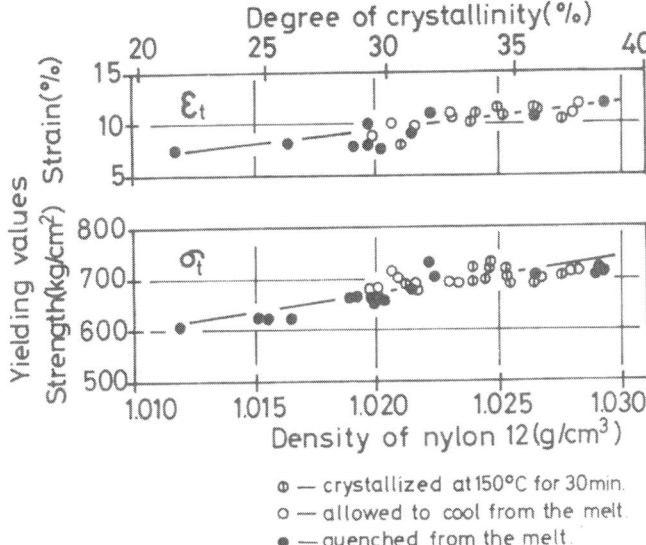

Fig. 10. The dependence of the yield strength and
          strain on the density or the degree of
          crystallinity. These quantities were ob-
          tained from uniaxial tension of nylon 12
          dumbbells annealed.

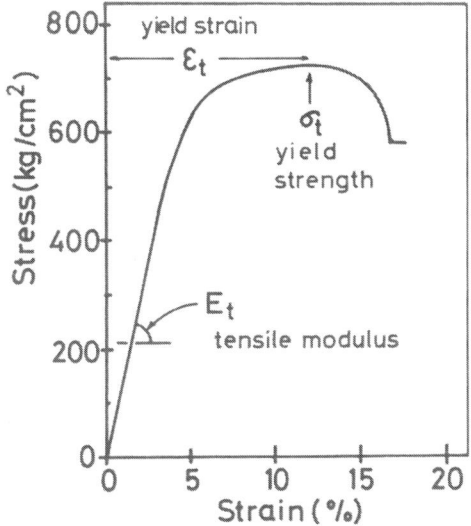

Fig. 11. An example of the conventional stress-strain
          curve of nylon 12 dumbbell and the definition
          of the mechanical properties adopted in this
          study.

$\frac{1}{2} \cdot \sigma_t \cdot \varepsilon_t$, on the degree of crystallinity or the density is given in Fig. 12. The density dependence of the former is similar to that of the tensile bond strength, as shown in Fig. 7. This similarity between Figs. 7 and 12 suggests that the tensile bond strength for the interfacial bond failure would be expected to reflect the yielding behavior of bulk polymer used as the adhesive. However, the tensile bond strength observed is very much lower than the tentative yield strength defined as $E_t \cdot \varepsilon_t$.

The cross lap joint has a large stress concentration, resulted from a bend of the adherend under applied tensile loading. From the other experiment in our laboratory, the stress concentration co-efficient of the cross lap joint adopted in this study was evaluated to be 6 to 10 times a butt joint. The coefficient of a butt joint is in the range of 1.0 to 1.2 as reported by Kobatake on the basis of a stress analysis (20). When the stress concentration is absent, the tensile bond strength will be expected to approach to the value of the product, $E_t \cdot \varepsilon_t$.

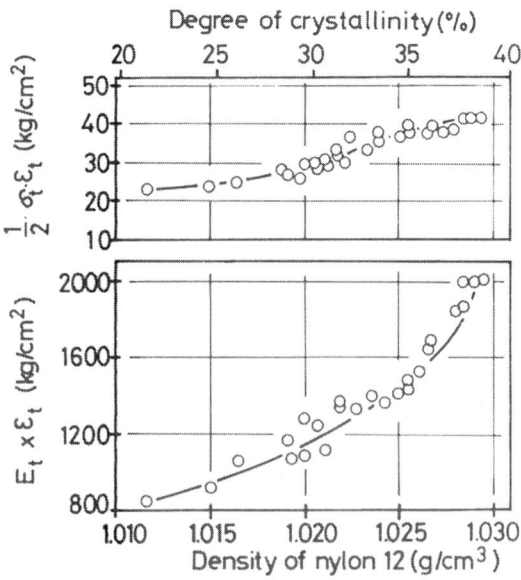

Fig. 12.   The dependence of the product of the tensile modulus and the yield strain (E_t x ε_t) and the half of the product of the yield strength and the yield strain ($\frac{1}{2}$ x σ_t x ε_t). These quan-tities were calculated by the use of the data shown in Figs. 9 and 10.

When the thin layer is bonded between rigid metal platens and tested under a uniaxial tension. Such specimen can not contract laterally and the triaxial stress state develops in the specimen. In accordance with our analysis (3,21), under triaxial stress state, the stress required to rupture the thin layer is always higher than the yield stress of the bulk polymer obtained from a uniaxial tension. When the interfacial bond failure occurs, the interfacial region of the adhesive would be subjected to the stress field equivalent to the yielding stress state under uniaxial tension. However, the observed bond strength will be lower than the yield strength resulting from a stress concentration.

In order to determine the fracture mechanism of the interfacial bond failure, the fracture surfaces were observed by the use of the reflecting microscope. The fractographic patterns are given in Fig. 13, which shows the change in the crack patterns with the density of the adhesive. The higher is the density, consequently, the higher is the tensile bond strength, the larger is the crack depth and width. These crack patterns of the boundary region is quite similar to the appearance observed on cup-bottom when the cup-and-cone fracture occurs in a specimen under tension. The fractography shown in Fig. 13 indicates that the interfacial fracture of the joint is attended by the ductile rupture of the interfacial region of the adhesive. In other words, the interfacial bond failure can be considered as such processes that the interfacial region of the adhesive is ruptured in a ductile manner and at the same time, the bond separation occurs at the interface. This feature on the interfacial bond failure is supported by the dependence of the ultimate properties of the joint on the time-temperature and on the degree of crystallinity. That is, this dependence results from the ductile deformation and fracture manner of the interfacial region of the adhesive.

In general, the ductile fracture will be caused by the shear component of a stress or strain energy induced in a specimen under tension. It suggests that the interfacial bond failure will occur when the shearing stress or the shearing strain energy operative at the interfacial region increases up to a critical value equivalent to the yielding point of the adhesive polymer under uniaxial tension. The mechanical behaviors of ductile materials are described quite well by the hypothesis of Tresca or Mises. They are employed frequently as the fracture criterion. In accordance with the Tresca condition, when the shear stress in a specimen reaches the maximum value equivalent to the yielding point of uniaxial tension and with Mises condition, when the shearing strain energy in a specimen reaches the maximum value equivalent to the yielding, the ductile fracture will occur, respectively. Both hyoptheses are valid criteria for the internal fracture of the butt joint as pointed out by the authors (3,21). We can also consider that the

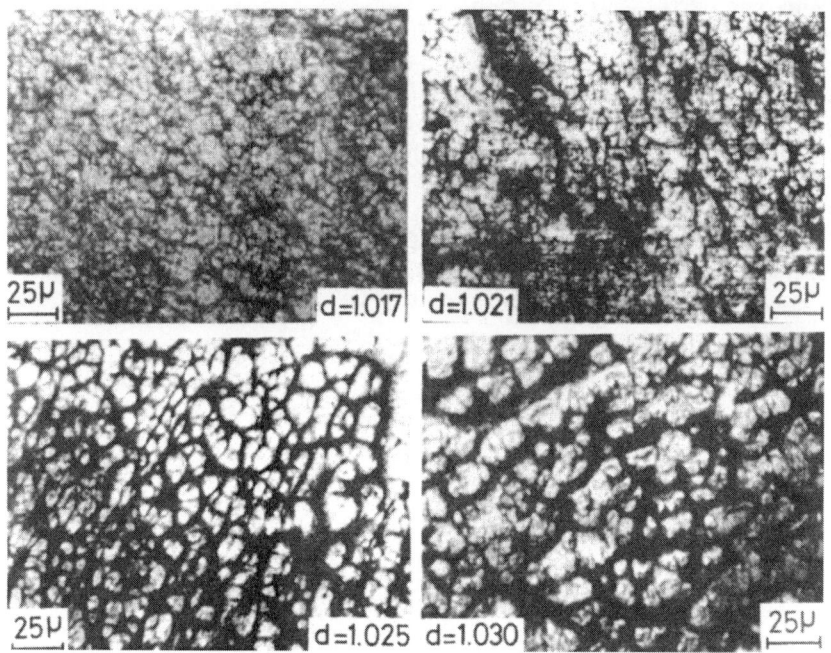

Fig. 13.   The change in the fractographic patterns
observed on an interfacial region of the
nylon 12 adhesive with the density (d).

Mises and Tresca hypotheses can be applied reasonably to the
interfacial fracture of the tensile adhesive joint as the fracture
criterion.

When a stress distribution in the adhesive could be obtained
for the cross lap joint under the condition of a bend of the
adherend, for example, by the use of the finite element method, an
analytical equation describing the interfacial region could be set
up and solved by combining the stress distribution with the hypotheses. It would be possible to establish the fracture criterion
for the interfacial bond failure. It has been pointed out by Gent
(1) that the bond failure, both for "cohesive" and "adhesive", is
governed by the energy criterion generalized from Griffith fracture
criterion. However, it is very difficult to determine which kind of
the hypotheses, either the shearing stress or the shearing strain
energy, will be predominant for the interfacial fracture of the
adhesive joint.

In the case of the internal fracture of aluminum-EVA-aluminum
butt joint having a rectangular bonded area (10mm x 32mm), an
interesting result was found as seen in Fig. 14, which shows the

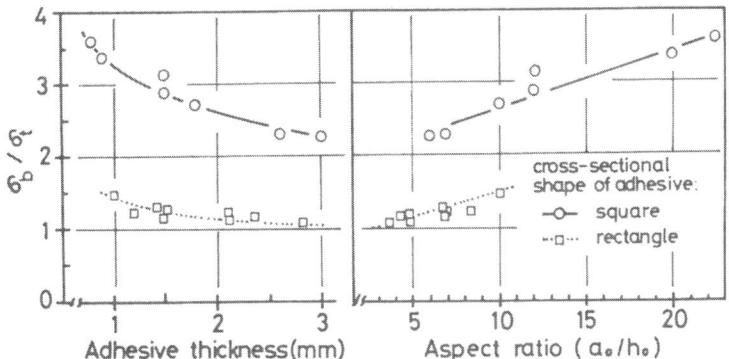

Adhesive: ethylene-vinylacetate copolymer (EVA)
            having the vinylacetate content of
            19 wt % and the melt index number of 20.

Fig. 14.   The lowering in the tensile bond strength
           with the anisotropic shape effect of a
           bonded area in the case of the internal
           fracture of butt joint.

dependence of the reduced tensile bond strength (σ_b/σ_t) on the
adhesive thickness or the aspect ratio of the adhesive, where σ_b and
σ_t mean the tensile bond strength and the yield strength of the EVA
obtained from uniaxial tension of dumbbell, respectively, and h_o is
the adhesive thickness. The observed tensile bond strength for the
rectangular joint decreases with the anisotropic shape effect to
about one-half that for the square joint even though both joints
have same bonded area. The effect of an anisotropic shape on the
tensile bond strength will be explained by the Mises hypothesis in
the same manner derived from Eq. (1) but not by the Tresca one. σ_x,
σ_y, and σ_z are the principal stresses. The x axis shows the tensile
direction. The y and z axis are the major and the minor one in
lateral direction, respectively. The Mises criterion is expressed
by the following equation.

$$(\sigma_x - \sigma_y)^2 + (\sigma_y - \sigma_z)^2 + (\sigma_z - \sigma_x)^2 = 2\sigma_t^2 \qquad (5)$$

Eq. (5) can be rewritten:

   For the square cross section (3);

$$\sigma_x - \sigma_y = \sigma_t \qquad (6)$$

and for the rectangular cross section;

$$\sigma_x - \sigma_z = (2/\sqrt{3})\sigma_t \qquad (7)$$

Eq. (6) shows the same form of the Tresca criterion. Eq. (7) is given by neglecting the lateral contract strain on the major axis. $(\sigma_b/\sigma_t)_{REC}$ and $(\sigma_b/\sigma_t)_{SQ}$ is the tensile bond strength of the rectangular and the square joint, respectively. The following relation is finally given by the use of the same manner derived from Eq. (1).

$$(\sigma_b/\sigma_t)_{REC} = (\sqrt{3}/2) \, (\sigma_b/\sigma_t)_{SQ} \tag{8}$$

That is, the tensile bond strength of the rectangular cross section lowers to about $\sqrt{3}/2$ that of the square one when both joints have the same bonded area. Eq. (8) agrees well with the result shown in Fig. 14. This anisotropic shape effect indicates that the Mises hypothesis is significant as the fracture criterion of the tensile adhesive joint. The analytical formulation of the anisotropic shape effect which has not been achieved in detail will be reported in a forthcoming paper.

CONCLUSIONS

     In the case of the interfacial fracture of steel-nylon 12-steel cross lap joint, the importance of the viscoelastic or mechanical responses of the adhesive polymer to the tensile bond strength is shown in this study. The temperature-time super-position, the failure envelope, the dependence of the tensile bond strength on the degree of crystallinity of the adhesive and the fractography of the interfacial region of the adhesive were inves-tigated to verify the fracture criterion of the tensile adhesive joint.

     It will be considered that the interfacial bond failure is identical with the ductile or brittle rupture of the interfacial region of the adhesive. Such rupture manner depends on the test conditions. Consequently, it is found to be reasonable that the observed ultimate tensile properties for the interfacial fracture of the adhesive joint are quite similar to those obtained from the tensile test of a ductile material. The activation energy of 51 kcal/mol was estimated by the Arrhenius plot of the shift factors. This value is much larger than that obtained from a rheological rate process of nylon 12, suggesting that a complicated fracture mechanism acts in the interfacial region.

     The dependence of the tensile bond strength on the degree of crystallinity is quite similar to that of the tentative yield strength of bulk nylon 12, defined by the product of the tensile modulus and the yield strain.

These facts observed in this study suggest that the Mises hypothesis is very significant for the interfacial and internal bond failure as the fracture criterion and is powerful especially to analyze the anisotropic shape effect on the tensile bond strength.

REFERENCES

1. A.N. Gent and A.J. Kinloch, "Adhesion of Viscoelastic Materials to Rigid Substrates. (III) Energy Criterion for Failure", J. Polymer Sci., A-2, 9, 659 (1971).
2. T. Hata, "Rheology of Adhesive Failure. (II) Rheological Consideration of the Transition between Interfacial and Cohesive Failure", J. Adhesion Soc., Japan, 8, 64 (1972) and "Mechanism of Adhesive Failure", J. Adhesion 4, 161 (1972).
3. M. Masuoka and K. Nakao, "Effect of Aspect Ratio on Tensile Bond Strength for Butt Joint of Internal Fracture" in "Adhesion Measurement of Thin Films, Thick Films and Bulk Coatings", ASTM STP-640, K.L. Mittal, ed., American Society for Testing and Materials, Philadelphia, (1978).
4. J.J. Bikerman, Chapter VI. "Final Strength of Adhints" in "The Science of Adhesive Joints", 2nd. Ed., Academic Press, New York, (1968).
5. L.H. Sharpe, "The Interphase in Adhesion", J. Adhesion 4, 51 (1972).
6. R.J. Good, "Theory of "Cohesive" vs "Adhesive" Separation in an Adhering System", J. Adhesion 4, 133 (1972) and "Locus of Failure and Its Implications for Adhesion Measurements" in "Adhesion Measurement of Thin Films, Thick Films and Bulk Coatings", ASTM STP-640, K.L. Mittal, ed., American Society for Testing and Materials, Philadelphia, (1978).
7. D.H. Kaelble, "Peel Adhesion: Influence of Surface Energies and Adhesive Rheology", J. Adhesion 1, 102 (1969).
8. M. Masuoka and K. Nakao, "Effect of Residual Stress on Tensile Bond Strength for Steel-Nylon 12-Steel Cross Lap Joint", J. Adhesion Soc., Japan, 12, 41 (1976).
9. A. Müller and R. Plüger, "Eigenschaften chemischer Aufbau und kristallinität von Polyamid-Kunststoffen", Kunststoffe 50, 203 (1960).
10. A.N. Gent and R.P. Petrich, "Adhesion of Viscoelastic Materials to Rigid Substrates", Proc. Roy. Soc., A, 310, 433(1969).
11. K. Fukuzawa, "Studies on Rheology of Pressure Sensitive Adhesive Tapes (V) Theoretical Equation for 180° Angle Peel Force at Constant Peel Rate", J. Adhesion Soc., Japan, 5, 301 (1969).

12. M. Masuoka and K. Nakao, "Temperature-Time Superposition and Failure Envelope for Interfacial Fracture of Cross Lap Joint", J. Adhesion Soc., Japan, 14, 341 (1979).

13. T.L. Smith, "Ultimate Tensile Properties of Elastomers (I) Characterization by a Time and Temperature Independent Failure Envelope", J. Polymer Sci., A, 1, 3597 (1963).

14. K.H. Illers, "Der Einfluss von Wasser auf die molekularen Beweglichkeiten von Polyamiden", Makromol. Chem., 38, 168 (1960).

15. A.N. Gent, "Adhesion of Viscoelastic Materials to Rigid Substrates, (II) Tensile Strength of Adhesive Joints", J. Polymer Sci., A-2, 9, 238 (1971).

16. K. Nakao, H. Endo, and M. Masuoka, "The Relation Between Tensile Bond Strength and Crystalline Properties of the Adhesive on the Steel-Nylon 12-Steel System", in "Mechanical Behavior of Materials", Proceedings of the 1971 International Conference on Mechanical Behavior of Materials, Vol. III, The Society of Materials Science, Kyoto, Japan, (1972).

17. H. Schonhorn and F.W. Ryan, "Effect of Morphology in the Surface Region of Polymers on Adhesion and Adhesive Joint Strength", J. Polymer Sci., A-2, 6, 231 (1968).

18. T.K. Kwei, H. Schonhorn, and H.L. Frisch, "Dynamic Mechanical Properties of the Transcrystalline Regions in Two Polymers", J. Appl. Phys., 38, 2512 (1967).

19. H.W. Starkweather, J.R., G.E. Moore, J.E. Hansen, T.M. Roder, and R.E. Brooks, "Effect of Crystallinity on the Properties of Nylons", J. Polymer Sci., 21, 189 (1956).

20. Y. Kobatake and Y. Inoue, "Theory of Adhesive Joint", Proceedings of the First Japan Congress on Testing Materials, 156 (1958).

21. M. Masuoka and K. Nakao, "Effect of Aspect Ratio on Tensile Bond Strength for Butt Joint of Internal Fracture", J. Adhesion Soc., Japan, 14, 125 (1978).

Deformation and Shear Strength of FRP Adhesive Joints

Hiroo Miyairi, Hideaki Fukuda and Atsuyoshi Muramatsu

Institute for Medical and Dental Engineering

Tokyo Medical and Dental Univ., 2-3-10, Surugadai,

Kanda, Chiyoda-Ku, Tokyo, Japan 101

ABSTRACT

The structural adhesives used in this study were a hard type polyester, a soft type polyester and an epoxy polyamide. The FRP adherend consisted of polyester matrix and reinforcement of glass mat/glass roving cloth combinations. The adhesive strength was obtained by using the tension shear test of single lap adhesive joints. The characteristics of adhesive strength were examined for the change in lap length, and the results were used to explain the deformation of adherend and the fracture pattern of layers. Furthermore, the peeling behavior of adhesive layers, the bending moment of FRP adherend and the influential factors on the adhesive strength of FRP adhesive joints were investigated.

INTRODUCTION

Recently, fiber reinforced plastics (FRP) are used for large-sized structural construction. Therefore, the adhesive construction is applied for the many purposes of practical use (1). FRP adhesive construction shows the complicated fracture pattern in comparison to other materials. In concerning with these problems, Hart-Smith (2) have reported the design matters of single, double and scarf joint, etc., for practical use. But in his paper, the adherend of lap joints was used for aircraft structural materials, boron/epoxy, glass/epoxy and aluminum, etc. And they were concerned with the adhesive layer to be separated.

However, the strength design for FRP adhesive construction has not been well established in engineering work. Under the present conditions the problem of adhesion is attracted by the attention to the development of new adhesives and their technology. For the isotropic or anisotropic adherend of single-lap joints, Renton and Vinson (3) studied the theoretical approaches to adhesive layers which are influenced by properties of adherend materials. But in their paper, the deformation of adherend parts and, for concrete example, FRP constitution of glass/polyester was not examined and discussed.

Therefore, aiming at the usefulness for the design of adhesive joints, we have investigated the deformation of the glass/polyester adherend used for single-lap joints and, in practice, used for long lap length of adhesive joints, and the displacement and stress distribution of a near-by adhesive single lap joint.

And so, this paper treated the single-lap joint as a basic adhesive joint. Consequently we are establishing the guides for the improvement in adhesive construction, and its technology and the adhesive strength for practical use, on the basis of the influences of their deformation and strength (4-6). The adhesives used in this paper are of three types: a hard type polyester, a soft type polyester and an epoxy-polyamide. A hard type polyester is the same as the matrix of FRP adherend. A soft type polyester is of the new type developed for FRP construction in this study. An epoxy-polyamide is widely used as the structural adhesive. FRP adherend is consisted of the combination of the glass mat and roving cloth reinforcements. The adhesive strength was obtained by using the tension-shear test. The characteristics of adhesive strength were examined for the change in lap length of joint, and the results were used to explain the deformation of adherend, the dimensional influence of lap length and the fracture pattern (7). Furthermore, the factors affecting the adhesive strength, such as the peeling from the single-lap ends, the bending moment at adherend, etc., and the stresses at the adhesive layer related to the deformation of adherend were investigated (8-10).

SPECIMEN

Constitution of FRP Adherend and Adhesives (11)

FRP adherend was composed of three layer reinforcements, such as glass-mat/glass-roving cloth/glass-mat, and polyester as matrix, and was made by hand layout method. The material description is shown in Table 1.

The adhesive, as shown in Table 2, is a hard and soft type polyester and epoxy-polyamide.

Table 1. Constitution of Glass Fiber Reinforced Plastics

Glass fiber reinforced materials	Matrix	Glass content (wt %)
Mat+Roving cloth+Mat (MC600 +WR570 +MC600)	Polyester resin (Ester R110A)	39.0

Table 2. Kinds of Adhesive and Treatment of these Specimens

Adhesives	Treatment
Hard type polyester (Ester R110A)	#450 chop mat 1/2 mat in Cured at room temperature
Soft type polyester (Ester AD5000)	Cured at room temperature
Epoxy-polyamide (Cemedine 1500)	Cured at room temperature

A hard type polyester is widely used as the so-called primary adhesive to FRP constructions. A soft type polyester was developed for the use of FRP adhesive structural construction as the secondary adhesive. An epoxy-polyamide is generally used as the so-called secondary adhesive.

Dimension

The lap length ℓ of the single-lap adhesive joint is 12.5, 25.0, 50.0, 100.0 and 200.0 mm, respectively. Here the lap length of 12.5 mm is based upon JIS (12) and ASTM standards; the width of adhesive joint b is all 15 mm.

The deformation of adhesive joint was measured with strain gages (Shinko Co., B-FAE-8-12T23) placed on both surfaces of adherend. The strain gage location is shown in Fig. 1. The gage length of all specimens is 280mm between both chucks, and both grip ends of specimen were fitted by the supporting plate of 25 mm to longitudinal direction because of the working of the pure shear stress at joint.

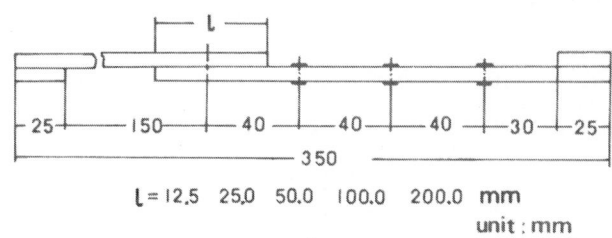

L= 12.5 25.0 50.0 100.0 200.0 mm

unit : mm

Strain gage position (Gage length L=280 mm)

Fig. 1. Dimension and shape of adhesive joint.

RESULTS AND DISCUSSION

Mean Shear Stress-Elongation Diagram

Figure 2 shows the mean shear stress-elongation diagrams (τ-λ diagrams) for the change in adhesive and lap length, when the crosshead speed of the tension-shear test was 1.5 mm/min.

The mean shear strength for each adhesive was different, and also the fractural elongation λ_B for each adhesive was different. Therefore, the value of λ_B of A-adhesive was relatively small with increasing lap length, while those of B and C-adhesives were relatively large. The apparent rigidity of joint is not explained from the value of $\Delta\tau/\Delta\lambda$ because τ decreases as lap length increases as shown in Fig. 2. But considering that the gage length L is a constant value of 280 mm the large value of $\Delta\tau/\Delta\lambda$, within the limits of small lap length, shows the elongation of ℓ to occupy a large percentage in comparison with the elongation of adherend, of total elongation of adhesive joints. And without regard to ℓ, the elongation at fracture of adhesive joints, λ_B, decreased from epoxy-polyamide, to soft and hard polyester adhesives in good order. Especially in the hard polyester the elongation at fracture λ_B was small in spite of the increase of lap length, and its mode of fracture was brittle differing from the modes of others.

Figure 3 shows the mean shear strength τ_B or P_B/b being the fracture load per unit width versus lap length ℓ.

Generally, τ_B decreases with increasing ℓ and in contrast to this, P_B/b increases, but in the case of GRP adherend, the increase of P_B/b is small in comparison with other materials. It is considered that the fracture of joint occurred at the mat/roving cloth interface of adherend because of the stress concentration at joint ends. For example, especially, the specimen with C-adhesive showed such a fracture. These results show that the fracture of

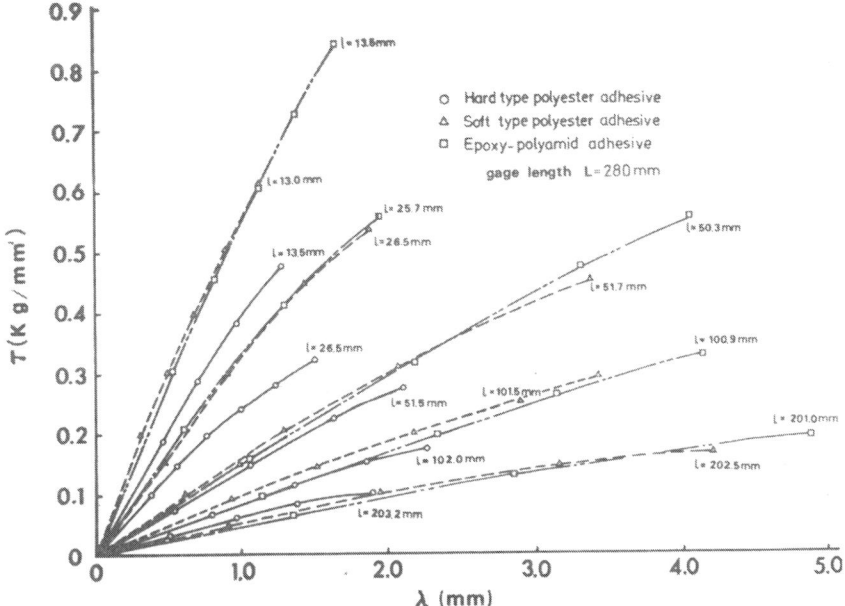

Fig. 2.   Shear stress-deflection diagram of some
          adhesive joints.

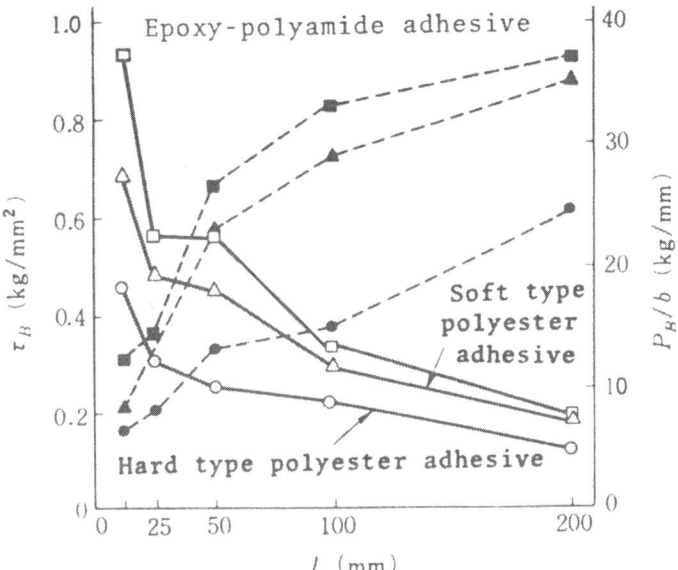

Fig. 3. τ_B, P_B/b - ℓ diagram of some adhesive joints.
          (τ_B: O△□, P_B/b: ●▲■)

joint is not caused only by the characteristics of adhesive but also by the characteristics of FRP adherend.

Classification of Fractured Bonding Surface and Fracture Pattern of Adhesive Joint

Figure 4 shows the five fracture modes of adhesive joint for change in adhesive and lap length. By using five symbols such as (a), (b), (c), (d) and (e) showing in Fig. 4, the fracture mode of each specimen was classified as shown in Table 3, where the numerical number before each symbol is the number of fractured specimen, and symbols a, b, c, ... are of fracture mdoe showing in Fig. 4.

As a result, in the cases of the joint with A-adhesive of low strength or the joint of short lap length, these fracture modes were displayed by the symbols (a), (b) and (c), in the consideration of the fracture all over the adhesive layer. Such a result is explained by the following matters. The shear stress of the joint with short lap length tends to distribute uniformly and the fracture load of the joint with A-adhesive was relatively small. Therefore, it is considered that, as these joints failed slowly through peeling from the ends of lap length, the fracture of FRP laminates was not observed. In the case of fracture mode (d) which the joint with C-adhesive showed, the fracture occurred at the glass-mat/roving cloth interface, for which the serious problem is the interfacial strength of adherend rather than the characteristics of adhesive.

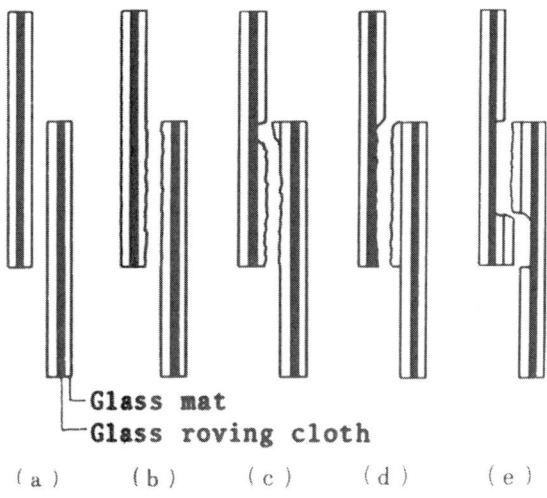

Glass mat
Glass roving cloth

(a)   (b)   (c)   (d)   (e)

Fig. 4. Fracture pattern of adhesive joints.

Table 3. Fracture Mode of Adhesive Joint

L / adhesive	12.5	25.0	50.0	100.0	200.0
A	5a	4a,c	5b	4b,c	5b
B	5a	5b	4b,c	4b,c	b,4c
C	b,4c	5b	4d,e	4d,e	4d,e

Therefore, to use the expensive, high strength adhesive such as an epoxy-polyamide to GRP adherend is not as efficient as other adhesives in improving lap lengths.

Adhesive Joint Rigidity and Apparent Elastic Modulus of Adherend

The adhesive joint part has the discontinuous stress distribution and the variation of rigidity in comparison with the flat adherend part. Therefore, in practical adhesive construction, there are many cases of the joint fracture resulting in the discontinuity of the joint deformation.

This tendency is especially seen on the flexible structure like GRP, and then the influence of lap length is an important problem to be associated with the joint rigidity. Here, this problem was investigated by analyzing the apparent elastic modulus of adherend, E'. Firstly, we divided the elongation of specimen, λ, into two parts: one with lap length ℓ and the other part L-ℓ (where L: gage length of 280 mm) when ignoring the shear deformation of adhesive layer. The equation is as follows:

$$\lambda = \frac{\tau \ell}{2E't} + \frac{\tau \ell}{E't} (L-\ell) = \frac{\tau \ell}{2E't} (2L-\ell)$$

$$\tau = \frac{2E't}{(2L-\ell)} \lambda \qquad\qquad\qquad (1)$$

Where τ: shear stress
       t: thickness of adherend
Hence E' is obtained by

$$E' = \frac{(2L-\ell)\ell}{2t} \frac{\Delta\tau}{\Delta\lambda} \qquad\qquad (2)$$

In Eq. (2) the shear deformation of adhesive layer and the flexural deformation of adherend are discounted; the influence of adhesive and the flexural load are examined.

Figure 5 shows E' versus lap length for some adhesives. In the case of A-adhesive, E' increases with increasing lap length, but in the case of B and C-adhesives it decreases.

So, comparing with the elastic modulus of GRP adherend (E=1120 Kg/mm^2), the values of E' obtained were small, especially in the case of A-adhesive, E' was influenced by lap length. Therefore, in the case of A-adhesive, the influence of shear deformation and flexural load gradually decreased as the lap length increased. And the decrease in the apparent elasticity E', with an increasing lap length, was accompanied by the decrease in rigidity of joints because of the joint's deformation. Therefore the increase of E' indicates that rigidity of joints according to large lap length starts to increase and it is expected that these influences give rise to the increase in the flexural rigidity.

Similarly, the apparent tension rigidity η of the adhesive joint is given as follows:

$$\eta = \frac{\Delta P/b}{\Delta \lambda} = \frac{\Delta P}{b \Delta \lambda} \qquad (3)$$

where $\Delta P/\Delta \lambda$ is the initial slope in P/b - λ diagram as shown in Fig. 6. Comparing the initial slope in P/b - λ diagram with that in $\tau - \lambda$ diagram, the value of $\Delta P/b \Delta \lambda$ is a small change with the variation in lap length ℓ. And the value of $\Delta P/b\Delta \lambda$ increases as the value of ℓ increases.

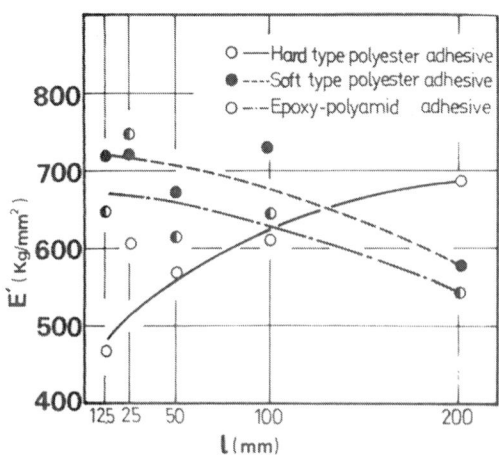

Fig. 5.   E' - ℓ diagram of some adhesive joints.

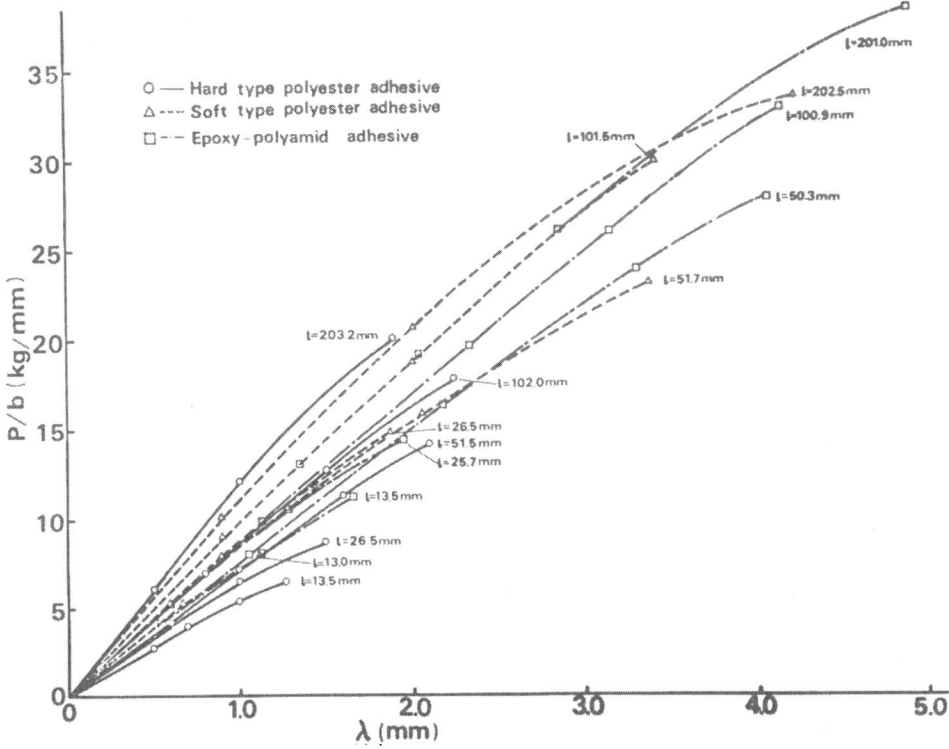

Fig. 6. P/b - λ diagram of some adhesive joints.

By using Eq. (3), the relation between η and ℓ is shown in Fig. 7. It is proposed that η is generally expressed with the increasing function for ℓ, but those behaviors were changed by adhesive, as shown in Fig. 7. The η of A-adhesive increased in comparison with other adhesives which showed respectively a similar pattern. In the case of B and C-adhesives, the increase in apparent tension rigidity was small. It is considered that the discontinuity of rigidity was weakened by the shear deformation of lap length. Therefore, the rigidity of adhesive joint depends on the characteristics of adhesive, it is the same effect influenced especially by FRP large structural construction.

Bending Moment Distribution of FRP Adherend

The test of single lap joint gives erroneous results to the occurrence of flexural load at adhesive position and adherend. In this paper, the flexural influences were investigated by measuring the strain at two sides of the adhesive joint. The different strain occurred at two sides of adherend results in the combination of

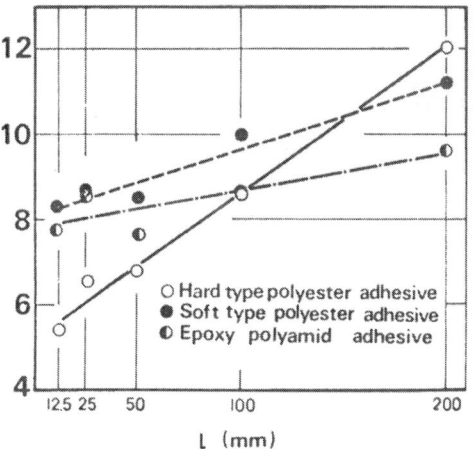

$$L \ (mm)$$

Fig. 7. η - ℓ diagram of some adhesive joints.

tensile and flexural loads. Therefore, the standing ε_t and ε_m for tensile and flexural strains, or the strains at two sides of adherend, respectively ε_1 and ε_2, are as follows:

$$\varepsilon_1 = \varepsilon_t + \varepsilon_m$$

$$\varepsilon_2 = \varepsilon_t - \varepsilon_m$$

$$(4)$$

And dividing the axial load acting on the adherend into tensile and flexural loads, F and M, these equations are given by using ε_t, ε_m and the elastic modulus of adherend E .

$$F = E \, \varepsilon_t \, b \, t$$

$$M = 2 \int_0^{1/2} \{ \frac{2E \, \varepsilon_m \, b}{t} \} \, y^2 dy = \{ \frac{E \, \varepsilon_m \, b \, t^2}{6} \} \qquad (5)$$

Here y is the distance from the neutral axis of the adherend. Finally, after eliminating ε_t and ε_m in Eq. (5) by using Eq. (4), Eq. (5) becomes:

$$F = \frac{E \, b \, t}{2} \, (\varepsilon_1 + \varepsilon_2)$$

$$(6)$$

$$M = \frac{E \, b \, t}{12} \, (\varepsilon_1 - \varepsilon_2)$$

Fig. 8 shows the normal and flexural forces, F and M, which were obtained from Eq. (6), when treating the case of a soft type polyester adhesive.

The bending moment near the joint increased with increasing load, but decreased as the debonding at adhesive joint progressed. And the maximum value of the bending moment increased with increasing lap length. The bending moment near the supporting point showed the same sign as that near adhesive joint in case of early loading and was relatively small, but its bending moment reversed the sign regardless of the lap length as load increased. Bending moment near the clumping parts tended to decrease along the adherend in comparison with that near the adhesive joint. Bending moment near the center of adherend tended to be influenced by lap length, and at its location the normal load increased. Thus, the bending moment near the supporting point and the adhesive joint changed its sign; therefore, this tendency was shown to be especially at the center of adherend.

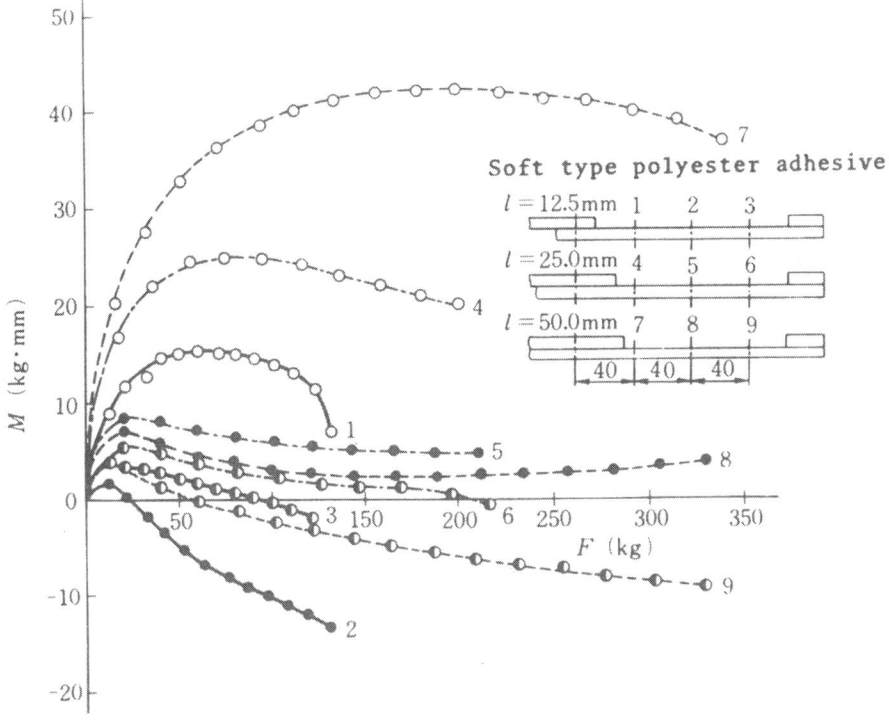

Fig. 8. M - F diagram of soft type polyester adhesive joint.

Figure 9 shows the bending moment and normal force along the longitudinal direction of joint. The normal force of the adherend increased with increasing lap length, but in the case of short lap length the curve tends to differ from others. And the normal force F occurred at adherend is smaller than the bending moment Pt (where t is thickness of adherend) as happened from the gap of both supports, and these values did not change with the location of adherend except on ℓ=12.5 mm.

Figure 10 shows the bending moment of adherend which changed with increasing load. Horizontal axis X, which started from the center of lap length, denotes the distance along the longitudinal direction of specimen. In the case of ℓ=12.5 mm, the bending moment reversed the sign in the region of X=40~80 mm and approached to zero near the supports. In the case of ℓ=50 mm, the bending moment reversed the sign in the region of X=80~120 mm; therefore in the case of single lap joints, the bending moment at adherend near adhesion with constant gage length increased as lap length increased and the sign of bending moment was reversed near the supports, and then the single lap joint was deformed to the shape of the capital letters. But in the case of ℓ=12.5 mm according to JIS and ASTM standards, the bending moment was reversed in the region of X=40~80 mm and X=120 mm, and the deformation pattern became very complicated in this case.

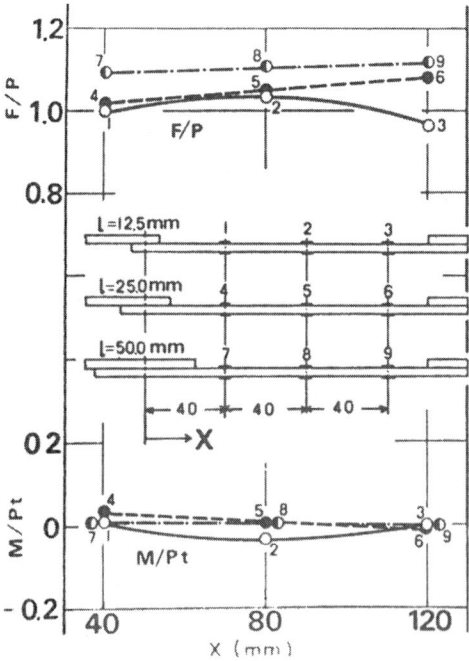

Fig. 9.   M/Pt - X Diagram of soft type polyester
          adhesive joint.

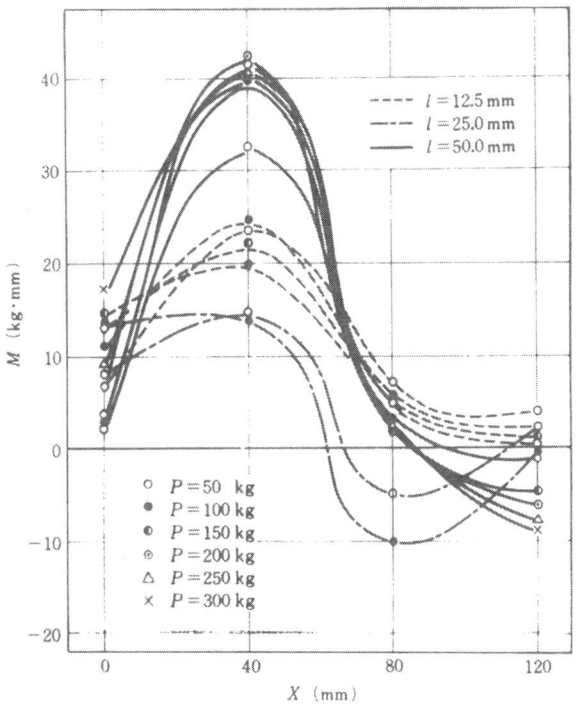

Fig. 10. M - X diagram of soft type polyester
adhesive joint.

Influence of the Scarf Angle of Adhesive Joint

In the single lap joint the stress concentration of adhesive
would be reduced to some degree by providing a scarf angle to the
joint. Therefore, in this study we examined the mechanical effects
of the epoxy-polyamide adhesive joint with a scarf angle $\theta = 35^{\circ}$.

Figure 11 shows the apparent shear stress - elongation diagram
(τ-X diagram) with the change in lap length.

The rigidity of the scarf joint is low in comparison with that
of the foregoing single lap joint, and particularly in the small
short lap length the fracture elongation λ_B became scarf joint. And
the scarf joint was known to be that the deformation of its adherend
was very much followed by the applied load to fracture.

Figure 12 shows the apparent shear strength τ_B or the adhesive
strength by unit width, P_B/b, in the change in lap length, and also
the results of a single lap joint. However, the difference of both
strength properties was not recognized.

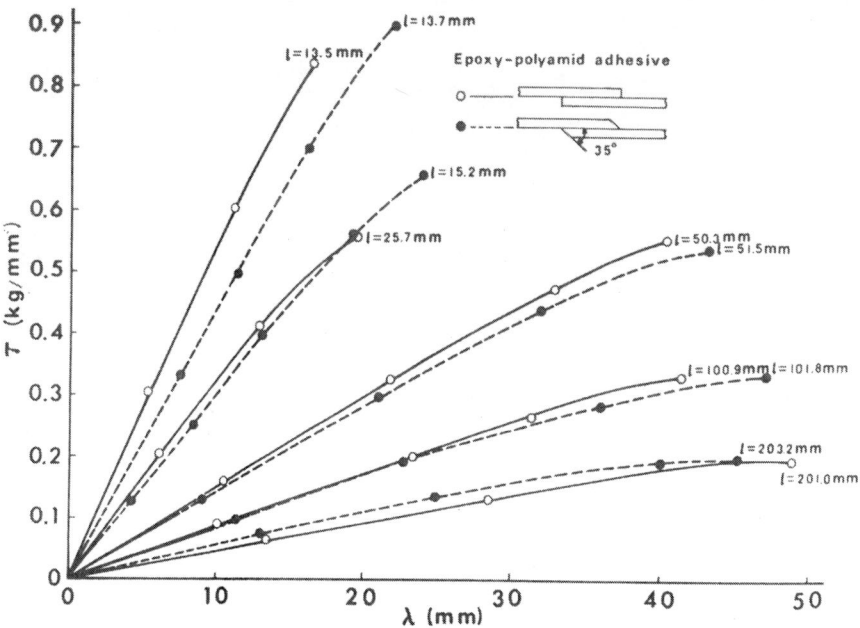

Fig. 11.   τ - λ diagram of epoxy polyamide adhe-
            sive joint.

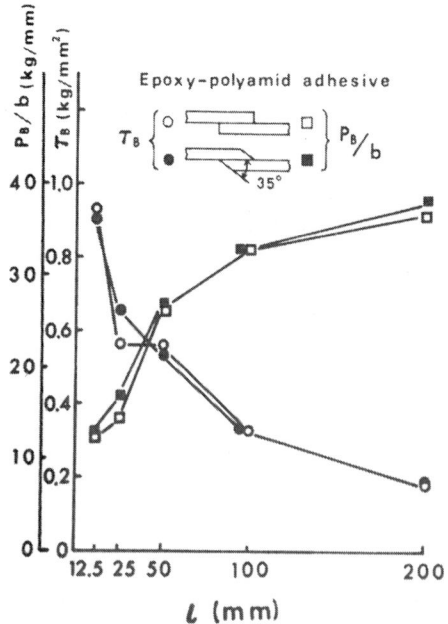

Fig. 12. $\tau_B, P_B/b$ - ℓ diagram of epoxy polyamide
            adhesive joint.

Figure 13 shows the distribution of bending moment of the scarf joint adherend in much the same way as the case of single lap joint. The distribution of the bending moment of both joints showed nearly a similar tendency, but these values of scarf joints in the center of lap length are larger than that of foregoing single lap joint. These data show that the decrease of rigidity of joint and of the flexural load apportionment at the center of adhesive layer were resulted by providing a scarf angle of joint. And in the large lap length (ℓ = 50 mm), the decrease of maximum bending moment at about X = 40 mm was not shown; therefore, the scarf angle provided to joint ends is not effective for the decrease of bending moment. But the slight increase of fracture strength is caused by the average of the distribution of bending moment.

Figure 14 shows the relation between the bending moment M and normal force F with an increasing load in the change of lap length.

In the case of the scarf joint dissimilar to foregoing single lap joint, its bending moment M does not become M<0 except the positions of #5 and #8. Therefore, it is characterized that the

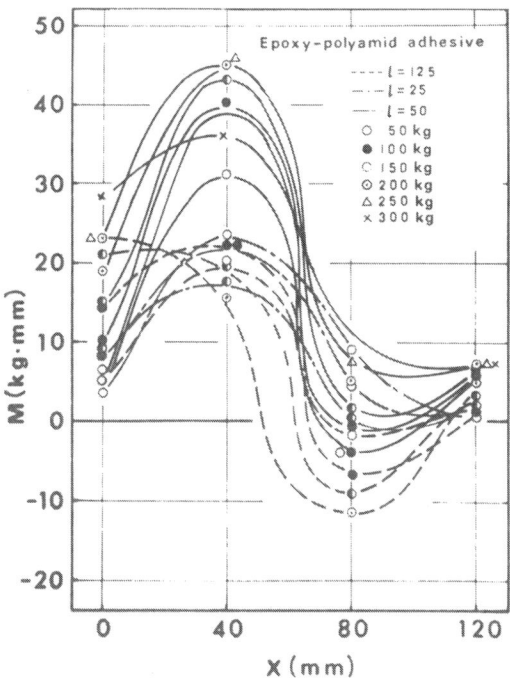

Fig. 13. M - X diagram of epoxy polyamide adhesive joint.

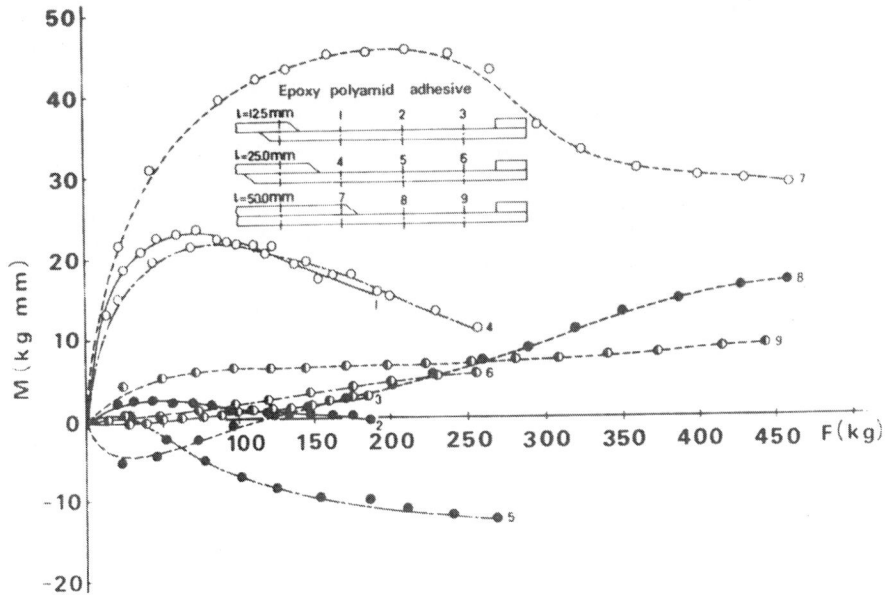

Fig. 14. M - F diagram of epoxy polyamide adhe-
          sive joint.

bending moment at the adherend of adhesive joints with each lap
length becomes M>0.

Fracture Pattern and Lap Length of Adhesive Joint

        Fracture pattern of adhesive joint is shown by the group of
Fig. 4 and the results of Table 3. And the state of fracture is
expressed by the initiation of the end of lap length to be caused by
stress concentration in spite of the increase of lap length. The
fracture mode (d) which epoxy-polyamide adhesive joint showed is
found as the mat/roving interfacial debonding, then the roving
layer is exposed at the fracture surface. But in the case of the
soft type polyester similar to the properties of the former adhesive, its fracture occurred at the adhesive interface and its
debonding surface did not progress to roving layer.

        Therefore, it is considered that the soft type polyester
exhibited maximum adhesive properties for the glass fiber reinforced plastics (GRP) adherend. In the scarf joint of epoxy-
polyamide adhesive, the apparent shear strength τ_B showed slightly
larger value than that of usual adhesive, but both fracture patterns
were not different. This result was caused by the large scarf angle
θ = 35°. Therefore, the fracture of adhesive joint started from the

lap length ends as the debonding and then progressed to the center of adhesive layer as load increased.

These states of fracture can be prevented with the decrease of stress concentration and bending moment of the adherend, but in this experiment these effectiveness were not revealed. So the scarf angle θ takes to be small and then the mechanical effects of adhesive joint will be improved. And if the good epoxy-polyamide adhesive was used for GRP adherend, we could not expect more than the adhesive properties because of the debonding fracture of adherend. And so it is important to examine the decrease of the stress concentration of the adhesive layer and the prevention of interface fracture of GRP adherend.

Therefore, it is considered that the possibility of the prevention of the joint fracture is to examine these fracture patterns at the lap length ends. And in the adhesive design of FRP it is pointed out that we must consider the adhesive work to be a part of engineering, including the decrease of stress concentration of the adhesive layer, and not only the improvement of the material for structural adhesive.

CONCLUSIONS

On the adhesive strength of glass-fiber reinforced plastics (GRP), the characteristics of tension-shear strength of single lap joint widely used for the secondary adhesive structure were examined practically in the consideration of lap length of the region in 12.5~200.0 mm. The results are summarized as follows:

(1) In the single lap joint of GRP adherend, the adhesive strength P_B/b (fracture load per unit width) did not increase in proportion to lap length, and even in the case of an epoxy-polyamide adhesive its strength was about 35.5 Kg/mm. Such results were caused by the fracture of the mat/roving cloth interface of GRP adherend. Therefore, even if a soft type polyester adhesive is used we will be able to show the same adhesive strength as in the case of an epoxy polyamide adhesive.

(2) Furthermore, such fracture is discussed reasonably in term of the stress concentration which increasingly occurs at the end of adhesive layer as the length increases, and the stress concentration is influenced by the bending moment occurred at the adherend. Therefore, in the case of such an adhesive joint as fractured at the mat/roving cloth interface of FRP adherend, the examination according to the remaking of joint in consideration of the relaxation of stress concentration and the adhesive technology is required for the adhesive structure.

(3) The apparent elastic modulus E' of GRP adherend decreased because of the bending moment and shear strain of the adhesive layer, when changed with lap length and adhesive. In the case of a hard type polyester, the value of E' increased as lap length increased, but in the cases of a soft type polyester and an epoxy-polyamide adhesive, it decreased. Consequently, the value of E' was in the region of about 1/2-2/3 of the actual elastic modulus for adherend.

(4) The reduction of stress concentration of adhesive layer ends with the scarf angle shows a little increased properties in the case of scarf angle $\theta=35^{\circ}$.

These situations are found in the fracture pattern and the deformation of adherend.   Therefore making the scarf angle to decrease, we would expect the improvement in the adhesive shear strength.

(5) Fracture modes of GRP joints were classified into five types depending on the adhesives and lap length. In the case of adhesive joint of short lap length showing low strength, the fracture occurred at adhesive layer, and in the case of adhesive joint of long lap length showing high strength it occurred at the interface of GRP adherend.

(6) In the case of adhesive joint according to tension-shear test with constant gage length, the bending moment occurred at the adherend depended on the lap length; therefore in the case of lap length 12.5 mm, it reversed the sign near the adhesive and supporting positions, but in the cases of lap lengths 25.0 mm and 50.0 mm it showed a maximum value near the adhesive position and decreased until reaching a minimum value near the supporting position.

REFERENCES

1. T. Hayashi, Fukugo Zairyo Kogaku (in Japanese), 840 (1971).
2. L.J. Hart-Smith, NASA Contractor Report, NASA CR-2218 (1974).
3. W.J. Renton and J.R. Vinson, J. Adhesion 7, 175 (1975).
4. C. Raphael, Appl. Polym. Sym. 3, 99 (1966).
5. H. Miyairi, Nippon Setchaku Kyokai-Shi (in Japanese) 7, 4, 217 (1971).
6. Fukugo Zairyo Gijutsu Shusei Henshu Iinkai, Fukugo Zairyo Gijutsu Shusei (in Japanese), 690 (1976).
7. FRP Adhesion Committee, Reinforced Plastics (in Japanese) 14, 6, 313 (1968).
8. L.M. Keer, J. Appl. Mech. 9, 697 (1974).
9. R.D. Adams and N.A. Peppiatt, J. Strain Anal. 9, 3, 185 (1974).
10. F. Erdogan and M. Ratwant, J. Comp. Mater. 5, 378 (1971).

11.   H. Miyairi, H. Fukuda and A. Muramatsu, _Zairyo_ (in Japanese)
         <u>26</u>, 282, 649 (1977).
12.   Japanese Industrial Standards K 6850 (1972).

Stress Distribution and Strength in Shear on the Adhesive Lap Joint Loading Bending Moment

Yukisaburo Yamaguchi and Susumu Amano

Kogakuin University

Nishishinjuku 1-24-2, Tokyo, Japan 160

ABSTRACT

The stress distribution and the maximum stress in shear at the circular area of adhesive lap joint were discussed theoretically and experimentally in this article. The joint carries the bending moment due to external force P which is parallel to lap surface and at a distance ℓ from the center of adhesive area. The following equation with respect to the maximum shear stress τ'_{max} was derived and verified experimentally.

$$\tau'_{max} = \frac{P}{\pi R^2} \left(\frac{2\ell}{R} \alpha + \beta\right)$$

where R is radius of adhesive circular area; α and β are stress concentration factors and are 1.2 and 2, respectively.

INTRODUCTION

There are two cases in which the bending moment acts on the adhesive lap joint; in one case, tensile and compressive stresses arise mainly from carrying the bending force perpendicular to the adhesive surface (1,2,3), and in the other, shear stress arises mainly from carrying the bending force parallel to the adhesive surface (4). The stress distribution and the maximum shear stress or the shear strength of adhesive lap joint in the latter case (Fig. 1) are treated theoretically and experimentally in this article.

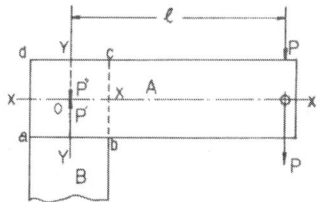

Fig. 1.   Adhesive Lap Joint Loading Bending
Moment $P\ell$

INDUCTION OF THEORETICAL FORMULA

Preliminary Experiment

In Fig. 1, the bending force P that is at a distance ℓ from the center 0 of the adhesive surface area and parallel to adhesive surface was applied to a lap joint with color paste between two transparent plates, and then it was found that the locus of relative movement of every adhesive point was circular around the center 0 of the adhesive area as shown in Fig. 2. Accordingly, we may assume that the relative movement or shear strain in the adhesive layer is directly proportional to the distance from the center 0, provided that the adherends are much higher in elastic modulus than that of adhesive. According to the total mechanical equilibrium system in Fig. 1, bending moment $P\ell$ should be balanced to both actions a couple P-P' and a compressive force P" to the center 0. Therefore, the balance actions to bending moment $P\ell$ in the adhesive lap joint may be divided into two reactions: the total shearing couples around the center 0 in the adhesive area and the shear stresses opposite to P".

Theoretical Formula

We assume that the adhesive area of lap joint is circular with radius R and similar to that of quadrangular abcd to make analysis simple as shown in Fig. 3, though the adhesive surface area is generally quadrangular. We may obtain the following relation from the above assumptions and the preliminary experiment:

$$\tau = \frac{r}{R} \tau_o \qquad\qquad\qquad (1)$$

where τ is shear adhesive stress at any point being r distant from 0; R is the outside radius; τ_o is the shear adhesive stress at the outside of the circular adhesive area.

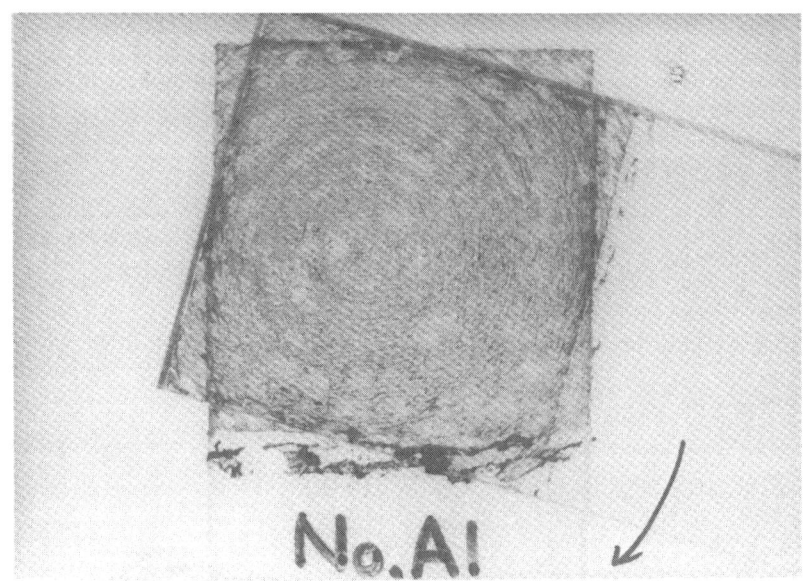

Fig. 2.   Photograph Showing the Locus of Movement
          of Adhesive Joints.

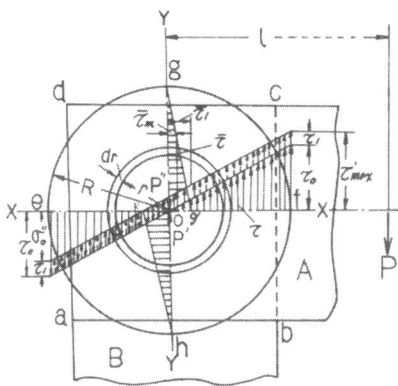

Fig. 3.   Stress Analyzing Diagram in Adhesive Lap
          Joint Due to Bending Moment.

The elementary moment dM of the circular elementary area dA being at a distance r from the center 0 and dr in width to be balanced to the corresponding couple $P\ell$ is presented in the following equation

$$dM = \tau \cdot dA \cdot r$$

$$= \tau \cdot 2\pi r \cdot dr \cdot r$$

$$= 2\pi r^2 \cdot \tau \cdot dr \tag{2}$$

From Eqs. (1) and (2)

$$dM = 2\pi r^2 \frac{r}{R} \tau_0 \cdot dr \tag{3}$$

$$M = P\ell = \int_{r=0}^{r=R} dM$$

$$= \int_{r=0}^{r=R} 2\pi r^3 \frac{\tau_0}{R} dr$$

$$= \frac{\pi}{2} R^3 \tau_0 \tag{4}$$

or

$$\tau_0 = \frac{2 \, P\ell}{\pi \, R^3} \tag{5}$$

The distribution of shear stress τ and the maximum shear stress τ_0 due to the coupling action are also shown by the diagram in Fig. 3.

The distribution of the shear adhesive stress $\bar{\tau}$ on the line gh due to compressive force P'' parallel to yy as an example is shown by the triangle gg'o and ohi in Fig. 3. Its maximum value $\bar{\tau}$, is as follows

$$\bar{\tau}_1 = \frac{P''}{A} \beta \tag{6}$$

where A is the adhesive area, πR^2, and β is the stress concentration factor concerning $\bar{\tau}_1$. The mean adhesive stress $\bar{\tau}_m$ due to P'' is to be P''/A, so that the value of β being the ratio of $\bar{\tau}_1$ to $\bar{\tau}_m$ should be 2, if the adhesive shear stress due to P'' $\bar{\tau}$ is directly proportional to the distance from the edge of the circle to the loading line XX as shown in Fig. 3.

Accordingly, the total shear stress τ' at any point on the adhesive layer must be the vector sum (↦) of vector τ and $\bar{\tau}$, so that the maximum total shear stress τ'_{max} may be obtained from Eqs. (5) and (6) as follows:

$$\tau'_{max} = \tau_0 \mapsto \bar{\tau}_1$$

$$= \frac{2\ P\ell}{\pi\ R^3} \mapsto \frac{P}{\pi R^2}\ \beta \tag{7}$$

It is clear in Fig. 3 that τ'_{max} arises at the point on the line XX as follows:

$$\tau'_{max} = \frac{2\ P\ell}{\pi\ R^3} + \frac{P}{\pi R^2}\ \beta$$

$$= \frac{P}{\pi R^2}\ [\frac{2\ell}{R}\ \alpha + \beta\ ] \tag{8}$$

where α is the stress concentration factor concerning τ_0 and theoretically equal to about 1.

EXPERIMENTAL DISCUSSIONS

On The Stress Distribution

The distribution of the total adhesive shear stress τ', $\tau + \bar{\tau}$, arose by the bending moment on the adhesive area, may be discussed experimentally by the test using five strain gauges, 1-5, on the line XX of the steel adherend surface, as an example, under various bending forces P as shown in Fig. 4. Compressive strain ε' and tensile strain ε'' at every strain gauge under various bending forces P are shown in Fig. 5. It is clear in Fig. 5 that compressive or tensile strain on the surface is directly proportional to the bending force and to the distance from the center 0. And it is also evident that the strain ε' at the right side from the center 0 is larger than ε'' of the left side, because the direction for the right side is similar to that of τ; however that for the left side is opposite to it. As the shear strain of adhesive layer may be directly proportional to the compressive or tensile strain of adherend, it may be recognized that the distribution of shear stress shown in Fig. 5 corresponds to that in Fig. 3 or Eq. (7).

Total Maximum Shear Stress and the Shear Strength of the Joint

The shear strength of the double lap joint bonded with an epoxy-poly-amide adhesive (BOND E and 0.05 mm in thickness) between

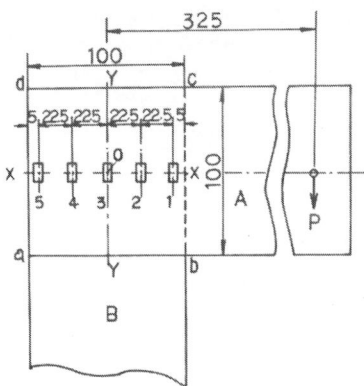

Fig. 4. Positions of Strain Gauges in Stress
         Distribution Measurement.

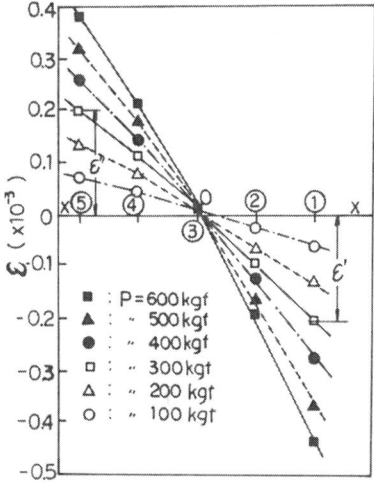

Fig. 5. Compressive and Tensile Strain on Adher-
         end Surface at the Position of Every
         Strain Gauge.

steel plates is about equal to the shear strength of the adhesive
itself τ_a. The morphology of fracture surface of the adhesive
bonding shows the cohesive break at the adhesive layer (5). And
then, if the value of τ'_{max} calculated from Eq. (8) in which the
values of P, ℓ, R, α and β were applied for the condition by which
the cohesive break of the adhesive bond due to bending moment Pℓ
happened is equal to the strength τ_a of the adhesive itself,
Equation (8) may be said to be reasonable.

Figure 6 shows the testing apparatus by which the actual shear strength of the adhesive lap joint was compared with the value of τ'_{max} in Eq. (8).

Figure 7 shows the values of τ'_{max} calculated from Eq. (8) applying the actual bending moment $P_{max}\ell_1$ for each five test pieces. In this experiment, R was 20 mm; ℓ_1 325 or 225 mm; P_{max} the maximum bending force to break the adhesive joint; α 1 or 1.2 and β 2. In Fig. 7, A and A' show the values in the cases of 325 mm of ℓ_1 to be 1 or 1.2 of α, and B and B' show those 225 mm of ℓ_1 to be 1 or 1.2 of α, respectively. Line C shows the shear strength τ_a of the adhesive itslef which was measured by using the method of ASTM D732.

We may recognize that the maximum total shear stress τ'_{max} at the breaking condition or the shear strength of lap joint due to bending moment is about equal to the shear strength τ_a of the adhesive itself, and becomes more closely to τ_a in the case of α being 1.2 than that being 1.0, as shown in Fig. 7.

CONCLUSIONS

The interrelations among the maximum shear stress τ'_{max} arising in the adhesive lap joint, the bending moment $P\ell$ loading parallel to the adhesive surface, the radius R of the circular adhesive area and the stress concentration factors α and β were derived as follows,

$$\tau'_{max} = \frac{P}{\pi R^2} \left[\frac{2\ell}{R}\alpha + \beta\right] \tag{8}$$

It is verified from two kinds of experiments, the stress distribution measurement and the breaking test of the adhesive lap

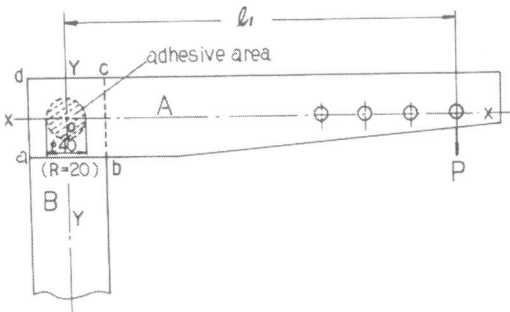

Fig. 6.   Breaking Test Apparatus of Adhesive Lap
          Joint.

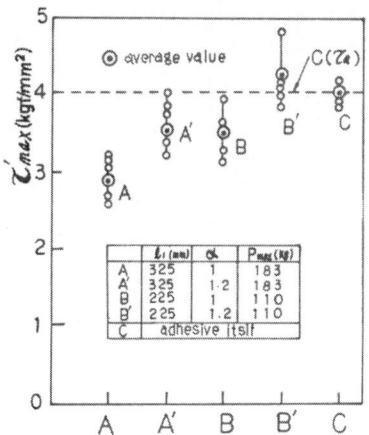

Fig. 7. Actual Shear Adhesive Strength Due to
         Bending Moment Under Various Conditions
         and Shear Strength of Adhesive Itself.

joint, that the distribution of shear stress at the adhesive layer
was similar to the assumption by which Eq. (8) was derived. The
actual shear strength of adhesive lap joint also corresponded
closely to the relation presented by Eq. (8).

REFERENCES

1.   MIL- A - 5090B, "Standard Temperature Bend".
2.   ASTM D 1184-69, "Strength of Adhesive on Flexural Loading".
3.   Y. Yamaguchi and S. Amano, Preprint of 14th Symposium on Adh.
         of Japan, "Testing Method of Tensile Adhesive Bonding
         Strength", 61 (1976).
4.   L.R. Lunsford, Applied Polymer Symposia No. 3, <u>Structural
         Adhesive Bonding</u>, "Stress Analysis of Bonded Joint",
         <u>Interscience</u> (1966).
5.   Y. Yamaguchi, S. Amano and K. Furukawa, Preprint of 15th
         Symposium on Adh. of Japan, "Testing method of the Adhe-
         sive Strength in Shear", 69 (1977).

Fracture Criteria on Peeling

Takashi Igarashi

Department of Technology, Gunma University

Kiryu, Japan

ABSTRACT

The mechanical conditions for the initiation and the propagation of fracture are considered in the case of trousers type of a specimen consisting of two flexible canvas bonded with rubbery adhesive. To reduce the difficulties in mathematical analysis a mechanical model for peeling is adopted. It was found that the condition for the initiation of peeling obeys the maximum stress criterion and that for the propagation of steady peeling obeys the minimum free energy criterion in thermodynamics. Here discussed are limits of validity of thermodynamics for the problem of fracture.

INTRODUCTION

When two canvas bonded with a rubbery adhesive are peeled off, trousers type peeling as shown in Fig. 1 occurs. This peeling has such an interesting character that there exists steady fracture: that is, the ultimate state of break is being realized continuously in this peeling. An external force required for its continuation remains constant in this steady peeling process. Since the fracture takes place continuously, measurement of the peeling force implies carrying out continuous and infinite measurement.

On the other hand, the initiation of peeling or tearing in a flawless sheet differs from the above mentioned steadily propagating fracture. Peeling or tearing starts immediately after the external force reaches some definite value. This feature, therefore, is thought to belong to an "ordinary" fracture like a tensile

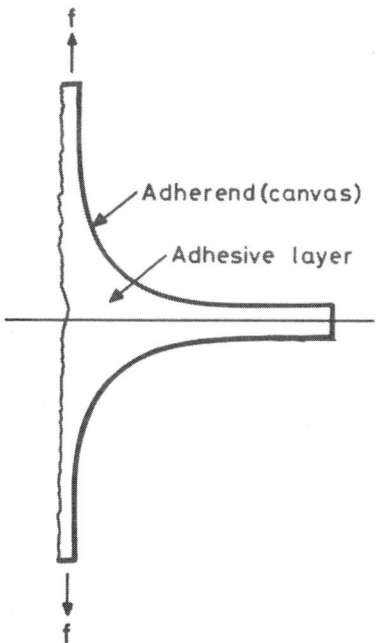

Fig. 1.   Trousers type peeling.

fracture of a wire. How does the "ordinary" fracture occur? When
tensile load is applied on a thin and long bar, tensile stress
occurs in it, and the breaking of the specimen takes place when the
stress reaches some definite value. It is well known that in some
materials yielding phenomenon is observed at a certain state of
stress before breaking of the material showing "necking" in its
appearance. The necking extends along its longitudinal direction
with the increase in elongation of the specimen, then the stress in
the necking part increases again to the definite level that leads to
breaking of the specimen. Here is a series of a process of
"ordinary" fracture without presenting steady feature from loading
to breaking. Then in the case of peeling or tearing, it may be
thought that the initiation and propagation of it belongs to
different kinds of fracture and that each criterion of the fracture
obeys different mechanical laws. It is a well known experimental
fact that the force required to start tearing or peeling is much
different from that for its propagation.

When we discuss the fracture phenomena theoretically with
regard to the above facts, it may be important to distinguish the
problem of the initiation of the fracture of the adhesive   that
completes after a series of successive stages from that on the
steadily continued fracture. As papers that discussed tearing or
peeling made no distinctions between the above two types of fracture

from this standpoint; some papers discussed either the problem of
the initiation of fracture or that of steady fracture and others
discussed both problems inconsistently (1).

A mechanical theory on the fracture of solids was proposed by
Griffith (2) in 1920, on which a number of studies have been
successfully developed. It is then interesting to consider what the
Griffith theory does or does not teach us about these problems of
two types of fracture. One of the purposes of this article is to
make the structure of the Griffith theory clear with applying it to
the fracture of rubbery materials, to show its limit of validity and
to find out the possibilities of its further development.

Griffith, who has studied effects of a surface finishing of a
metal on its mechanical strength, found that the decrease in
material strength caused from existence of cracks was much greater
than the value expected from the calculated stress value based on
the elastic theory. This meant that the generally accepted hypothesis at that time, i.e., fracture takes place when the maximum
stress in materials reaches a definite value, was not true.

Then, Griffith proposed a new version that the condition for
the fracture might be derived from minimizing the thermodynamic
potential (free energy) of the system. Thermodynamic potential of a
system consists of the free energy of the elastic deformation of a
solid caused by the external force and the surface free energy of
cracks existing in a solid. When work is done by the external force
on the solid, there occurs isothermal, reversible, elastic deformation of the solid storing the potential energy as free energy. In
the case that the external force exceeds a definite level, it leads
to lowering the increase in the total free energy to transform the
work done into surface free energy with opening the cracks rather
than to transform it into free energy of further deformation. This
condition is analogical to the gas-liquid equilibrium where evaporation leads to lowering the increase in the total free energy at the
boiling point.

Two facts can be pointed out concerning the Griffith theory.
First, the theory is for the case that there exist cracks in the
solid from the beginning; second, it is the principle of minimum
free energy, which is expressed

$$\delta F \equiv \delta F_e + \delta F_s = 0 \qquad\qquad (1)$$

where δF is the variation of the total free energy; δF_e is that of
the free energy of the deformation and δF_s is that of the surface
free energy.

Thus, Griffith discussed tensile strength of a plate which was in a two-dimensional plane stress state (Fig. 2). In this case, the variation of free energy of the elastic deformation for change in the crack length had been obtained from the stress analysis by Inglis (3) as

$$\delta F_e = -(2\pi\sigma^2 L\delta L)/E \tag{2}$$

where σ is tensile stress, E is Young's modulus. On the other hand, the corresponding variation of surface energy is given as

$$\delta F_s = 4\gamma\delta L \tag{3}$$

where γ is the surface free energy. Then, from Eqs. (1), (2) and (3), one obtains a well known result for fracture stress σ_b

$$\sigma_b = (2E\gamma/\pi L)^{\frac{1}{2}} \tag{4}$$

It is noted that the free energy F_e has to be given in analytic form. Significance should be emphasized on the fact that with introduction of thermodynamics into Inglis' stress analysis the condition of fracture was given not as a consequence of a hypothesis but as that of laws of nature; here exists the merit of Griffith theory. Thus, the ultimate property of fracture strength could be expressed in terms of physical quantities of the substance, i.e., so-called material constants, such as elastic moduli and surface energy. However, from the facts that the Griffith theory is the equilibrium theory of cracks and that Eq. (4) includes length of the

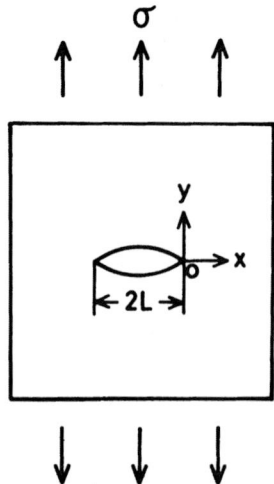

Fig. 2. Griffith crack.

crack L, the problem of the strength of materials had to deal with
the substantial problem of the crack. In addition, the problem of
estimation of the surface free energy of solids is of importance
both from the viewpoint of strength of materials and as a link
between molecular theory and the Griffith theory.

Griffith's theory, i.e., principle of minimum free energy, was
extended to the studies on materials with large deformation by
Rivlin et al. (4) who discussed tearing of rubbery materials. On
the other hand, Hata (5) in Japan made a study on peeling a few
years earlier than Rivlin et al. For the peeling of a flexible and
inextensible tape, he proposed a simple and elegant relation,

$$f = W/(1-\cos \theta) \qquad\qquad\qquad\qquad (5)$$

where f is the peeling force per unit width of the tape; W is the
work of adhesion, and θ is the peeling angle (Fig. 3). Though he
did not refer to the Griffith theory or thermodynamics, his study
included an idea of energy balance thought to be equivalent to
Griffith theory. Here, the work of adhesion W had the same
dimension (force/length = energy/area) as that of the surface
energy and evalulated to the order of 10^3 erg/cm^2 for vinylite(PE-
PVAc copolymer)-glass interface (6).

Since the Griffith idea came to be applied to the problems of
tearing or peeling, problems on the influence of a cut made in a
rubbery sheet on the steady fracture strength have been mainly
treated. The Griffith theory, however, should be of no validity in
discussing the problem of initiation of a crack because it is
essentially the equilibrium theory of a crack.

In this paper, first, the condition of initiation and the
following propagation of peeling are considered with the use of a

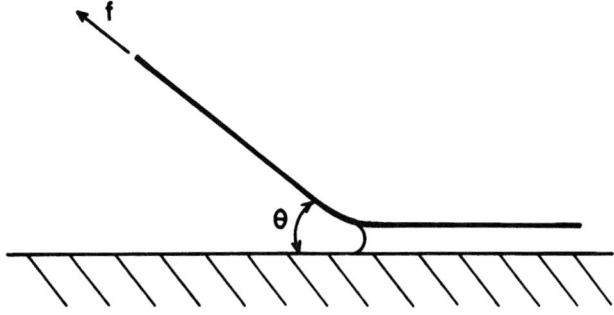

Fig. 3. Peeling of flexible tape from a rigid
         plane.

mechanical model to reduce the difficulties in the stress analysis. It is valuable not only for simplifying the analysis but for bringing the structure of the theory into relief.   Second, the influence of relaxation on the steady peel strength is discussed and the effect of energy dissipation in materials is made clear.

EQUILIBRIUM THEORY ON PEELING (7)

The specimen, for which the peel strength is to be calculated, is shown in Fig. 1.   The experimental peeling force obtained for such specimens is in the order of 10 kg/cm; therefore it can be assumed that the adherends of canvas are flexible but inextensible. It follows from this assumption that no work is done on the adherends by external forces during bending and also that no work is done when a tensile force is applied during peeling.   The rubbery adhesive is made from unfilled synthetic rubber and thus can be regarded as sufficiently elastic.   Therefore energy dissipation during deformation can be neglected.

Because of the above assumption a simple Hooke model for peeling, as shown in Fig. 4, can be adopted.   In order to simplify the calculations, it is assumed that symmetry exists with respect to the x axis and only the upper half needs to be considered.   In Fig. 4, the adhesive layer has unit width, and y_o is the half thickness of the adhesive.

Thus, one can obtain the following basic equations of mechanical equilibrium with respect to x and y directions respectively:

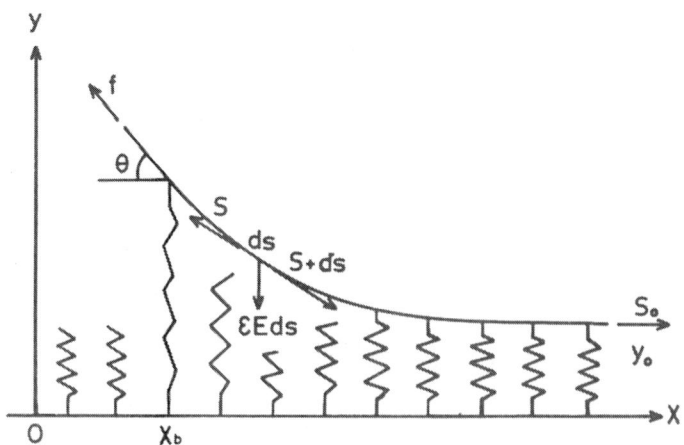

Fig. 4.   Mechanical model of peeling.

$$\frac{d}{dS}\left(S\frac{dx}{ds}\right) = 0 \tag{6}$$

$$\frac{d}{dS}\left(S\frac{dy}{ds}\right) - \varepsilon E = 0 \tag{7}$$

where ε is the strain of adhesive layer directed to the y axis and E is its Young's modulus. As the specimen has unit width, S is the tension in the adherend for unit width and s is the length along the adherend. From Eq. (2),

$$S\frac{dx}{ds} = \text{const.} \equiv S_o \tag{8}$$

It follows from Eq. (8) that the x component of the tensile force is constant and is equal to S_o everywhere along the adherend. Eqs. (6) and (7) can be rewritten as

$$S\frac{d^2x}{ds^2} + \frac{dSdx}{dsds} = 0 \tag{6'}$$

$$S\frac{d^2y}{ds^2} + \frac{dSdy}{dsds} - \varepsilon E = 0 \tag{7'}$$

Using the relation $(dx/ds)^2 + (dy/ds)^2 = 1$, these equations can be combined to obtain

$$\frac{dS}{ds} - \varepsilon E \frac{dy}{ds} = 0 \tag{9}$$

Since $\varepsilon = (y-y_o)/y_o$ and $S=S_o$ at $y=y_o$, integration of Eq. (9) leads to

$$S = Ey_o\varepsilon^2/2 + S_o \tag{10}$$

From Eqs. (3) and (6),

$$\left(\frac{Ey_o}{2}\varepsilon^2 + S_o\right)\frac{dx}{ds} = S_o \tag{11}$$

Using the relations

$$\frac{dx}{ds} = \frac{dy/dx}{[1+(dx/dy)^2]^{1/2}}$$

and $dy=y_o d\varepsilon$, Eq. (11) leads to

$$\frac{dx}{d\epsilon} = \mp \frac{2S_o}{E} \left[\epsilon(\epsilon^2 + \frac{4S_o}{y_o E})^{1/2} \right]^{-1} \tag{12}$$

in which the minus sign is adopted so that for $x \to 0$, ϵ goes toward $+\infty$, and thus the origin of the x axis corresponds to infinite values. Integration of Eq. (12) gives

$$\epsilon = 4(\frac{S_o}{y_o E})^{1/2} \left\{ \frac{\exp[-(E/y_o S_o)^{1/2}x]}{1 - \exp[-2(E/y_o S_o)^{1/2}x]} \right\} \tag{13}$$

Since $\epsilon = (y/y_o)-1$, the shape of the adherend under the action of peeling force is given by

$$y = 4(\frac{y_o S_o}{E})^{1/2} \left\{ \frac{\exp[-(E/y_o S_o)^{1/2}x]}{1 - \exp[-2(E/y_o S_o)^{1/2}x]} \right\} + y_o \tag{13'}$$

It should be noted that the relation can be derived without introducing any criterion for fracture.

The variation of the free energy of the deformation, δF, with respect to the difference of the crack length can be expressed:

$$\delta F_e = -y_o E \epsilon_b^2 \delta x \tag{14}$$

where ϵ_b is the ϵ at the breaking point. The variation of the surface free energy, δF_s, is given like as Eq. (3):

$$\delta F_s = \Gamma \delta x \tag{15}$$

where Γ is twice the surface free energy.

The second law of thermodynamics requires the relation (Griffith criterion):

$$\delta F \equiv \delta F_e + \delta F_s = 0 \tag{16}$$

and this leads to the results with Eqs. (13) and (15):

$$\epsilon_b = (\Gamma/y_o E)^{1/2} \tag{17}$$

or

$$\sigma_b = (E\Gamma/y_o)^{1/2} \tag{18}$$

On the other hand, the value of S_o is obtained for a given peeling angle θ using the relation

$$- (\frac{d\epsilon}{dx})_{x_b} = \frac{1}{y} \tan \theta \tag{19}$$

in Eq. (12):

$$\frac{y}{\tan \theta} = \frac{2S_o}{E} [\epsilon_b (\epsilon_b^2 + \frac{4S_o}{y_o E})^{1/2}]^{-1} \tag{20}$$

Then

$$S_o = y_o E \epsilon_b^2 \{\frac{1 + \sec \theta}{2\tan \theta}\} \tag{21}$$

From Eqs. (10) and (17), the steady peeling strength f_s, which corresponds to the stress S at the point of peeling, is derived as

$$2f_s = \Gamma + 2S_o \tag{22}$$

Combining Eqs. (17), (21) and (22), one obtains

$$f_s = \Gamma/(1-\cos \theta) \tag{23}$$

For the 90° peeling, Eq. (23) becomes:

$$f_s = \Gamma \tag{24}$$

Equation (23) shows that the steady peeling force is determined only from the surface energy and the peeling angle in the case of this peeling.   Stress distribution obtained with much hard working and expressed analytically does not appear explicitly in the result (though in the case of Griffith too!   cf. Eq. (4)). Further, Eq. (23) contains no Young's modulus nor the thickness of the specimen that determines the strain or the stress at the peeling point.   Eq. (24) merely tells us that the work done by the external force f_s is equal to the work to produce the surface, that is:

$$f_s ds = \Gamma dx_b \tag{25}$$

Let us consider this fact in other thermodynamic form.   As the work done by the external force is stored as the potential energy F_e, the following relation is obtained:

$$-F_e = \int_o^{(\epsilon_b)} fds \tag{26}$$

where the integration with ds is taken to the displacement at which
the elongation of the adhesive layer reaches the breaking point.
The increase in the surface free energy has been given by Eq. (15).
The condition of equilibrium is given:

$$\delta F_e + \Gamma \, dx_b = 0 \qquad (27)$$

From Eqs. (26) and (27),

$$\int^{(\varepsilon_b)} \frac{\partial f_s}{\partial x_b} \, ds = \Gamma \qquad (28)$$

From the geometrical relations, one obtains:

$$ds = (1 - \cos \theta) dx_b \qquad (29)$$

Then, from Eqs. (28) and (29), integrating df to the peeling force
f_s, one obtains the same result as Eq. (23),

$$\int^{(f_s)} df = f_s = \Gamma / (1 - \cos \theta) \qquad (30)$$

The above treatment is the feature of the Griffith theory
applied to the simplest example of the steady peeling. As for the
equilibrium or the stability of the crack or the cut in this
experiment, the peeling point is almost completely elongated along
the tensile direction and no shape of crack tip can be found as
shown in Fig. 1.

To compare this behavior with that of the initiation of the
peeling may be valuable.

THE CONDITION FOR INITIATION OF PEELING (7)

As mentioned above, the condition for the formation of the
initial crack leading to the fracture may not be derived thermodynamically. Then we assume that peeling starts when the extension
of the spring reaches the breaking elongation ε_b or the breaking
stress σ_b (these two are different from the values given by Eqs.
(17) and (18)) at the peeling point $x = x_b$ in Fig. 4. This assumption
is a working hypothesis which may not be derived from thermodynamics
or other mechanical laws. However, adopting this hypothesis instead of Griffith condition, one can obtain the initial peeling
force, f_i, as follows.

In this case, Eq. (10) gives

$$2f_i = y_o E \epsilon_b^2 + 2S_o \qquad (31)$$

From Eqs. (29) and (31) the force required to initiate peeling according to the maximum stress criterion is derived as

$$f_i = \frac{1}{2} y_o E \epsilon_b^2 / (1 - \cos \theta) \qquad (32)$$

Since $\epsilon_b E = \sigma_b$, Eq. (32) can be rewritten as

$$f_i = \frac{1}{2} y_o \epsilon_b \sigma_b / (1 - \cos \theta) \qquad (33)$$

For the 90° peeling, Eq. (33) becomes

$$f_i = \frac{1}{2} \epsilon_b \sigma_b y_o \qquad (34)$$

Γ in Eq. (23) led from Griffith condition is replaced by $\epsilon_b \sigma_y y_o / 2$ in these results. As $\epsilon_b \sigma_y y_o / 2$ is the work required for elongation of unit volume of Hookean materials to the breaking point, Eq. (33) shows that the fracture takes place when the external force f_i has done the work required for breaking a unit length of the adhesive layer. However, the value of "fracture energy" obtained here is generally not equal to the value of free energy in elastic deformation by which the condition of the steady peeling is led; nor has it any relation with the free energy within this treatment thus far. While f_i is proportional to the thickness of the adhesive layer obviously from Eq. (33), f_s is independent of the thickness. Therefore, in the case of a peeling experiment using a specimen (Fig. 5) with thickness more than $2\Gamma / \epsilon_b \sigma_b$, the value of load continues to rise until the start of peeling, then falls down to the definite value of Γ at the same time as an initial crack appears, and remains constant during steady peeling. The peak value is proportional to the thickness.

The results of the experiment are shown in Fig. 6, which suggests the validity of the maximum stress criterion for the initiation and of the thermodynamic theory for the propagation of peeling. Though the effectiveness of Griffith theory for the continuation of peeling is thought to be elucidated from the consideration above, we have no theoretical foundation why the start of peeling is subject to the conditions of the maximum stress. This problem is out of the frame of thermodynamics.

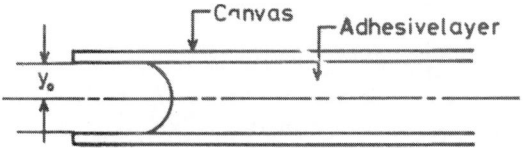

Fig. 5.   Peeling specimen.

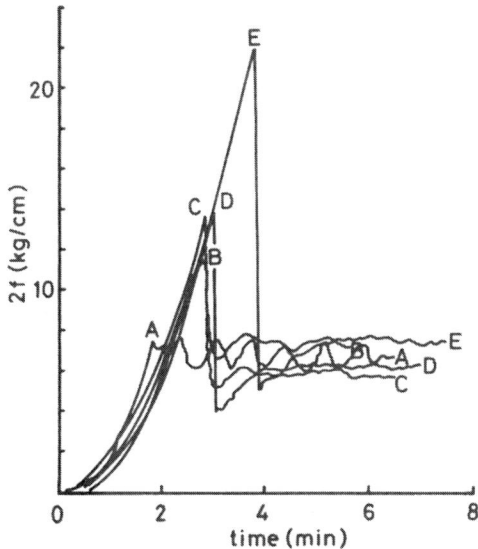

Fig. 6.   Peeling force - time curves for constant
          rate of extension. y_o(mm):   A,0.3; B,0.8;
          C,1.0; D,1.6; E,3.5.

PEELING AND ENERGY DISSIPATION (8)

One of the problems which are required to solve within the
frame of thermodynamics is the treatment of relaxation phenomena in
fracture.   Different from the thermodynamic equilibrium state, the
free energy cannot provide a condition for steady state in irreversible process and it indicates only the direction of change in the
system.   In the case that the steady state of a thermodynamic system
is realized, the system, being worked by an external force, converts
a part of the work done into heat.   In such a system the free energy
is maintained at a higher level than the equilibrium state.   It is
known that the condition for the steady state is determined by the
principle of minimum entropy production (9).   Therefore, the

criterion for the steady peeling of dissipative system should be derived from this principle. However, it may be allowed to assume that the condition for the steady state becomes near the condition of minimum free energy with a small correction in the case of the system close to the equilibrium.

From the above consideration, let us adopt the following expression from the condition for the steady peeling of dissipative materials:

$$\delta F = \delta F_e + \Gamma \, dx + \delta U' = 0 \qquad\qquad (35)$$

where U' is the energy dissipation during the deformation of the adhesive in peeling. Therefore, in this expression the dissipative energy is also to be formalistically regarded as a sort of free energy equivalent to surface free energy.

Provided that u' is the dissipation of energy per unit volume of the adhesive during the deformation to fracture and subsequent recovery (Fig. 7), $\delta U'$ becomes as:

$$\delta U' = hu'\delta x \qquad\qquad (36)$$

where h is the thickness of the adhesive layer. For $90°$ peeling, the following relation can be derived from Eq. (36) by the same treatment as deriving Eq. (24) from Eq. (16):

$$f_s = \Gamma + hu' \qquad\qquad (37)$$

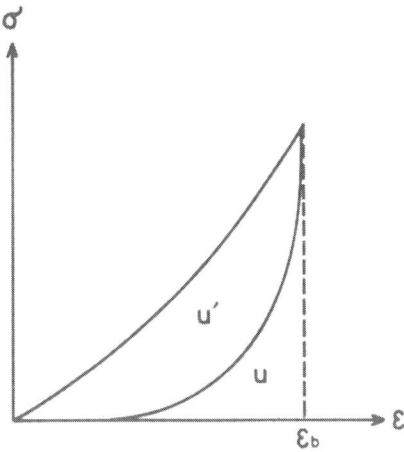

Fig. 7.   Stress-Strain curve.

Comparing Eq. (37) with Eq. (24), one can find that the peeling strength is to be increased by the second term. In the peeling, this additive term due to the energy dissipation represents the dependence of the peel strength on the thickness of the deformed layer.

The results of peeling experiment for the specimens with the various values of u' are shown in Fig. 8 and in Table 1.

Concerning evaluation of the energy dissipation, a question arises. As the energy dissipation takes place in the adhesive layer, the value of u' can be determined from the area enclosed by the hysteresis (Stress-Strain) curve in a simple extension test of the adhesive. However, the extent of the elongation (i.e., ε_b or σ_b) cannot be determined. This value is theoretically given by Eq. (17) or (18) in the case of elastic materials. Irreversible thermodynamics, i.e., the principle of minimum entropy production, must tell us a way to solve this problem. It is noted that assuming some fracture condition in such a steady fracture will be always tentative and conventional. Agreement of the values of u' obtained from S-S curve elongated to the breaking point of simple tensile test with those obtained by Eq. (37) from the dependence of peeling strength on the thickness is rather satisfactory as in the case of this kind of experiment.

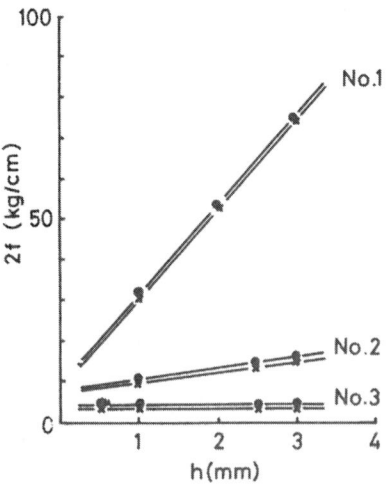

Fig. 8. Dependence of peeling force on thickness of the adhesive layer. ($\bullet$),$\dot{\varepsilon}$=50min^{-1}, (x),$\dot{\varepsilon}$=5min^{-1}.

Table 1. Results of Tensile and Peeling Tests

Specimen No.	$\dot{\varepsilon}$, min^{-1}	σ_b, kg/cm^2	ε_b	u', kg/cm^2	Γ, kg/cm	u', kg/cm^2
		Tensile			Peel	
1	5	150	7.0	320	9.0	220
	50	155	7.0	413	9.0	220
2	5	21.2	6.2	29	7.5	31
	50	25.6	7.3	29	8.5	27
3	5	15.5	0.75	0	3.5	0
	50	20	1.0	0	4.0	0

DISCUSSION

As shown in Fig. 6, the initial peeling force increases with the thickness of the adhesive layer. On the other hand, the force during the subsequent steady peeling process is practically independent of the adhesive thickness. These results correspond to those given by Eqs. (32) and (24), respectively. This is the experimental verification for the fracture criteria on peeling of elastic materials as mentioned above.

It may be interesting to classify the various formulas which have been proposed for the peel strength according to the results of this investigation and to examine whether they relate to the initiation or to the steady propagation of peeling. The studies made by Rivlin et al.(4) may be considered to refer to the propagation where the tearing energy governed the peeling or tearing strength. On the other hand, the treatment of peeling for flexible ribbons made by Bikerman (10) may concern the initiation of fracture, where the tensile strength of the cement is the major factor in the peel strength.

As for the effect of energy dissipation on the peeling, the relation between peel strength and thickness of the adhesive is found experimentally to be linear as expected from Eq. (37). For the adhesive with no energy dissipation, the strength is independent of the thickness of the adhesive as expected from Eq. (37) by setting u'=0.

Values of Γ, however, are of order of 10^6 erg/cm^2, which is much greater than the values for a usual solid surface (10^2 to 10^3 erg/cm^2). This may be explained by the energy of rupture and by the "draw out" of the polymer chains (11). This implies that Γ treated as free energy formalistically has the character of the dissipation energy. This consideration, however, may not require to rewrite the minimum relations represented by Eqs. (16) and (35).

Since the Griffith theory and its extension have the generality as the phenomenological theory from thermodynamical essence, it may be true that the theory needs not be modified in the future. In other words, the theory can be considered as only presenting the framework in which various phenomena can take place according to the laws of nature. Cracks in a solid, for example, govern the strength of the materials whose behaviors can be considered within this framework.

REFERENCES

1. T. Igarashi, J. Polym. Sci. Phys. Ed. 16, 407 (1978).

2. A.A. Griffith, Phl. Trans. Roy. Soc. (London) A 221, 163 (1920).
3. C.E. Inglis, Proc. Inst. Naval Architects, 14, March (1913).
4. R.S. Rivlin and A.G. Thomas, J. Polym. Sci. 10, 291 (1952); H.W. Greensmith, J. Appl. Polym. Sci. 3, 183 (1960); H.W. Greensmith, L. Mullins and A.G. Thomas, Trans. Soc. Rheol. 4, 179 (1960).
5. T. Hata, Kobunshi Kagaku (Polym. Chem.) 467, 72,77 (1947).
6. T. Hata, Zairyo (Materials) 13, 341 (1964).
7. T. Igarashi, op. cit.
8. T. Igarashi, J. Polym. Sci. Phys. Ed. 13, 2129 (1975).
9. I. Prigogine, "Étude Thermodynamique des Phénomenès Irréversibles", Desoer, Liège (1947).
10. J.J. Bikerman, J. Appl. Phys. 28, 1484 (1950).
11. T. Igarashi, J. of The Adhesion Soc. of Japan 8, 5 (1972).

Superposition of Peel Rate, Temperature and Molecular Weight for T Peel Strength of Polyisobutylene

Toshiya Tsuji, Mineo Masuoka and Kazumune Nakao

Osaka Prefectural Industrial Research Institute

2-1-53, Enokojima, Nishi-ku, Osaka, Japan 550

ABSTRACT

Effects of molecular weight of adhesive polymer on T peel strength were investigated over a wide range of temperature and peel rate. Polyisobutylene fractions having narrow molecular weight distribution were used as the adhesive, and low-density polyethylene films as the substrate. The peel strengths were found to yield a single master curve with respect to the peel rate and the temperature for both the cohesive and the interfacial failures. It was found that the shift factor a_T agrees with that of the Williams-Landel-Ferry (WLF) equation and is almost independent of the molecular weight of the adhesive. Furthermore, the master curves related to the rate-temperature are superimposed by the use of the shift factor a_M to yield a single master curve relation with respect to the molecular weight of the adhesive. In the case of the cohesive bond failure, a_M is related to the ratio of $(M/M_0)^{3.4}$.

INTRODUCTION

Many studies on peel strength have shown that the time-temperature reduced variable treatment can be applied to superimposed peel data in the similar way as the application to the viscoelastic response of adhesive polymers (1-5).

It is well known that adhesion properties are greatly influenced by the molecular weight of the adhesive polymer (6-9).

However, little has been reported in detail on the relationship
between them by the use of the adhesive polymer having narrow
molecular weight distribution. This will be attributed primarily
to the fact that the preparation of a large quantity of polymer
fraction having narrow molecular weight distribution is very
laborious. We were able to obtain a large quantity of polyisobutylene (PIB) fractions by the use of a large column fractionation
method.

The purpose of this study is to investigate the relationship
between the molecular weight of the adhesive polymer and the peel
strength in terms of the superposition principle by the use of PIB
fractions as the adhesive. T peel strength was measured in a wide
range of molecular weight of the adhesive between 2.0×10^4 and
90.2×10^4 over a wide range of temperature and peel rate.

The peel strength obtained for each molecular weight of the
adhesive was found to yield a single master curve with respect to
peel rate, temperature, and molecular weight of the adhesive.

EXPERIMENTAL PROCEDURES

Two kinds of PIB were used as the starting materials for
fractionation, VISTANEX MML-80(M_v:60×10^4) and VISTANEX LMMH(M_v:
5×10^4) both supplied by ESSO CHEMICALS Co. The fractionation of
VISTANEX MML-80 was accomplished by the use of the theta elution
column method using benzene as the solvent and changing elution
temperature. In the case of LMMH, benzene-acetone mixtures of
increasing solvent power were used as the eluting solvent at
constant temperature, $25^{\circ}C$. The load of PIB was about 80g per batch
and the same fractionation procedure was repeated several times.
The results of fractionation are summarized in Table 1. The ratio
of the weight- to number-average molecular weight, ($\bar{M}_w/\bar{M}_n$), indicates fairly a narrow molecular weight distribution of the fractions obtained. The viscosity-average molecular weights, $\bar{M}_v$, of
eight fractions selected as the adhesive from these fractions for
the present study were 2.0×10^4, 3.7×10^4, 6.7×10^4, 10.6×10^4,
21.8×10^4, 36.0×10^4, 46.6×10^4 and 90.2×10^4, respectively.

The low-density polyethylene film of 0.13 mm thick, supplied
by MITSUBISHI JUSHI Co., was used as a substrate and, prior to use,
wiped with distilled acetone and dried in atmosphere for two to
three hours.

Test specimens were prepared as follows: Each substrate was
coated with a n-hexane solution of the adhesive, and after dried
overnight in vacuo at about $40^{\circ}C$, one film was placed over another

Table 1.   Fractions of Polyisobutylene

LMMH fra.No.	$[\eta]$ [d)] dl/g	$\bar{M}_v$ ($\times 10^{-4}$)	$\bar{M}_w/\bar{M}_n$ [a)]	$\bar{M}_w$ [b)] ($\times 10^{-4}$)	$\bar{M}_n$ [c)] ($\times 10^{-4}$)	$\bar{M}_w/\bar{M}_n$
2	0.097	0.8	1.28			
3	0.127	1.3	1.19			
4	0.158	2.0	1.17			
5	0.187	2.7	1.17			
6	0.220	3.7	1.18			
7	0.236	4.3	1.17			
8	0.265	5.2	1.17	5.6	5.4	1.0_4
9	0.300	6.7	1.16			
10	0.362	9.6	1.17			
11	0.382	10.6	1.18	11.7	11.5	1.0_2
MML-80 fra.No.						
2	0.560	21.8	1.41			
3	0.730	36.0	1.34			
4	0.838	46.6	1.33			
5	0.950	59.0	1.33			
6	1.111	79.4	1.31			
7	1.190	90.2	1.31			

a) GPC
b) light scattering
c) osmotic pressure
d) benzene, 25°C

with a spacer of 100μ thickness between them and then they were pressed at 60°C for 10 min. Prior to testing, the bonded specimens were cut by 25 mm in width and conditioned overnight at 23°C and at RH 60%.

T peel strength was measured at five temperatures in the range from -30 to 40°C and at seven or eight rates in the range from 0.5 to 100 cm/min at each temperature by the use of Shimadzu Autograph DSS-2000 tensile tester and a temperature control chamber. T peel strength was averaged over three specimens.

RESULTS AND DISCUSSION

Relationship between Molecular Weight of Adhesive and T Peel
Strength

    The dependence of the peel strength on the molecular weight of
the adhesive is given in Figs. 1 and 2, in the case of the test
temperatures of 40 and -20°C, respectively.

    As the molecular weight increases, T peel strength for any
rate, except with 50, 20 and 10 cm/min at -20°C, rises steadily up
to a maximum strength and the mode of bond failure in this region
was cohesive. Since the cohesive force of adhesive polymer increases generally with the molecular weight in the region of low
molecular weight, the peel strength can be expected to increase with
the molecular weight. Beyond a maximum value, the peel strength
decreases abruptly and the mode of bond failure changed from
cohesive to interfacial failure, which had no visible adhesive
remaining on one side of the substrate, because the cohesive force
of the adhesive is expected to exceed the interfacial adhesive force
between the adhesive and the substrate. Finally, the peel strength
reaches an approximately constant value with increasing molecular
weight. The peel strength is altered with the molecular weight even
slightly at lower peel rate and markedly at higher rate. With any
molecular weight of the adhesives, the peel strength increases with
the peel rate for both modes of bond failure.

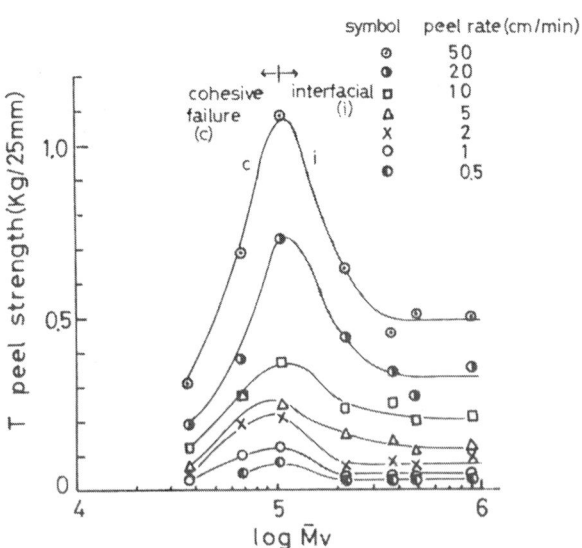

Fig. 1.   Effect of molecular weight of polyiso-
         butylene on T peel strength (40°C).

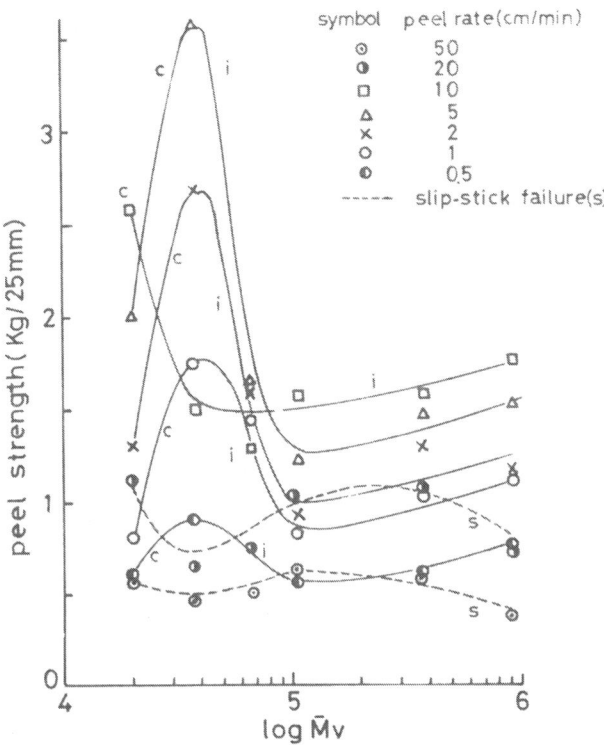

Fig. 2.   Effect of molecular weight of polyisobutylene on T peel strength (-20°C).

At the peel rates of 50 and 20 cm/min at -20°C, the slip-stick
failure was observed over the range of the molecular weight studied,
where the peel strength fluctuated widely with time as shown in Fig.
3. Therefore, the mean value of each peak strength was taken as the
peel strength. The schematic representation of slip-stick failure
is shown in Fig. 4. The slip-stick phenomenon would be associated
with the onset of the transition from rubberlike to glasslike
behavior of the adhesive. The peel strength was almost independent
of the molecular weight of the adhesive in this slip-stick region.

Results similar to those shown in Fig. 1 were obtained at the
test temperatures of 20 and 0°C. In the case of the adhesive having
$\bar{M}_v$ of 10.6×10^4, a mixed failure of both cohesive and interfacial was
observed at 20°C for the peel rates of 5, 10, 20, 40 and 100 cm/min,
and at 0°C, 0.5, 1 and 2 cm/min. The dependence of the peel
strength on the molecular weight obtained at -30°C was similar to
that obtained at -20°C. The maximum peel strength shifted to higher
molecular weight side with increasing temperature from 2.7×10^4 at
-30°C to 10.6×10^4 at 40°C.

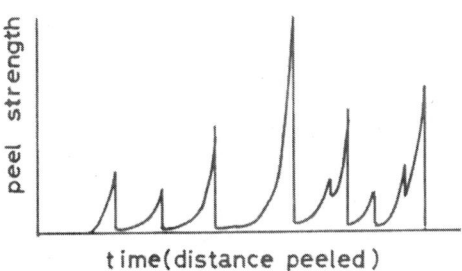

Fig. 3. Example of pattern of slip-stick failure.

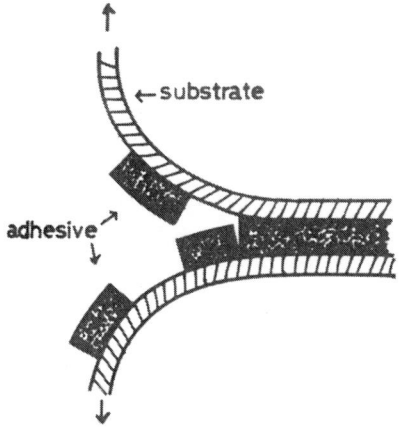

Fig. 4.   Schematic representation of slip-stick
          failure.

Superposition of Peel Strength

The dependences of the peel strength on both the peel rate and the temperature, and the peel rate-temperature superposition of the peel strength are given in Figs. 5 and 6 in the case of molecular weight of 3.7×10^4 and 90.2×10^4, respectively. In these and later figures, the peel strength was reduced by the ratio of 293/T, where $T(^{\circ}K)$ is the test temperature.

The solid line and the dashed line in Fig. 5 indicate the cohesive and the slip-stick failure, respectively. On the left side, the peel strength is related to the peel rate, R, for five test temperatures. In the region of the cohesive failure, the peel strength increases with the peel rate, but in the range of the slip-stick failure, the peel strength decreases with the peel rate. A single master curve obtained is shown on the right side, where the

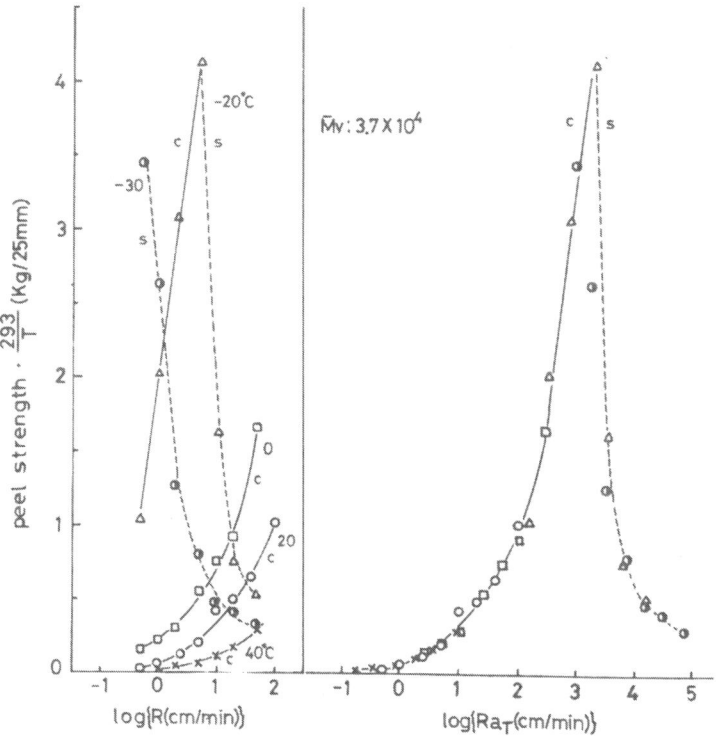

Fig. 5. Master relation for T peel strength
against peel rate, reduced to 20°C.

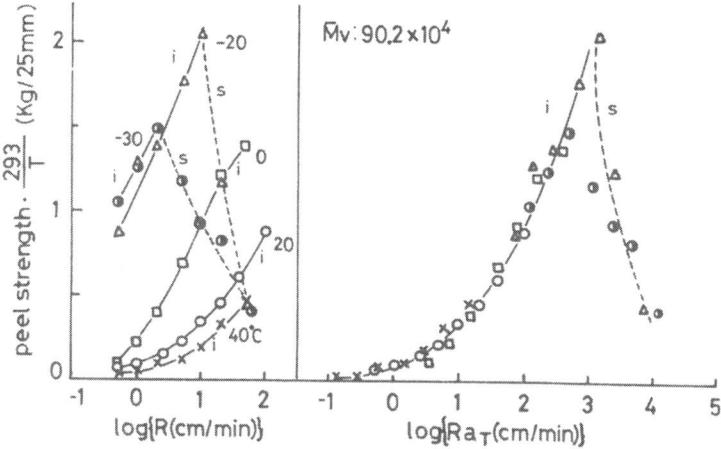

Fig. 6. Master relation for T peel strength
against peel rate, reduced to 20°C.

factor a_T relating equivalent rates at a temperature T and a reference temperature (293°K) was estimated by superposing the strength versus peel rate relation at any temperature upon that at a reference temperature. The peel strength drops abruptly at the transition region from cohesive to slip-stick failure.

Though not shown here, the results similar to that in Fig. 5 were obtained for the adhesives having $\bar{M}_v$ of 2.0×10^4 and 6.7×10^4, respectively.

The solid line and the dashed line in Fig. 6 refer to the interfacial and the slip-stick failure, respectively. As shown on the left side, in the region of the interfacial failure the peel strength increases with the peel rate, but in the region of the slip-stick failure the peel strength decreases with the peel rate. A single master curve obtained is shown on the right side, where the peel strength falls abruptly at the transition region from interfacial to slip-stick failure. Though not shown here, similar master curves were obtained for the adhesives having $\bar{M}_v$ of 36.0×10^4 and 46.6×10^4.

It is noted that the single master curve relation is obtained through both the cohesive and the slip-stick failure (Fig. 5) or both the interfacial and the slip-stick failure (Fig. 6), respectively.

For the adhesive having $\bar{M}_v$ of 10.6×10^4, however, data points were divided into two regions, one consisting of the cohesive failure and the other consisting of the interfacial one. The results obtained are shown in Fig. 7. The solid line refers to interfacial failure, the dotted line cohesive one, and the dashed line slip-stick one, respectively.

The good superposition obtained under different test conditions indicates that the dependence of the peel strength on the temperature and peel rate is due to a response of viscoelastic properties of the adhesive polymer for any failure mode.

Figure 8 shows the shift factor a_T versus temperature relation reduced to 20°C. It can be seen that a_T is almost independent of the molecular weight of the adhesive polymers. This fact would be explained from the rheological rate process, that is, a_T is approximated as

$$a_T = \eta_T / \eta_{T_0} \qquad\qquad (1)$$

where η_{T_0} and η_T are the viscosities measured at the standard

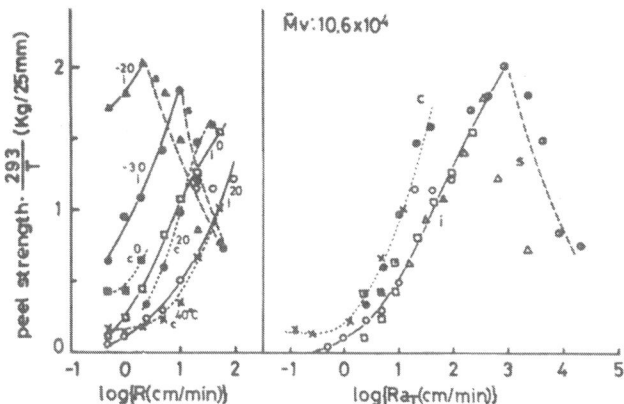

Fig. 7. Master relation for T peel strength
against peel rate, reduced to 20°C.

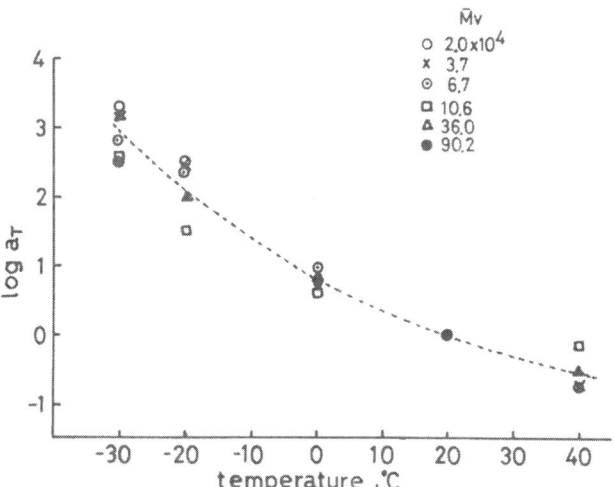

Fig. 8. Dependence of a_T on temperature, reduced
to 20°C.

temperature, T_0, and the test temperature, T, respectively. On the
other hand, the relationship between viscosity, η, and molecular
weight, $\bar{M}$, is expressed by

$$\eta_T = K_T \, M^{\alpha} \qquad\qquad (2)$$

for the temperature T and

$$\eta_{T_0} = K_{T_0} M^{\alpha} \tag{3}$$

for the temperature T_0, where K_T and K_{T_0} are constants for each temperature and α is equal to 3.4 above the critical molecular weight, M_c (M_c of PIB:1.7×10^4 (10)).

Substituting Eqs. (2) and (3) into Eq. (1), we obtain

$$a_T = K_T / K_{T_0} \tag{4}$$

The ratio K_T / K_{T_0} is independent of molecular weight and dependent exclusively on temperature. It is found, therefore, from Eq. (4) that the shift factor a_T is independent of the molecular weight as shown in Fig. 8.

In order to compare a_T value in this study with that of the literature, a_T was re-estimated with respect to the reference temperature of $243^{\circ}K(T_s)$ (11). The results obtained are given in Fig. 9. The solid line indicates the Williams-Landel-Ferry equation (12) defined by the following:

$$\log a_T = \frac{-8.86(T-T_s)}{101.6 + T-T_s} \tag{5}$$

T_s is the temperature at which a_T is unity. It can be seen that the W-L-F equation is superposed on our data.

We attempted to superpose all the above master curves with respect to molecular weight of adhesive. As the reference molecular weight, 2.0×10^4 was adopted, Fig. 10 shows the results obtained. It is found that a high portion of the data are superposed well on a master curve.

The symbol a_M means the shift factor for the molecular weight, by which the curves are shifted along the log Ra_T axis to superpose the reference curve.

It is quite interesting that the dependence of the peel strength on the time, the temperature and the molecular weight can be unified by the use of the shift factors a_T and a_M as shown in Fig. 10. That is, the principle of the superposition based on the W-L-F theory is extended to the effect of the molecular weight on the peel strength in this study. The dependence of a_M on the molecular weight is shown in Fig. 11. The symbol o[*] indicates the datum which

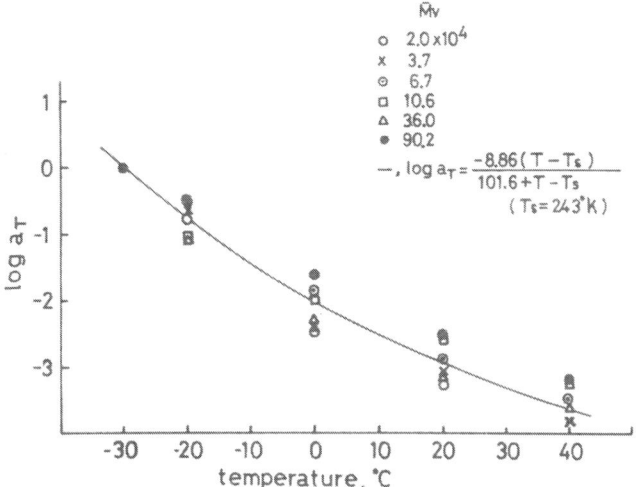

Fig. 9. Dependence of a_T on temperature, reduced to $-30°C$.

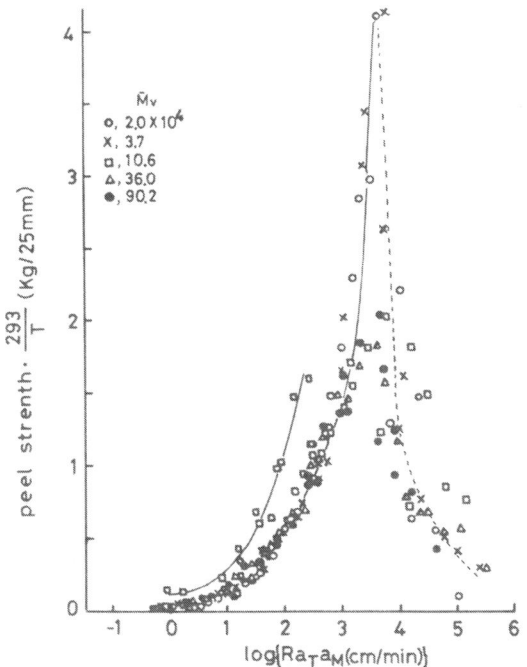

Fig. 10. Superposition of peel rate, temperature and molecular weight for T peel strength of polyisobutylene.

corresponds to the master curve consisting of the cohesive failure
for the adhesive having $\bar{M}_V$ of 10.6×10^4.

In accordance with our consideration, the shift factor a_T
includes all of the rate-temperature effect on the peel strength and
on the other hand, the shift factor a_M has the effect of the
molecular weight on the peel rate. From the viewpoint of the
rheological rate process, a_M is expected to be equal to η_M/η_{M_0},
where M_0 is the standard molecular weight, in the similar manner to
Eq. (1). The relationship between the viscosity and the molecular
weight is expressed by

$$\eta_M = K M^\alpha \tag{6}$$

and

$$\eta_{M_0} = KM_0^\alpha \tag{7}$$

Thus, the relationship between a_M and molecular weight is given by
the following:

$$a_M = (M/M_0)^\alpha \tag{8}$$

or

$$\log a_M = \alpha \log(M/M_0) \tag{9}$$

That is, the plot of $\log a_M$ against $\log(M/M_0)$ must show a straight
line having a slope of α. In general, α is equal to 3.4. The

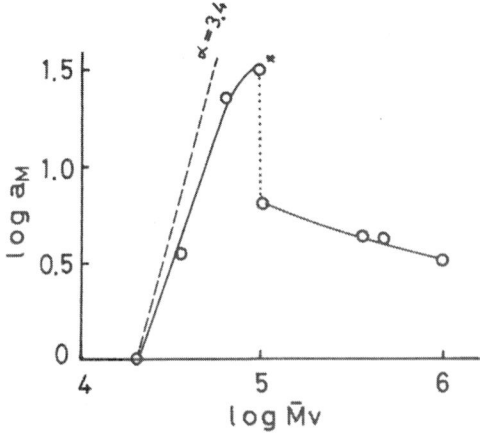

Fig. 11.   Dependence of a_M on molecular weight
(Reference molecular weight: 2.0×10^4).

relation of $\log a_M = 3.4 \log(M/M_0)$ is given by the dashed line in Fig. 11. As shown in Fig. 11, in the region of the molecular weight below 10^5 where the cohesive bond failure occurred mainly, the dependence of the shift factor a_M on the molecular weight agrees well with that predicted by Eq. (9), which was derived from the 3.4 power's law between the viscosity and the molecular weight. On the other hand, in the range of the higher molecular weight where the interfacial failure occurred, the shift factor a_M decreases abruptly to much smaller value than that predicted from Eq. (9). It is concluded, therefore, that the shift factor a_M can be related to the ratio of $(M/M_0)^{3.4}$ under the test conditions where the bond specimen shows the cohesive bond failure. The rate process on the bond fracture consists of many elemental steps and the rate-determining step is still uncertain. Consequently, it is very difficult to give a reasonable clarification of the shift factor a_M for the interfacial bond failure.

It will be noted from Fig. 10 that the time-temperature dependence of the peel strength can be further superimposed on the molecular weight over a wide range of molecular weight by the use of the shift factor a_M, in other words, the principle of superposition is extended so as to include an effect of the molecular weight on the peel strength in terms of the shift factor a_M defined in this study.

CONCLUSIONS

The dependence of the T peel strength on the peel rate, the temperature and the molecular weight of the adhesive polymer was investigated by the use of the principle of the superposition. A new parameter of the superposition is introduced by the rheological rate process based on the W-L-F theory. This parameter is the shift factor, a_M, for the molecular weight, by which the master curves related to the rate-temperature can be shifted along the axis of log Ra_T to superimpose the reference molecular weight. The fractionated polyisobutylene was used as the adhesive.

As the molecular weight of the adhesive increases, the peel strength shows a maximum where the mode of the bond failure changed from cohesive to interfacial failure, and the maximum point is shifted to higher molecular weight side with increasing temperature.

The peel strength was found to yield a single master curve with respect to the peel rate and the temperature for both the cohesive and the interfacial failures. The shift factor a_T agrees entirely with that of W-L-F equation and is almost independent of the molecular weight of the adhesive. The master curves related to the rate-temperature are further superposed by the use of the shift factor a_M to get a single master curve relation with respect to the molecular weight of the adhesive. In the case of the cohesive bond failure, the shift factor a_M is related to the ratio of $(M/M_0)^{3.4}$. On the other hand, in the case of the interfacial bond failure, a_M is much smaller than the theoretical value. The shift factor a_M defined in this study on the molecular weight was found to be explained qualitatively by the W-L-F theory in the range of relatively low molecular weight.

REFERENCES

1. T. Hata, M. Gamo, K. Kojima and T. Nakamura, "Mechanics of Peeling. II. Time-Temperature Superposition Principle in Adhesive Failure". Chemistry of High Polymers (Japan), 22, 160 (1965).
2. D.H. Kaelble, "Theory and Analysis of Peel Adhesion: Rate-Temperature Dependence of Viscoelastic Interlayers". J. Colloid Sci., 19, 413 (1964).
3. A.N. Gent and R.P. Petrich, "Adhesion of Viscoelastic Materials to Rigid Substrates", Proc. Roy. Soc. (London), A310, 433 (1969).
4. Y. Nonaka, "Mechanics of Peeling. 5. Temperature Dependence of Peeling Strength of Polyisobutylene", J. Adhesion Soc. Japan, 4, 207 (1968).
5. D.H. Kaelble, "Peel Adhesion: Influence of Surface Energies and Adhesive Rheology", J. Adhesion, 1, 102 (1969).
6. K. Nakao, M. Matsunaga and H. Shakutsui, "Studies on Acrylic Adhesives. Part 1. Bond Strength and Molecular Weight of Polyethylacrylate", J. Adhesion Soc. Japan, 1, 15 (1965).
7. S.W. Lasoski and Jr.G. Kraus, "Polymer to Metal. The System Polyvinyl Acetate-Steel", J. Polymer Sci., 18, 359 (1955).
8. S.S. Voyutskii, "Adhesion and Autohesion of Polymers", Adhesives Age, 5, (4), 30 (1962).
9. A.D. McLaren and C.J. Seiler, "Adhesion. (III) Adhesion of Polymers to Cellulose and Alumina", J. Polymer Sci., 4, 63 (1949).
10. T.G Fox and S. Loshaek, "Isothermal Viscosity-Molecular Weight Dependence for Long Polymer Chains", J. Appl. Phys., 26, 1080 (1955).
11. M.L. Williams, R.F. Landel and J.D. Ferry, "The Temperature Dependence of Relaxation Mechanisms in Amorphous Polymers and Other Glass-forming Liquids", J. Am. Chem. Soc., 77, 3701 (1955).

12. M.L. Williams, "The Temperature Dependence of Mechanical and
     Electrical Relaxations in Polymers", J. Phys. Chem., 59,
     95 (1955).

Discussion

A. Silberberg (Weizmann Institute, Israel): I find it very interesting that your failure mechanism seems to be viscosity related. Have you checked this by going below M_c into the region where your α should be about 1. Can you suggest why interfacial failure should depend upon the bulk disentanglement mechanism which accounts for the viscosity molecular weight dependence?

T. Tsuji, M. Masuoka and K. Nakao (Osaka Prefectural Industrial Research Institute, Japan): Your questions are of great importance from the viewpoint of the fracture mechanism of the adhesive joint. (1) The polyisobutylene fractions below $M_c (=1.7 \times 10^4)$ have very low viscosity, therefore, the appropriate specimen for peel test was not prepared. (2) The temperature dependence of the shift factor, a_T, for both cohesive and interfacial failures fits the WLF equation exactly as obtained in our results. It was found from the fractography of the interfacial failure that the adhesive was deformed and ruptured in ductile manner at the interfacial region. These facts show that the viscosity of PIB adhesive plays a very important role in the interfacial failure. In order to verify the shift factor, a_M, an application of the viscosity-molecular weight relation was attempted according to the WLF theory. This attempt is successful for the cohesive failure but not for the interfacial one. The rate-temperature-molecular weight superposition gave a fairly good idea about what the interfacial failure means.

On the Paper by T. Igarashi

R. Evans (Mameco): What were the criteria for an adhesive that had (or did not have) thickness dependence for peeling forces?

T. Igarashi (Gunma University, Japan): In both cases that the adhesive had and had no energy dissipation, the steady peeling force could be derived formalistically from principle of minimum free energy. Since the dissipation energy shall be the quantity different from the free energy, the criterion in the case of dissipative adhesive must be the others, say, may be the minimum entropy production principle. Because of the difficulty in theoretical treatments, the dissipation energy in this paper was regarded as a kind of free energy and was "renormalized" in it as shown in Eq. (35).

R. Evans: What adhesive had no thickness dependence? What adhesive did?

T. Igarashi: The adhesives that have no energy dissipation do not show the dependence of peeling force on the thickness. Such adhesives behave quite elastically and do not show hysteresis in the stress-strain curve; they, however, do not always obey Hooke's law. In our experiment, the adhesive layers consisted of rubber vulcanizates which contain BR, SBR and no reinforcing fillers. To promote adhesion the surface of adherend canvas was treated with RFL.

AUTHOR INDEX

(The numbers in parentheses are references for discussion participants. Pp. 1-456 are found in Volume 12A, pp. 457-858 in Volume 12B.)

A

Abraham, M.H., 825
Abricosova, I.I., 75
Adam, N.K., 854
Adams, A.L., 682, 854
Adams, D.F., 472
Adams, M.J., 101
Adams, R.D., 193, 410
Adamson, A.W., 174
Aggarwal, S.L., 362
Ah-Sing, E., 825
Ahlquist, C.N., 362
Ahn, J., 519, 538
Ainbinder, S.B., 75
Akhmatov, A.S., 74
Aleksandrova, N.A., 627, 641
Alexandrowicz, Z., 748
Allara, D.L., 549, 689, 751,
         756, (794)
Alred, R.E., 163
Amano, S., 413, 420
Ananthanarayanan, S., 845
Anderson, H.R., Jr., 519
Andrews, E.H., 163, 274, 501
Arakawa, H., 40
Arakawa, M., 41
Aris, R., 664
Armstrong, D.E., 823
Armstrong, R.C., 664
Ashbaugh, N., 472

Avotin'sh, Y.Y., 549

B

Baakker, G., 13
Barich, V., 627
Bach, D., 823
Bafr, Y., 122
Bagley, E.B., 52
Baier, R.E., 681, 822
Baker, D.W., 790
Bale, W.F., 682, 835
Barash, Y.S., 75
Barber, M.N., 14
Barquins, M., 203, 274, 275
Barrick, N.J., 824
Bascom, W.D., 39, 162, 173, 193,
         321, 338, 773
Basin, V.E., 75
Bates, R., 497
Bates, R.J., 501
Baun, W.L., 121, 122
Baxt, L., 575
Beaumont, P.W.R., 472
Beerbower, A., 52
Bell, A.T., 538
Bell, J.P., 122, 162
Bellamy, L.J., 756
Bellemans, A., 5, 13, 14
Belyi, V.A., 67, 74, 75, 539,

549, 550
Benko, B., 824
Benninghoven, A., 122
Benoit, H., 600
Bergh, A.A., 520
Bergmark, T., 122
Berlin, A.A., 75
Berry, J.P., 274
Bettelheim, F.A., 844
Betz, G., 122
Bigelow, C.C., 825
Biggs, H.G., 682
Bijsterbosch, B.H., 825
Bikerman, J.J., 391, 437,
        549, 627
Binford, J.S., Jr., 640
Birch, D.A., 275
Bird, R.B., 664
Birdi, K.S., 824
Birkle, D.L., 320
Bisch, P.M., 643, 664, 665
Bitner, J.L., 321
Bjeletich, J.G., 193
Blanchard, A.F., 362
Blank, R.E., 101
Blumenthal, R., 664
Bockris, J.O.M., 825
Bodnar, M.J., 312, 495
Boerio, F.J., 773
Bogie, K., 338
Bolger, J.C., 312
Bonekamp, B., 824
Boone, C.W., 673
Borisevich, N.A., 75
Botham, R.A., 641, 728
Boussinesq, J., 274
Bowden, F.P., 75
Bradley, A., 537
Brakere, R.R., 549
Brandts, J.F., 824
Brash, J.L., 682, 823, 824, 835
Braun, P., 122
Breece, J.M., 537
Brewis, D.M., 495
Briggs, D., 495
Briggs, G.A.D., 275
Brinson, H.F., 141, 162, 163,
        279, 298, 299,
        (363), (364)

Briscoe, B.J., 275
Brodkey, R.S., 338
Brody, H., 362
Bronsveld, W., 822
Brooks, R.E., 392
Brownell, R.E., 824
Brunt, N., 275
Brzozowski, K.J., 64
Bucci, C., 173, 193
Bucknall, C.B., 338
Bueche, A.M., 362
Buff, F.P., 13
Bui, H.D., 274
Bulas, B., 747
Bull, H.B., 824, 854
Bullett, T.R., 519
Burhnam, R.K., (196)
Burns, H., 640
Busse, W.F., 501
Butler, D.W., 519, 538

C

Cagle, C.V., 320
Callen, H.B., 274
Cannon, S.L., 362
Cantow, H., 626
Carlson, D.E., 537
Carlson, E., 575
Carpenter, D.K., 640
Carr, S., 575
Cartner, J.S., Jr., 299
Chan, D., 600
Chan, M.G., 549
Chan, R.K.S., 40
Chandrasekhar, S., 664
Chang, C.C., 538
Changeux, J.P., 664
Chapman, B.N., 538
Chase, K.W., 298
Chasovnikova, L., 847, 854
Chattoraj, D.K., 854
Chen, Y.D.M., 362
Chesters, G., 823
Chia, C.Y., 472
Chiang, T.C., 362
Chiu, T.H., 825
Choo, Y.H., 121

Chowols-Englert, A., 13
Chu, B., 627
Chumaevskii, A.N., 75
Cirlin, E.H., 362
Clark, A.T., 14, 626, 749
Clark, D.T., 575, 706
Clark, E.S., 362
Cleary, E.G., 845
Coburn, J.W., 121, 538
Cohen, J.B., 790
Cohen, R.E., 338, 362
Colson, J.P., 748
Comper, W.D., 845
Conley, R.T., 773
Cottington, R.L., 39, 338
Cotton, J.P., 600, 627
Counsell, P.J.C., 501
Counts, M.E., 162
Courtel, R., 274, 275
Coutts, M.D., 537
Crane, L.W., 537
Crochet,, M.J., 162, 298
Crooks, H.N., 538
Crowley, J.L., 165, 193, (197)
Cuthrell, R.E., 174
Cuypers, P.A., 824, 825, 854

D

Dahlquist, C.A., 501
Dahmen, E.A.M.F., 320
Dakin, T.W., 575
Dance, W.E., 163
Danielli, J.F., 822
Danil De Namor, A.F., 825
Dann, J.R., 52, 65
Daoud, M., 600
Dasgupta, A.A., 298
Datta, P., (78), (197), (364),
        (579), (581), 521
Dautzenberg, H., 627
Davidov, R.M., 824
Davidson, R.G., 313
Davidson, T., 775, 790
Davis, H.T., 13
Davis, L.A., 362
Davis, W.M., 338
Davson, H., 822

Davy, T.J., 706
Debenedictis, T., 495
Debye, P., 627
Deckert, C.A., 519, 520
Defay, R., 13
Deferran, E., 472
Degennes, P.G., 600
Dehl, R.E., 827, 835
Dejardin, P., 747
Dell'Oca, C.J., 122
Delollis, N.J., 495
Derjaguin, B.V., 276
Deryagin, B.V., 74, 75, 275
Desalos-Andarelli, G., 275
Dettre, R.H., 65
Devanathan, M.A.V., 825
Devereux, O.F., 121
Devine, A.T., 495
Devore, D.I., 749
Devries, K.L., 362
Dewitt, T.W., 338
Dezelic, G., 823, 824
Dezelic, N., 823, 824
Diggle, J.U., 122
Dijstra, R., 320
Dillman, W.J., 823
Dimarzio, E.A., 599, 640
Dini, J.W., 519
Divens, W.G., 575
Doi, T., 728
Doner, D.R., 472
Doumas, B.T., 682
Downie, T.C., 122
Drago, R.S., 52
Drutowski, R.C., 275
Dryfuss, P., (365)
Drylov, O.V., 824
Drzal, L.T., 121
Dubrovina, L.V., 75
Dudek, L.P., 101
Duplessix. R., 600
Dwight, D.W., (78), 121, 141,
        162, (195), (197),
        (198), (364), (577),
        (580), (581), (794)
Dyatlovitskaya, E.V., 673
Dynes, P.J., 312
Dzyaloshinskii, I.E., 75

E

Easterling, K.E., 275
Edelman, G.M., 854
Egorenkov, N.I., 75, 539, 549, 550
Eguchi, M., 575
Eirich, F.R., 52, 639, 640, 747, 748
Eisner, Y., 627
Elbing, E., 65
Elhorst, J.K., 824
Ellerstein, S., 640
Ellison, H.A., 538
Eltekov, A., 641
Eltekov, Y.A., 640
Endo, H., 392
Ensminger, D., (198)
Epand, R.M., 41
Erdogan, F., 410
Erman, V.Y., 627, 641
Eskin, V., 627
Etzkorn, H.W., 122
Evans, C.A., 575
Evans, J.T., 275
Evans, R., (455), (456)
Everett, D.H., 13
Ewing, P.D., 193
Eyring, E.M., (196)

F

Faber, W., 122
Fabulyiak, F., 627
Fahey, J.D., 321
Fairen, V., 665
Farnoux, B., 600, 627
Farrow, M.M., (196)
Feichtmayr, F., 575
Feinerman, A., 627
Felbeck, D.K., 472
Feng, M.Y., 472
Fenstermaker, C.A., 748, 824, 835, 854
Ferry, J.D., 274, 337, 452, 501
Feyen, J., 824
Fiande, E., 338

Fieschi, R., 173, 193
Fisher, M.E., 13
Fixman, M., 749
Flaggs, D.L., 193
Fleer, G.J., 823
Flory, P.J., 641, 748
Folkman, J., 673
Fontana, B.J., 52, 641, 747
Forest, T.W., 706
Foster, F.J., 824
Fowkes, F.M., 40, 43, 52, 65, (78), (79), (80), 728, 756
Fox, H.W., 64
Fox, K.K., 641
Fox, T.G., 452
Francis, S.A., 538
Frank, H.S., 825
Fridkin, V., 575
Frisch, H.L., 392, 639, 640
Frojmovic, M.M., 844
Fromageot, H.P.M., 825
Frommer, M.A., 748
Fuks, S., 13
Fukuda, H., 393, 411
Fukuzawa, K., 391
Fuller, K.N.G., 275
Furasawa, K., 825
Furneau, R.C., 122
Furukawa, K., 420

G

Gabel, D., 824
Gaever, I., 673
Gail, M.H., 673
Gaines, G.L., Jr., 39, 65
Gall, W.F., 854
Gamo, M., 452
Gardon, J.L., 52
Garrett, C.F., 121
Garroway, A.N., 339
Gebhard, H., 747
Gelius, U., 122
Gemant, A., 575
Genkin, M.V., 824
Gent, A.N., 274, 391, 392, 452, 773

Gerald, J., 627
Gergelson, L.D., 673
Gershman, H., 825
Gersho, A., 101
Gerstlf, F.P., Jr., 163
Gessler, A.M., 640
Gettings, M., 773
Gibbs, J.W., 13
Gilliland, E.R., 640
Gilman, A.B., 537
Gilula, N.B., 673
Ginsburg, V.L., 75
Glansdorff, P., 665
Gledhill, R.A., 163
Gleiter, H., 361
Goan, J.C., 472
Goddard, R.L., 682
Godde, J.S., 122
Goldsmith, W., 298
Golovachev, V., 627
Good, E.F., 121
Good, R.J., 65, 162, 275,
        391, 501
Goodland, R.L., 835
Goodman, A.M., 537
Goody, W., 52
Goring, D., 575
Goulding, C.W., 122
Gounder, R.J., 775
Grabovac, I., 320
Graham, D.E., 822, 823
Grant, W.H., 587, 641, 799,
        823, 824, 827, 835,
        (855), (856), (857)
Grayson, M.A., (366)
Greeland, J., 626
Greenler, R.E., 773
Greensmith, H.W., 437
Gregg, R.A., (363)
Gregor, L.U., 537
Gregory, H.J., 495
Gribova, I.A., 75
Grievenkemp, J.E., 773
Griffith, A.A., 437
Griffiths, P.R., 538
Grinnel, F., 673
Gross, B., 575
Gross, J., 844
Grossberg, A.L., 854

Grossman, R.F., 574
Gueleva, S., 362
Guess, T.R., 163
Guggenheim, E.A., 13
Guidi, G., 173, 193
Gusev, S., 627
Gutmacher, R.G., 640
Gutmann, W.H., 122
Gutoff, E.B., 640
Guttmann, A.J., 14

                            H

Haager, O., 854
Hafner, J., 472
Hahn, H.T., 463, 472
Hall, J.R., 495
Hall, J.T., 756
Hammermesh, C.L., 537
Hammes, J.P., 537
Hamrin, K., 122
Han, C.C., 823
Hanlin, D., 103
Hanna, G.L., 472
Hansen, C.M., 52
Hansen, F.Y., 664
Hansen, J.E., 392
Hansen, R.H., 495, 575
Hansma, P.K., 756
Hara, K., 640, 641, 728
Harey, E.N., 728
Harper, R.C., Jr., 338
Harrington, U., 122
Harris, B., 472
Harris, J., 338
Harris, L.A., 538
Harshbarger, R., 100
Hart-Smith, L.J., 163, 410
Hartog, J.J., 174, 193
Harvey, E., 627
Harvey, E.N.J., 641
Hashimoto, N., 520
Hassager, O., 664
Hata, T., 15, 39, 40, 41,
        (77), 391, 437, 452,
        496
Hauschka, S., 681
Hauser, E.A., 41

Hayama, H., 728
Hayashi, T., 410
Heaviside, O., 575
Hedeman, J., 122
Heden, P.F., 122
Hedvig, P., 173
Heimbuch, R.A., 472
Heimenz, P.C., 65
Helfand, E., 65, 600
Hellman, M.Y., 640
Helmkamp, R.W., 682, 835
Hemker, H.C., 824, 825, 854
Hennenberg, M., 664, 665
Herakovich, C.T., 162, 298
Herbst, M., 854
Hermens, W.T., 824, 9825, 854
Herzlinger, G.A., 825
Hesselink, F.T., 599, 747,
        748, 823
Heurtel, A., 275
Hill, T., 825
Hino, K., 501
Ho, T.C., 338
Hochmuth, R.M., 682
Hock, C.W., 501
Hoeve, C.A.J., 599, 626, 640,
        747, 748, 823
Hoffman, A.S., 673, 681, 682,
        691, 706, 822
Hohnson, W., 575
Holly, F.J., 706
Honig, R., 122
Horbett, T.A., 677, 681, 682,
        (684), (685), 706,
        (855), (857)
Horh, K., 575
Howard, G.J., 52, 640, 641
Howell, H.B., 163
Hoy, K.L., 52
Hsu, E.C., 773
Huft, C., 274
Huneke, J.T., (581)
Hunston, D.L., 321, 339
Hyndman, D., 790
Hyodo, S., 275

            I

Ibanez, J.L., 665

Igarashi, T., 421, 436, 437
            (456)
Ignatova, T., 627
Ikeda, M., 561
Illers, K.H., 392
Imoto, T., 640, 641, 728
Inagaki, H., 641
Inglis, C.E., 437
Inoue, Y., 392
Inubushi, H., 728
Iosipescu, N., 193
Ishida, H., 773
Ishikawa, H., 362
Israelachili, J., (683), (792)
Ito, T., 501

            J

Jackson, D.S., 845
Jannink, G., 600
Jay, R.R., 320
Jenkins, T.R., 174
Jennings, L.W., 162
Johansson, G., 122
Johnson, D.J., 101
Johnson, H.R., 519
Johnson, K.L., 274, 275
Johnson, R.E., Jr., 65
Jonath, A.D., (78), 165, 175,
            193, (195), (197),
            (198), (364)
Jones, L.G., 338
Jones, R.L., 173, 193
Jones, R.N., 773
Joppien, G.R., 641
Josefowitz, D., 501
Jud, K., 362

            K

Kaasteleyn, P.W., 13
Kaelble, D.H., 40, 52, 64,
            201, 274, 301,
            312, 363, (364),
            (365), 391, 452,
            537, 538
Kaganowicz, G., 521
Kagawa, I., 748, 749

Kahn, A., 640
Kalnin, M.M., 549
Kamagata, K., 501
Kambe, H., 501
Kammer, H.W., 65
Kanazashi, M., 575
Kaneko, M., 41
Kang, A.H., 844
Kanig, G., 39
Kanstad, S.O., 101
Kaplan, D., 338
Karasaki, H., (580)
Karasek, F.W., 100
Karlivan, V.P., 549
Kasemura, T., 15, 39, 40
Katagai, R., 41
Katchalski, E., 824
Katchalsky, A., 748
Kato, T., 729, 748, 749
Kato, Y., 495
Kausch, H.H., 362
Kawaguchi, M., 729, 748
Kay, C., 121
Kayayama, H., 728
Keer, L.M., 410
Kelly, A., 472
Kelly, M.A., 706
Kendall, K., 274, 275
Kerchner, J., 122
Khomutov, V.A., 75
Khopina, V.V., 640
Khramova, T., 626
Kikuchi, K., 41
Killmann, E., 640, 728, 747
Killon, M., 575
Kim, C.Y., 575
Kim, S.W., 823, 854
King, R.T., 538
Kinloch, A.J., 163, 274, 391,
        501, 773
Kirkbright, G.F., 101
Kirkwood, J.G., 13
Kiselev, A.V., 640, 641
Kitazaki, Y., 39, 40, 496
Klappratt, D.K., 122
Klein, F., 822
Klotz, I.M., 825
Kloubek, J., 52
Knibbs, R.W., 501

Kniseley, R.N., 100, 101
Knollman, G.C., 174, 193
Kobatake, Y., 392
Kobayashi, H., 538
Kobayashi, T., 551
Kochwa, S., 824
Koenig, J.L., 773
Kohno, A., 275
Kojima, K., 452
Kolotyrkin, V.M., 537
Kolthoff, I.M., 640
Kondo, T., 39, 40
Konieczo, M.B., 495
Kopfle, J.T., 320
Koral, J., 52, 640
Korshak, V.V., 75
Kosaka, H., 501
Koutsky, J., 338
Kovaleva, N.V., 640
Krasnov, A.P., 75
Kratky, O., 854
Kraus, G., Jr., 452
Kresheck, G.C., 825
Krieger, I.M., 338
Krimm, S., 790
Krotova, N.A., 74, 275
Krugers Dagneaux, P.G.L.C.,
        824
Kubacki, R., 537
Kuno, M., 748
Kurotsu, K., 41
Kutscha, D., 163
Kuzavkov, A.I., 539, 549, 550
Kuznezova, D., 626
Kwei, T.K., 392

                    L

Lal, M., 14, 626, 749
Laland, M., 13
Lambert, J.M., 501
Lampert, M., 575
Landau, L.D., 75
Landel, R.F., 452, 501
Landolt, D., 121
Langley, P.G., 501
Laplace, P.S., 13
Lasoski, S.W., 452
Lavrentev, V.V., 847

Leblanc, J.L., 362
Lecomber, P.G., 537
Lee, E.S., 854
Lee, L.H., 3, 4, 75, (77),
        85, 86, 87, (195),
        (196), (197), (363),
        549, (582), (684),
        (858)
Lee, R.G., 823
Lee, R.H., 338
Lee, R.L., 854
Lee, S.H., 757, 773
Lefever, R., 664
Legrand, D.G., 39, 65
Lehmann, S.L., 122
Leo, F., (78), (79)
Levi, D.W., 312, 495
Levin, M.A., 75
Lewis, G., 501
Lifshits, E.M., 75
Lin, C.J., 122, 162
Lin, D.G., 549, 550
Lin, J.W., (795)
Lin, J.W.P., 101
Lipatov, Y.S., 549, 601, 626,
              627, 728, 756
Loeb, G.I., 824
Logan, M.A., 845
Loginova, A.Y., 75
Lohse, M.I., 361
Lorensen, L.G., (794)
Loshaek, S., 452
Lott, A.D., 561
Low, M.J.D., 101
Ludwik, P.G., 162
Lundberg, J.L., 640
Lunsford, L.R., 420
Lussow, R.O., 503, 520
Lygin, V.I., 641
Lyklema, J., 823, 824, 825, 835
Lyman, D.J., 823, 824, 835

                    M

Mackor, E.L., 641
Macosko, C.W., 338
MacQueen, R.J., 519
MacRitchie, F., 823, 835

Mahanty, J., 4
Mahoney, C.L., 122
Malcolm, B.R., 854
Malers, A.Y., 549
Malvern, L.E., 298
Manne, R., 122
Manning, G.S., 749
Marceau, J.A., 122
Margis, D., 275
Margolis, L.B., 673
Maricic, J., 824
Mark, H., 501
Markovitz, H., 338
Markus, G., 854
Maroudas, N.G., 673
Marsden, J.G., 773
Martin, R.J., 537
Martin, T.W., 472
Maruchi, S., 52
Masares, M., 122
Mason, J.H., 575
Mason, W.P., (198)
Masters, L., (366)
Masuoka, M., 371, 391, 392, 439,
              (455)
Matduura, R., 749
Mathiew, H.J., 121
Mathot, C., 854
Matsumae, K., 495
Matsumoto, S., 748
Matsumoto, T., 707, 728, (793)
Matsunaga, M., 452
Matsuzawa, T., 520
Matthies, D., 575
Mattson, J.S., 854
Matuura, R., 599
Maugis, D., 203, 274
McCafferty, E., 728
McClellan, A.L., 52
McClelland, J.F., 100, 101
McConnell, P., 640
McCrackin, F.L., 538, 640, 748
McDevitt, N.T., 103, 121, 122
McGinniss, V.D., 561
McGrath, M.J., 52
McKenna, G.B., 362
McLaren, A.D., 452, 823
McMillin, C.R., 824
Means, A.M., 537

Memgeres, G., 627
Meyer, E.F., 52
Meyer, F.A., 844
Michaels, A.S., 312
Middlemiss, K.M., 14
Mihairi, H., 393
Milan, M., 673
Miles, M., 361
Miller, I.F., 823
Miller, I.R., 748
Mitchell, D.J., 600
Mittal, K.L., 496, 503, 519,
        538, (580), (581),
Miyairi, H., 410, 411
Miyamoto, T., 626, 641
Miyoshi, S., 728
Mizuhara, K., 640, 641, 728
Moddeman, W.E., 122
Mofnigman, J.R., 122
Mohandas, N., 682
Moniz, W.G., 320
Mons, J., 627
Montoya, O., 495
Moore, G.E., 392
Moreno, R., 575
Moriguchi, H., 501
Morimoto, T., 728
Moritz, A.G., 313
Morris, C.E.M., 313, 320, (365),
        (366)
Morrissey, B.W., 823, 824, 835
Morton, M., 362
Mosback, K., 823
Moscarello, M.A., 41
Moseona, A., 673
Mostafa, M.A., 52, 728, 756
Mostovoy, S., 162, 338
Motomura, K., 599, 749
Moynihan, R.E., 790
Moysya, E., 626, 627
Mukai, S., 728
Muller, A., 391
Muller, K., 825
Muller, V.M., 276
Mullins, L., 362, 437
Muramatsu, A., 393, 411
Murch, S.A., 162, 298
Murphy, M.C., 472
Mutschler, L.E., 682, 835

N

Nagao, M., 728
Nagasawa, M., 748, 749
Nagdi, P.M., 298
Naghdi, P.M., 162
Nagura, M., 362
Nakamae, K., 707, 728, (793)
Nakamura, T., 452
Nakanishi, K., 773
Nakao, K., 369, 371, 391, 392,
        439, 452, (455)
Naono, H., 728
Nasehzadeh, A., 825
Natarajan, R.T., 790
Nate, K., 551, (580), (581),
        (582)
Needham, T.E., 52
Nemethy, G., 825
Nenow, D., 362
Nesterov, A., 627
Neuman, R.E., 845
Neumann, A.W., 41
Nicolis, G., 665
Nicolson, G.L., 664
Niinomi, M., 538
Ninham, B.W., 4, 600
Nishikawa, A., 823
Niu, T.F., 338
Noda, I., 749
Noether, H.D., 362
Nonaka, K., 728
Nonaka, Y., 452
Nordal, P.E., 101
Norde, W., (684), (685), 801,
        822, 823, 824, 825,
        (855), (856)
Nordling, C., 122
Nose, T., 40
Numa, H., 362
Nyilas, E., 825

O

O'Boyle, D.R., 362
O'Connor, D.M., 361
O'Sullivan, J.P., 122
Okamoto, H., 575

Okawa, T., 728
Onsager, L., 13
Oreskes, J., 823
Origlio, G.F., 790
Orofino, T.A., 748
Othman, A.B., 274
Otsuki, K., 728
Outwater, J.O., 472
Owen, M.J., 472
Owens, D.K., 40, 52, 64, 575
Oya, M., 41

P

Padday, J.F., 65
Paganelli, J., 122
Paghupathi, N., 338
Palm, W., 854
Palmberg, P.W., 121
Pappillo, C.A., 362
Pandalai, K., 338
Paolini, G., 122
Papahadjopoulos, G., 664
Park, I.K., 362
Park, J.B., 362
Park, S.M., 362
Parker, J.G., 101
Parkinson, D., 362
Parodi, G.A., 101
Pascalf, J.V., 495
Pasechnik, N., 627
Passaglia, E., 538, 640, 747
Patat, F., 640
Patterson, D., 64
Pavinich, W., 362
Pavlov, S.A., 75
Pawel, R.E., 121
Pademonte, E., 362
Pefferkorn, E., 747
Peilman, M., 575
Penler, J.P., 121
Peppiatt, N.A., 193, 410
Peria, W.T., 538
Perkel, R., 640
Perkins, J., (365)
Perleberg, C.R., 538
Perzyna, P., 298
Petermann, J., 361
Peters, D.A., 520

Petersen, D.H., 163
Petrich, R.P., 274, 391, 452
Petrokovets, M.I., 74
Peyser, P., 640, 728
Phillips, D.C., 472
Phillips, M.C., 822, 823
Philofshy, H.M., 575
Picot, C., 600, 627
Piez, K.A., 844
Pilz, I., 854
Pimental, G.C., 52
Plazek, D.J., 338,339
Pleskachevskii, Y.M., 75, 549
Plueddemann, E.P., 773
Pluger, R., 391
Pollock, H.M., 275
Poranski, C.F., 320
Porter, A.W., 39
Poste, B.E., 664
Powers, W.J., 495
Prager, W., 298
Prescott, R., 472
Pressman, D., 854
Priel, Z., 844
Prigogine, I., 5, 13, 437, 665
Prosser, J.L., 519
Puchwein, G., 854
Pye, E.K., 824

R

Raj, R., 362
Rao, C.H.R., 773
Raphael, C., 410
Rappaport, C., 673
Rastogi, A.K., 39, 40, 64
Ratner, B.D., (195), (196),
                  673, 681, 691,
                  706, (791), (792)
                  822
Ratwant, M., 410
Rauhut, H.W., 495
Read, B.E., 790
Reeves, O.R., 673
Refojo, M.F., 706
Regan, R.J., 706
Reichard, H.S., 101
Reid, J.C., 338
Reiner, M., 298

Renieri, M.P., 162, 298

Renton, W.J., 163, 193, 299, 410

Rentzepis, P.M., 495

Rice, J.R., 274

Rice, K.K., 122

Richards, F.M., 824

Richardson, J.A., 122

Rivlin, R.S., 275, 437

Robb, I.D., 641

Roberts, A.D., 274, 275

Roberts, R.F., 519

Robin, M.B., 100

Robinson, C., 41

Roder, T.M., 392

Roe, R.J., 65, 599, 600, 629, 640, 749, 790, 823

Rohrer, J., (582)

Rosen, J.J., 667

Rosencwaig, A., 86, 100, 101

Rosenfield, R.E., 824

Ross, KD.L., 537

Ross, R.C., 312

Rossmann, K., 495, 575

Rothen, A., 854

Rothstein, E., 747

Rowland, F., 747

Rowland, T.J., 361

Roylance, D.K., 362

Rubin, R.J., 599

Ryan, F.W., 392

Rytov, S.M., 75

S

Sacchi, F., 122

Sadakne, G.S., 641

Sakai, I., 728

Samak, Q.M., 823

Samaras, N.N.T., 13

Samueld, R.J., 790

Sancaktar, E., 141, 163, 279, 299

Sandorfy, C., 773

Sanfeld, A., 643, 664, 665, (683)

Sapse, A.M., 65

Saraga, L., 13

Sarma, G., 600

Sarolea, L., 13

Sasaki, T., 673, 706

Sato, T., 640

Savkin, V.G., 74

Savkoor, A.R., 275

Sawada, Y., 338

Schaeffer, B.E., 664

Scharpen, L.H., 706

Scheraga, H.A., 825

Scheutjens, J.M.H.M., 823

Schich, M.J., 627, 641, 728

Schildenecht, C.E., 501

Schlag, J., 575

Schliebener, C., 640

Schonhorn, H., 392, 495, 519, 575

Schrader, W.H., 495

Schrag, J.L., 338

Schulman, F., 65

Schultz, J., 274

Schulz, R.A., 825

Schumaker, R.V., 854

Schwartz, H.S., 122

Schwarz, E., 854

Schway, M.B., 667

Scigliano, J.M., 52

Scott, J.R., 501

Scriven, L.E., 664

Searingen, L.E., 41

Seiler, C.J., 452

Semenovich, G., 626, 627

Sergeeva, L.M., 626, 627, 728, 756

Shafrin, E.G., 65

Shakutsui, H., 452

Shapira, R., 823

Shapiro, A.B., 824

Sharp, R.S., (198)

Sharpe, L.H., 391, 459

Shem, M., 538

Sherriff, M., 501

Shifrin, V., 627

Shinomiya, J., 41

Shunji, O., 549

Siegbahn, K., 122

Silberberg, A., 13, (77), (82), (195), (363), (455), 591, 599,

600, 640, 641,
(683), (684),
747, (791),
(792), (793),
(794), 823,
(856), (858)
Simha, R., 627, 639, 640
Singer, J.M., 822, 823
Singer, S.J., 664
Singleberry, C.R., 39
Siow, K.S., 64
Skolnick, J., 749
Small, P.A., 52
Smilga, V.P., 74
Smith, C.A., 854
Smith, L.E., 640, 728, 823,
824, 835, 854
Smith, O.E., 641
Smith, P., 123
Smith, R., 641
Smith, T., 123, 139, 312, 537
Smith, T.L., 392
Smolders, C.A., 824
Smolinsky, G., 537
Smurugov, V.A., 67, 75
Soderquist, M.E., 825
Sojka, S.A., 320
Sokolniloff, I.S., 338
Soloman, P.H., 773
Solomon, J.S., 103, 121, 122,
(196)
Solomonoas, I.N., 641
Somoano, R.B., 100
Sonnonstine, T., 575
Sorensen, T.S., 664, 665
Sorokin, J., 847
Sowell, R.R., 495
Spaeth, E.E., 682
Spar, I.L., 682, 835
Spear, W.E., 537
Sprague, B.S., 362
Springer, J., 162
Srere, P.A., 673
St. Pierre, L.E., 39, 40
Standefer, J.C., 682
Stansbarger, D.L., 163
Starkweather, H.W., 392
Statton, W.O., 361, 362
Stedman, M., 748

Stefan, J., 501
Stein, R.S., 790
Steinbach, A., 673
Steinberg, G., 728
Steinberg, I.Z., 824
Steinburg, H.L., 538
Steinchen, A., 643, 664, 665
Steingiser, S., 472
Steppo, R.F.T., 14, 626
Sterman, S., 773
Sternling, C.V., 664
Sternstein, S.S., 338
Stevens, W.W., 338
Stewart, C.W., 65
Stoddart, C.T.H., 519
Stradal, M., 575
Strasser, H.J., 640
Stromberg, R.R., 538, 640, 641,
747, 823, 824,
835, 854
Strong, J., 538
Strongberg, R.R., 728
Struik, L.C.E., 339
Stuart, P.R., 519
Sudo, M., 40
Sumiya, T., 728
Sung, C.S.P., 757, 773
Sung, N.H., 757, (795)
Suzuki, K., 39, 40
Sviridyonok, A.I., 67, 74, 75
Swalen, J.D., 519

T

Tabor, D., 75, 275, 277,
Tadros, T.F., 835
Takahashi, A., 729, 748, 749,
(793), (794)
Tanaka, T., 600, 640
Tanford, C., 824, 825
Tanner, W.C., 312
Telisman, Z., 823
Teodorovich, E.V., 75
Ter Minassian-Sarada, L., 854
Tesapnk, A.K., 537
Thies, C., 627, 641
728
Tholen, A.R., 275

Thomas, A.G., 275, 437
Thomas, C., 854
Thomas, H.R., 706
Thomas, J.R., 52
Thomas, R.L., 472
Thompson, G.E., 122
Tien, R.K., 537
Timmons, C.O., 173, 193, 338
Tishkov, N.I., 550
Tobin, N.R., 362
Tobolsky, A.V., 299
Todosyichuk, T., 626, 627
Tomoshige, S., 641
Tompkins, H.G., 756
Toriyama, Y., 575
Torrie, G.M., 14
Toyama, T., 501
Tozaki, H., 39, 40
Trantina, G.G., 193
Trapesnikov, A., 854
Tsai, S.W., 463, 472, (577),
          (579)
Tschoegl, N.W., 338
Tsuge, T., 749
Tsuji, T., 439, (455)
Tsutsumi, A., 41
Tsyganok, V.N., 549
Turpin, M., 626
Tutas, D.J., 640, 747
          728

U

Ueberreiter, K., 39, 41
Uhlmann, D.R., 362
Ullman, R., 52, 640
Uniyal, S., 823
Upadyayula, S.K., 362
Uraki, H., 40
Usmanova, D.O., 641
Uvarov, A.V., 627, 641
Uy, K.C., 40
Uzgiris, E.E., 825
Uzi, F., 39, 40

V

Vail, W.J., 664

Valasek, J., 748
Validov, M.A., 75
Van Craen, J., 13
Van Der Linden, R., 563
Van Der Waals, J.H., 641
Van Lamsweerde-Gallez, D., 664
Van Romunde, L.K.J., 822
Van Turnhout, J., 575
Vanderscheer, A., 823, 824
Vanderjagt, D., 682
Varoqui, ., 747
Vedove, W.D., 643, 665
Veis, A., 845
Velarde, M.G., 665
Vendery, R.B., 501
Vereshchagin, V.G., 75
Viehbock, F.P., 122
Vignes-Adler, M., 664
Viksne, V.A., 549
Vincent, B., (793), 825
Vinogradov, A.V., 75
Vinson, J.R., 163, 299, 410
Vogel, G.C., 52
Von Frankenburg, C.A., 65
Voyutskii, S.S., 452
Vroman, L., 682, 854
Vuk-Pavlovic, S., 824

W

Wagner, R.E., 52
Wakamatsu, T., 728
Wake, W.C., 501
Wakefield, T.D., 101
Walton, A.G., 824, 825, (856),
          (857)
Ward, C.A., 706
Ward, E., 673
Warner, C., 854
Wasserman, J., 824
Watatani, K., 728
Watson, W.A., 682
Way, S., 274
Weathersby, P.K., 681, 682, 706
Weber, R.E., 538
Weber, T.A., 600

Wecker, S.M., 790
Wegman, R.F., 312
Wehner, G.K., 122
Wehs, E.E., 321
Weinberg, Z., 575
Weisman, Z., 844
Welgand, H., 728
Wells, E.E., 339
Wen, W.Y., 825
Wendt, R.C., 40, 52, 64
Werme, L.O., 122
Westerdahl, C.A.L., 495
Wetzel, F.M., 501
White, J.L., 641
White, J.R., 275
White, L.R., 600
White, M., 537
Whitehouse, R.S., 501
Whitney, J.M., 163
Whitney, W., 362
Whittington, S.G., 14
Wielogorski, J.W., 706
Wightman, J.P., 162
Wilcox, D.L., 538
Williams, J.G., 193
Williams, M.L., 452, 453, 501
Winn, M.L., 854
Winters, H.F., 538
Wisniewski, S., 854
Wittberg, T.N., 122
Woermann, D., 627
Wolman, M., 822
Wood, G.C., 122
Wood, S.J., 641
Wool, R.P., 341, 361, 362
Wronski, C.R., 537
Wu, S., 39, 40, 52, 53, 64,
        65, (80), (81), (82)
Wurstein, F., 575
Wyckoff, H.W., 824
Wydeven, T., 537

                    Y

Yamaguchi, Y., 413, 420
Yamakawa, H., 627
Yamakawa, S., 471, 495, (579),
        (580), (791)

Yamamoto, F., 495
Yamashita, N., 39, 40
Yamashita, S., 728
Yamori, S., 749
Yanazawa, H., 520
Yavorsky, P.M., 338
Yeow, Y.T., 299
Yoshida, T., 640
Young, T., 13

                    Z

Zapas, L.J., 338
Zbinden, R., 790
Zettlemoyer, A.C., 728
Zheludev, I., 575
Zisman, W.A., 52, 64, 65, 501

SUBJECT INDEX

(Pp 1-458 are found in Volume 12A, Pp 459-866 in Volume 12B)

A

Abrasion
  of specimen, 625
Absorption spectra, 68
Acid-base interaction, 43, 78
Acrylate
  oligomer, 558
  resins, 582
Activation energy, 321
Adaptation
  conformational-, 591
Additivity law, 21
Adherence
  of polyurethane, 204
  of punch, 204
  of spheres, 204, 236
Adherend
  aluminum, 165
  bondline thickness, 144
  shear stress, 144
  surface, 758
  surface preparation, 142
  thickness, 144
Adhesion
  cell, 667, 681
  measurements, 503
  mechanism, 491
  of filled PE, 539
  of polymer, 97
  of thin film, 521
  photoacoustic spectro-
    scopy, 87
Adhesive

bonding, 473
bulk properties, 141
characteristics, 551
debonding, 500
failure, 123, 124
failure energy, 143
interaction in polymer, 67
interface, 175
metal structure, 301
penetration, 113
structural reliability, 301
tack, 497
Adhesive-adherend
  by Auger spectroscopy, 103
  by ESCA, 103
  interface, 103, 124, 165
Adhesive bond
  by dielectric relaxation
    gradient, 165
Adhesive joint
  FRP, 393
  tensile, 371
Adsorbance
  of NaBr, 738
  of train, 744
  ratio, 639
Adsorbed
  monolayer, 751
  trains, 592
Adsorbent
  ammonium chloride, 618
  solution ratio, 602
Adsorption
  affinity, 639

degree of, 605
degree of selectivity, 629
enthalpy of, 816
kinetics, 736
of albumin, 684
of dyes, 87
of gamma-globulin, 816
of hemoglobin, 677
of plasma, 681
of polymer mixture, 618
of polymer on metal
    oxide, 707
on polystyrene, 684
preferential, 638
selective, 629
Adsorption isotherm, 724, 736
binary, 635
of protein, 804
of water, 710
plateau, 814
Aerosil, 603
Ageing
of adhesive, 316
Aggregate
fractionation of, 610
Aggregative
character of
    adsorption, 602, 603, 608
Albumin, 807
crosslinked, 809
radiolabeled, 804
Alkyl diphenyls, 17
Alkyl naphthalenes, 17
Alpha-metal free
phthalocyanine, 93
Aluminum, 105, 301
alloy, 105, 301
-aluminum oxide, 111
-epoxy system, 188
-ethylene vinyl
    acetate, 372
film (anodized), 106
-PE joint, 544
substrate, 751
Aluminum oxide, 757
as filler, 539
plate, 553
Aminosilane, 757
γ-Aminopropyltriethoxysilane,
                              759

Angle-resolved
    photoemission, 797
Anisotropic shape
    effect, 390
Anodic oxide, 106
Antioxidant, 540
Anti-reflection
    coating, 521
Arrhenius
    rate constant, 355
    relation, 166
Aspect ratio, 372
Athermal solution, 11
ATP, 644
Auger
    electron spectroscopy, 103,
                    129, 521, 528
    thin film analyzer, 528
Average absorption, 780
Azide
    as buffer, 685

                    B

Barry's representation, 234
Bending moment, 393, 419
    distribution, 401
Benzene, 714
Benzil, 553
Benzoin, 553
    isopropyl ether, 553
Benzophenone, 553
Beta-counter, 829
Beta-ray scattering, 123
Beta-relaxation, 321, 335
Bimolecular
    inhibition, 659
Bingham model
    modified, 281
Biocompatibility, 677, 801
Biomaterial, 827
Biomembrane, 643
Biopolymeric layer, 646
Bisphenol
    diglycidyl ether, 331
Blister method, 505
Block copolymer, 28, 341, 350
Blood

platelet adhesion, 837
Boltzmanns constant, 45, 166, 355, 744
Bond
    durability, 491
    exposure time, 301
    joint failure, 176
    strength, 131
Bondability, 473
Bonding force, 70
Boundary layer
    structure of, 601
Bovine
    albumin, 678
    gamma-globulin, 828
    pancreas
        ribonuclease
        (RNase), 802
Brittle
    failure, 466
    fracture, 124
Broadline
    NMR, 775, 785
Bromocresol, 678
Brønsted acid, 44
Brush-like failure, 466
Bulk shear, 150
Butt joint, 372
    coefficient, 386

                C

Cadmium telluride
    detector, 752
Calcium carbonate, 49
Calf serum, 667
Carbon black, 352
Carbon tetrachloride, 714
Carboxylic acid
    absorption of, 542
CASING, 474, 565
Castigliano theorem, 215, 237, 267
Cathetometer, 16
Cationic polymer, 718
Cavitation, 645
Cell
    adhesion, 667
    deformation, 643

    membrane, 663
    mobility, 667
    model, 8
Cellulose fiber, 91
Chain
    conformation, 612
    length, 633
    molecule, 10
    segment, 591
Chandrasekhar principle, 594
Chaotropic ion, 816
Characterization
    of adsorbed polymer, 707
    of hydrophilic-hydrophobic
        polymer, 691
Charge
    density, 563, 573, 811
    distribution, 810
    repulsion, 834
Chemical bonding, 118
Chemisorption
    of acetic acid, 752
Chip-coat, 582
Chlorinated
    polyvinyl chloride, 49
Chondroitin sulfate, 838
Chopping frequency, 89
Chromatogram, 317
Chromium
    oxide, 809
    plate, 793
Chromophores, 195
Circular dichroism, 35
Circumferential stress, 464
Coating glass, 691
Cobalt-60
    radiation, 678
Cohesive
    energy density, 25, 602
    failure, 123, 124
    force, 442
Coil conformation, 596
Collagen, 837
    fiber, 838
    multimer, 837
    soluble molecule, 858
Collagenase, 839
Complementary
    energy, 214
Complex

shear modulus, 323
Compliance, 239
Composite strength, 463
Computer maping, 126
Concentration
    gradient, 636
    of damage, 354
    of polymer segment, 637
Conformational
    change, 847
    constraint, 591
    entropy, 634
    transition, 35
Conservation
    equation, 648
Constitive equation, 281
Contact angle, 25, 55
Contact area, 212, 218
Contrast microscopy, 693
Cooperative
    transconformation, 647
Cooperativity constant, 647
Copper
    as filler, 539
Corona discharge, 563
Corresponding states, theory, 10
Corrosion
    inhibiting primer, 115
Coulomb interaction, 805
Counterions, 805
Coupling
    agent, 794, 795
    relation, 650
Covalenties, 847
Crack
    extension, 210, 213, 468
    healing of polymer, 341
    initiation, 304
    length, 145, 428, 468
    tip, 131
    velocity, 216, 234, 260
Crack propagation, 144, 211
    kinetics of, 204
    with fixed cross-head
        velocity, 260
    with fixed grip, 260
Creep stress, 283
Critical
    deformation velocity, 376
    fiber length, 470

opalescence phenomena, 609
    strain rate, 376
    stress, 134, 304
    stress intensity factor, 179
    surface tension, 53
    wavenumber, 662
Crochet's
    delayed yield equation, 283
Crosshead
    displacement, 212
    velocity, 204
Cross lap joint, 373
Crosslink density, 757, 795
Crosslinking, 564, 768
    reactoin, 551, 557
Cross-section
    circular, 330, 331
    rectangular, 330, 331
Crystalline band
    of PTFE, 778
Crystallinity
    index of, 783
Cyclohexane
    for polycarbonate, 603
    solution, 770
Cytoplasmic spreading, 671

D

Debye length, 744
Deformation
    field, 363
    of adherend, 393
    ratio, 778
    velocity, 218
Degradation, 564
Degree
    of crystallinity, 371, 374
    of hydrolysis, 478
Dehydration, 702
Density
    of gas, 18
    of liquid, 18
    of packing, 540
Deprotonation, 807
Depth of penetration, 612
Desorption, 722
    from hydrophobic
        surface, 803

Dextran
    crosslinked, 685
Dichroic orientation
    function, 780
Dicyandiamide, 313
Dielectric
    constant, 10, 745, 811
    depolarization spectro-
        scopy, 165
    layer, 521
    permeability, 68, 70
    relaxation of
        epoxy, 165
Differential
    IR spectroscopy, 809
    scanning calorimeter, 552
Diffusion
    convection, 645
    demixing, 349
    migration, 647
Dimer, 11
Dipole, 165
    -dipole interaction, 45
    moment, 45, 167
    relaxation, thermally
        stimulated, 175
Dispersion equation, 651
Dispersion force, 45
Displacement
    equation, 651
    factor, 216
    transducer, 327
Dissipative
    energy, 433
    function, 229
Dividing surface, 7
Donnan
    equilibrium, 733
    exclusion, 738, 745
Donor-acceptor
    complex, 536
    interaction, 43
Double lap joint, 144
Drago relation, 44
Draw direction, 776
Ductile
    brittle transition, 150
    fracture, 372, 387
Du Nouy ring, 703, 793
Dupre's

energy of adhesion, 208

E

Eastman 910
    adhesive, 146
Egg albumin, 37
Elastic
    deformation, 222
    displacement, 239
    energy, 215, 222, 239, 263
    modulus, 145, 218, 399
    shear strain rate, 149
    wave, 181
Electret formation, 566
Electric field, 811
    d.c., 166
    overlap, 817, 818
    polarizing, 189
Electrical
    double layer, 811
Electrokinetic
    phenomena, 171
    potential, 805
Electrolyte technique, 580
Electromagnetic field, 68
Electron
    gun, 107
    micrograph, 319, 521
    microscopy, 498, 780
    probe microanalysis, 144
Electroneutrality, 811
Electrophoresis, 807
Electrophoretic analysis, 828
Electrostatic repulsion, 742
Ellipsoid
    end-on, 831
    side-on, 831
Ellipsometer, 829
Ellipsometry, 521, 524, 729,
                        848, 856
    of protein, 809
End-to-end distance, 742
Energy
    balance theory, 204
    dispersive analysis, 157
    of adsorption, 814
    of rupture, 436
Enthalpy, 209, 224

Entropy, 206
    conformational, 630
    of mixing, 630
Epoxy
    adhesive, 106, 141
    aluminum bond, 175
    equivalent weight, 316, 320
    phenolic adhesive, 301
    polyamide adhesive, 417
    resin, 313, 618, 621
    rubber toughened, 157
    tetraglycidyl
        diaminomethane, 366
    thermosetting, 322
Epoxylite, 191
Equation
    of state, 55
Equilibrium theory
    of crack, 425
    of peeling, 426
ESCA, 509, 693
Etchant, 511
Ethyl methacrylate, 191
Ethylene
    acrylic acid copolymer, 79
    glycol condensate, 17
    oxide-propylene oxide
        copolymer, 27
    propylene copolymer, 22
    vinyl acetate, 27, 29
Evaporative analyzer, 314
Excluded volume, 10, 598
    function, 745
Expansion factor
    of loop, 742
External
    force, 429
    polarization, 566
    reflection IR
        spectroscopy, 751
Extinction coefficient, 524,
    751, 754, 755, 761, 778

                    F

Failure
    envelop, 375
    mode, 165, 166, 463

Fiber
    debonding, 466
    efficiency, 470
    -matrix bond, 466
    -matrix debond energy, 469
    -matrix failure, 150
    reinforced
        plastics (FRP), 393
    stiffness, 150
    volume, 464
Fibrillar texture
    in polypropylene, 783
Fibrinogen, 685, 687, 807
    on polyethylene, 857
Fickian diffusion, 646
Filled elastomer, 341
Film
    deposition, 522
    plasma-polymerized, 524
    thickness, 857
    thickness monitor, 109
Fixation, 647
Flory
    equation, 742
    -Huggins term, 632, 744
    -Huggins theory, 11
Fluorescence
    spectroscopy, 809
Forced oscillation, 323
Fourier transform
    IR spectrometer, 752, 759, 760
    IR spectroscopy, 752, 757
Fowkes equation, 54
Fractograph, 124
Fractography, 152, 371, 377
Fracture
    criteria of peeling, 421
    energy, 148, 431
    mechancis, 203
    mode, 150
    pattern, 408
    stress, 424
    toughness measurement, 176
Free energy
    criterion, 421
    of deformation, 428
    of interaction, 632
Free volume, 22
Friction

of irradiated film, 73
stress, 469

G

Gadolinium, 146
Gamma-counter, 829
Gamma-globulin, 807, 857
Gauss theorem, 811
Geiger Mueller
    counter, 126
Gelatin, 91
Geometric
    mean equation, 53
    mean expression, 45
    parameter, 217
Gibbs
    adsorption formula, 7
    -Duhem equation, 7
    energy, 814
    free energy, 209
    surface, 5
Glass, 758
    mat, 394
    -polymer joint, 544
    roving, 394
    -rubber contact, 245
    support, 314
Glass temperature, 19, 217, 335
    of adsorbed layer, 610
    of UV resins, 555
Globular protein, 801
Glueline inhomogeneities, 146
Gold-palladium alloy
    film, 148
Grafted
    polyethylene, 791
    silicone, 791
Grafting
    radiation induced, 474
Graphite epoxy
    composite, 464, 577, 579
Grating spectrometer, 778
Griffith
    criterion, 207, 209, 247
    -Irwin theory, 145
    theory, 423

H

Hamaker constant, 44
Hard and soft acids and
    bases, 79
Harmonic mean equation, 53
Healing
    line mode, 354
Heat
    of mixing, 46
Helical content, 36
Helium
    neon laser, 829
    plasma, 473
Helmholz
    free energy, 207
    plane, 805, 818
Hemostasis, 837
Hertz
    load, 246
    problem, 236
Hertzian contact, 245
Hooke's Law, 149, 287
Huggin's
    coefficient, 840
    constant, 843
Human
    plasma albumin, 802
    serum albumin, 827
Humidity
    absorption, 579
Hydrodynamic
    boundary condition, 650
    equation, 683
    technique, 730
    velocity, 644
Hydrogen
    abstraction, 553
    bonding, 25, 45, 565, 710, 794
    ion transfer, 817
Hydrolysis
    of silane monomer, 716
    polymerization, 767, 795
Hydrophile
    -lyophile balance, 26
Hydrophilic
    -hydrophobic interaction, 668
    surface, 667, 851
Hydrophilicity, 857

Hydrophobic
    bond, 804
    dehydration, 815
    interaction, 817
    surface, 827
Hydrophobicity
    of protein, 804
    of sorbent surface, 803
Hydroxyproline, 838
2-Hydroxyethyl
    methacrylate, 668, 677, 691

                    I

Immobilization of
    protein, 801
Immunoglobulin, 681, 847
Incident
    sound speed, 185
Inconel, 528
Inelastic electron tunneling, 752
Infrared
    dichroism, 775
    spectroscopy, 521, 525, 541,
        567, 617
Inorganic filler, 539
Instron test machine, 147, 522
Intensity
    of molecular interaction,
        ratio, 72, 766
Interaction parameter, 55, 132
Intercellular
    communication, 668
Interface
    energy, 207
    roll of, 467
Interfacial
    accomodation zone, 167
    bond failure, 371
    failure, 126, 446
    fracture of joint, 371
    free energy, 704, 855
    strength, 466
Interfacial tension, 6, 16
    of polymer melt, 15
    of polymer solution, 15
    time-dependence, 792
    water-oil, 707

Interlaminar
    shear strength, 470
Intermolecular
    interaction, 16, 70
Internal
    energy, 225
    slip, 788
    stress, 561
Interphase, 104
Intrinsic viscosity, 714, 840,
    843
Iodine, 677, 827
Ion
    beam etch, 105
    co-adsorption, 856
    migration, 509
    pair, 817
Ionic strength, 722, 729, 847,
    851
Iosipescu specimen, 180
Iron oxide, 758
    localization of, 707
Irwin relation, 271
Isoelectric point, 37, 805, 853,
    856
Isotropic material, 177

                    J

Joint strength, 150

                    K

Kaolin
    as filler, 539
Kelvin-Zisman probe, 574
Kinetics
    of adsorption, 677
    of cell adhesion, 667
    of detachment, 204
    of healing of polymer, 353
Kramers-Kronig
    relation, 68

L

Labeled protein, 677
Lag-phase, 673
Lagrangian
    multiplier, 595, 633
Lamellar crystal, 780
Laminate, 577
Langmuir
    adsorption, 829
    equation, 804
    isotherm, 638
Lap
    joint, 414
    joint deformation, 148
    length, 395
    shear joint, 141, 302
    shear strength, 124
    shear test, 475
Laplace
    equation, 6
    operator, 649
    transform, 287
Lateral stress, 645
Lauryl methacrylate, 555
Legendre
    theorem, 244
    transformation, 207
Length
    of train, 593
Limit
    of stability, 212
Linear
    elasticity, 205
    electron accelerator, 475
    viscoelasticity, 281
Lipid
    of membrane, 644
Liquid
    crystal, 36
    state, 8
Locus
    of failure, 127, 143,
        165, 482
London dispersion
    force, 43
Longitudinal
    shear, 464
    strength, 464

Loop, 593
    partition function, 743
    -train-conformation, 744
    -train-structure, 742
    -train-tail conformation,
        729
Lorenz-Lorentz
    equation, 732
Loss factor, 218
Ludwik's equation, 149, 283
Lyophilicity, 34

M

Macleod's exponent, 18
Macromolecule
    nonadsorbed, 593
Macrophage, 643
Magnesium
    X-ray source, 107
Magnetic
    recording tape, 724
Matrix
    -primer interfacial
        failure, 152
Maximum
    strain, 346
    stress, 421
Maxwell
    model, 282
    relations, 209
Mechanical
    adhesion, 580
    loss, 624
    properties gradient, 621
    recovery, 345, 354
Melt index, 540
Mercaptoethanol, 851
Mercury lamp, 552
Metal
    carboxylate, 754
    oxide interface, 112, 196
Methacrylate
    resins, 582
Methyl acrylate, 473
    graft, 474
    2-Methylanthraquinone,
        553, 561

Methyl group
    effect, 22
Methyl violet 2B, 91
Methylol acrylamide
    graft, 474
Michaelis-Menten
    desorption, 652
Microvoid
    formation, 341
Migration
    site of, 514
Minimum entropy
    principle of, 434, 456
Mises hypothesis, 371
Mixing condensation
    critical-, 609
Model biomembranes, 643
Modified epoxy, 126
Moffitt's equation, 35
Moisture degradation, 310
Molar
    enthalpy, 816
    refraction, 732
Molecular
    forces in adhesive, 67
    mobility of adsorbed
      layer, 607
    rearrangement, 855
Molecular weight
    critical-, 448
    distribution, 440, 630
    number average, 21, 440
    of polymer, 683
    viscosity average, 440,
      455
    weight average, 440
Monochromator, 778
Monolayer
    adsorption isotherm, 852
Monosulphide
    as antioxidant, 541
Monte Carlo method, 11
Multicomponent system, 601
Multilayer film, 97
Multiple internal reflection,
    759
Muscle-like
    contraction, 644

N

Navier-Stokes
    equation, 644
Necking, 422
Neumann equation, 6
Neutron
    flux, 146
    radiography, 141, 146
Newtonian
    fluid, 644
    liquid, 498
Nitrogen
    reaction on surface, 521
Non-crystalline phase
    PTFE, 775
Non-linear
    viscoelastic deformation,
      292
Normal stress, 216
Notched
    strength, 468
Novolac resin, 167, 177
Nuclear magnetic resonance,
    high resolution, 603
Nylon-12, 374

O

Optical
    absorption, 88
    coating, 521
    density, 541, 766, 848
    rotary dispersion, 35
Organic-metal
    interface, 751
Orientation function, 778
Oscillation regime, 658
Oxidation, 565
    induction period, 546
Oxide
    etch rate, 509
    thickness, 137
Ozonisation, 565

P

Packing density, 602, 617

Packing perfection
    degree of, 617
Papain, 848
Parachor, 18, 78
Partition function, 12, 594
Peel
    rate, 439
    test, 475, 505
Peeling, 263
    angle, 429
    bend energy, 263
    force, 455
    initiation of, 421, 430
Peel Strength (T), 439
    of PE-epoxy, 473
    superposition, 444
    Wet, 482
Penetration depth, 769
Perfect solution, 9
Perfluorodecanoic acid, 54
Permittivity of medium, 811
Phagocytosis, 643
Phase
    angle, 329
    reversal, 185
Phosphate buffer 841
Phosphoric acid, 109
Photoacoustic
    spectrometer, 89
    spectroscopy, 87, 196
Photodegradative
    etching, 486
Photoelectron spectroscopy,
    103
Photoelasticity, 218
Photoinitiator, 552
Photopolymerization, 553
Photoresist, 503
    coater, 668, 693
    KTFR, 510, 580
    negative, 518
    positive, 510
    process, 516
Photovoltaic cell, 522
Phthalic acid, 109
Piperidine, 331
Plasma, 685
    protein, 827, 838
    reactor, 475
Plateau adsorbance, 821

Platelet
    adhesion, 827
    aggregation, 837
    baboon, 792
    human, 792
    -platelet contact, 838
Platinum surface, 729
Poisson's ratio, 181, 221,
    283
Polarity
    of polymer, 712
Polarizability
    of molecule, 10, 621
Polarized radiation, 776
Polarizer, 778
Polyacrylamide, 792
    gel, 685
Polyacrylates, 25
Polyacrylic acid, 751, 794
Poly-dl-alanine, 37
Poly-alpha-methylstyrene, 17
Polybutadiene, 718
    ageing, 365
    carboxylate rubber, 618
1,2-Polybutadiene
    dimethacrylate, 551
Polycaproamide, 70
Polycarbonate, 91, 603
Polychloroprene, 499
Polyelectrolyte, 729
    adsorption of, 794
polyester, 91
    soft, 393
    hard, 393
Polyethylene, 22, 56, 59, 70
    adhesion of, 539
    graft on, 677, 692
    high density, 540
    linear, 828
    low density, 440, 475
    medium density, 475
    oxidation of, 541
    surface treatment of, 563
Polyethylene glycol, 21
Polyethylene oxide
    oligomer, 375
Polyethylene terphthalate,
    54, 70
Poly-2-ethylhexyl acrylate,
    25

Polyethyl methacrylate, 667,
    677
Poly-gamma-benzyl-l-glutamate,
    35
Polyhydroxyethyl methacrylate,
    667
Polyhydroxyproline, 838
Polyion, 729
    coil, 742
Polyisobutylene, 439
    fractions, 441, 455
Polyisobutylene glycol, 22
Polyisoprene, 499
Poly-L-lysine, 37
Polymer
    chain flexibility, 602
    melt, 30
    on solid surfaces, 601
    solution, 33
    solvation, 714
    solvent interaction, 610
Polymethyl methacrylate, 49,
    70, 91
    film, 612
    isotactic, 632
    solution, 631
    syndiotactic, 632
Polypeptide
    chain, 804
    fragment, 829
    solution, 35
Polyproline, 838
Polypropylene, 22, 539
    isotactic, 540
Polypropylene glycol, 21
Polysiloxane network, 757
Polystyrene, 30, 91
    in DCE, 603
    interfacial tension, 716
    latices, 802
    solution, 631
Polytetrafluoroethylene, 56,
    59, 70
    grafted surface, 476
    morphology, 775
    orientation, 775
    semicrystalline, 780
Polyurethane, 218
Polyvinyl acetate

    interfacial tension, 716
Polyvinyl alkylate, 25
Polyvinyl caprylate, 25
Polyvinyl chloride
    Inconel interface, 528
    metal, 529
Porter's equation, 16
Post-treatment, 569
Potassium dichromate, 93
Practical adhesion, 475, 504
Preferential localization, 712
Prestrain, 266
Prestress, 266
Pressure, 216
    sensitive adhesive, 497
Profile analysis, 105
Profilometry, 693, 697
Proline, 838
Propylene glycol
    condensate, 17
Protein
    adsorption, 677, 801, 855
    dissociation, 794
    globular, 856
    I-125 labeled, 685
    iodination, 678
    sorbent contact, 817
Pull test, 505, 534

                        Q

Quasi-isotropic
    elastic moduli, 577
    strength, 578
Quasi-lattice model, 592
Quasistatic force
    of adherence, 203, 212,
        240, 265
    of punch, 234

                        R

Radial stress, 464
Radius
    of contact (actural), 239
    of gyration, 714, 742
Rail

shear test, 146
test, 363
Random walk
    two choice, 594
Randomization
    of fiber orientation, 577
Rate
    of change of damage, 344
    of change of strain energy,
        344
    of dissipation, 344
    of healing, 344
    of working, 344
Rayleigh
    angle, 185
    stability of jet, 683
    -Taylor problem, 651
Reactive monomer, 551
    methacrylate, 553
Recovery half life, 356
Reduced parameter, 217
Reflecting microscope, 375
Reflectivity, 89
Refractive index, 524, 612,
    732, 794
    complex, 732
    of saphire, 759
Regular solution, 9
Relative humidity, 221
Relaxation time, 166, 283
Resin
    -metal interface, 143
    -rubber binary mixture,
        498
Reverse osmosis
    membrane, 521
Rheovibron, 169
Rhodamine B, 97
Rigid punch
    adherence, 221
Rigidity
    of joint, 396
Rolling resistance
    of glass cylinder, 268
Root mean square distance,
    742
Rotational
    freedom, 820
    mobility, 817
Routh-Hurwitz stability

conditions, 656
Rubber friction, 218

S

Sapphire, 757
Saturation
    magnetic flux, 725
    point, 743
    problem of PAS, 88
Scanning electron
    microscope, 106
    microscopy, 141, 143, 693
Scarf
    angle, 405
    joint, 177
Scintillation
    counting, 667
Scratch test, 506, 533
Screen printing, 582
Second viral
    coefficient, 745
Secondary ion mass
    spectra, 117
    spectrometry, 758
Secular equation, 660
Segmental mobility, 623
Selective adsorption, 27, 77,
    619, 633
Semicrystalline
    polymer, 341
Serological test, 801
Serum albumin, 837, 839
Serum-free media, 669
Sessile drop method, 15
Shaft-angle
    encoder, 327
Shear
    distribution, 144
    modulus, 172, 185, 624
    speed, 172
    strain rate, 364
    strength, 150, 396
    stress, 372, 413, 464
    stress (maximum), 467
    wave speed, 185
    (metal), 198
Shift factor, 335, 379

Short fiber
   composite, 577
   strength, 578
Silane, 758
   coupling agent, 557, 757
   primer, 115
Silica, 49, 758, 794
   adsorption on, 630
   gel as filler, 539
   surface, 581
Silicon oxide, 510, 511, 831
   hydrophilic, 833
   pretreatment, 515
   photoresist interface, 514
   plate, 848
Silicone
   dimethacrylate, 558
   rubber (graft), 691
Single lap joint, 394
Sliding
   friction, 281
   speed, 72
Slip-stick
   failure, 443
Small angle
   X-ray scattering
   (SAXS), 342
Sodium bromide, 730
Sodium chloride
   solution, 847
Sodium polyacrylate, 729
Solubility parameter, 45, 709
Solvent power, 714
Sorbent surface
   degradation, 819
Span-to-depth ratio, 467
Specific
   refraction, 616
   volume, 19
Spontaneous rupture, 245
Spreading pressure, 55
Sputter ion gun, 107
Sputtering
   rate, 111
   time, 111
Squareness ratio
   of magnetization, 724
Stability of equilibrium, 210
Statistical mechanics
   of adsorption, 5

   of surface tension, 5
Stearic acid
   contamination, 129
Steel
   Nylon-12-steel, 381
Steric stabilization, 793
Stern
   layer, 805
   potential, 805
Stiffness, 150, 210, 497, 577
   specific, 578
Stored
   elastic energy, 345
Strain
   energy release rate, 203,
     208, 239, 250
   rate, 150, 282
Stress
   concentration factor, 413
   distribution, 372, 413, 417
   distribution of punch, 270
   elongation diagram, 396
   intensity approach, 216
   intensity factor, 204,
     216, 270
   relaxation, 778
   softening effect, 351
   -strain response, 291
Strontium Chromate, 117
Structure
   of adsorbed protein, 808
Structural adhesive, 177
   failure characterization,
     141
   viscoelastic characteriza-
     tion, 321
   viscoelastic shear
     behavior, 279
Structural rearrangement
   in protein, 807
Structurization
   degree of, 608
styrene, 521
   -butadiene rubber, 499
   -butadiene-styrene block,
     copolymer, 342
   -dimethylaminoethyl
     methacrylate copolymer,
     718

-maleic anhydride
    copolymer, 34
Sudgen's equation, 15
Surface
    ageing, 301
    analysis, 148
    area, 852
    charge density, 818
    contamination, 133
    elasticity, 684
    energetics of metal
        structures, 301
    energy, 77, 301, 567
    entropy, 77
    generation mode, 185
    hydrophilicity, 692
    instability, 643
    modification of PE, 473
    potential, 574, 807
    pressure, 847
    roughness, 143, 705, 732
    site fraction, 638
    topography, 705
    turbulence, 643
    viscosity, 684
    wettability, 563, 567
Surface free energy, 423, 428,
    521
    dispersive component, 531
    measurement, 529
    polar component, 531
Surface tension, 15, 16, 644,
    709
    dispersion component, 54
    M.W. dependence, 17
    of polymer melts, 15
    of polymer solutions, 15
    of solid, 53
    polar component, 54
    statistical mechanics, 5
Surfactant, 28
Symmetric
    lap shear, 286
    rail shear, 286

                    T

Tack

adhesive, 497
Tack force, 498
Tackification, 499
Tackiness, 233
Tacticity
    of polymer, 22
Talc
    as filler, 539
Talysurf profile, 148
Tape test, 505
Tapered
    double cantilever beam,
        177
Tearing energy, 436
Teflon plate, 17
Temperature, 206
    post bake, 516
Tensile
    delayed failure, 285
    force, 263
    modulus, 374
    stress, 424
    stress (maximum), 467
    tester, 441
Tensile strength, 372
    of UV resins, 552
    reduced, 379, 389
Tetrafluoroethylene
    tensile deformation, 788
Tetrahydrofuran
    -ionen block copolymer, 730
    -propylene oxide copolymer,
        27
Tetramethylbenzidine, 678
Texture transition, 775
Thermal
    damping constant, 89
    diffusion length, 88
    diffusivity, 89
    stimulated current, 197
    stimulated discharge, 574
Thermodynamic potential, 203,
    207, 423
Thermosetting resin, 551
Theta
    point, 742
    solvent, 608
Thick film of A-1100, 765
Thickness

dimension of adhesive, 132
of adhesive layer, 431
of adsorbed layer, 738
of oxide film, 509
of polymeric layer, 540
of silane film, 761
sample, 88
Thin film analyzer, 107
Thrombus, 827, 838
formation, 792, 831
Thumb tack test, 498
H^3-Thymidine, 667
Time-temperature
superposition, 321, 322,
359, 375, 371
Titanium
alloy, 308
dioxide, 794
Toluene, 714
Torsion bar, 323
Torsional
elastic constant, 323
elasticity, 329
Total internal
reflection (ATIR), 612
Toughness
of adhesive, 150
Transconformation, 646
Transcrystalline
region, 383
Transferred ions, 817
Translational entropy, 803
Transmission spectrum, 753
Transverse
failure strain, 578
tension, 464
Tresca
failure criteria, 149
hypothesis, 372
Trialkoxysilane, 758
Trichloroethylene, 145, 714
Triglycidyl p-aminophenol, 315
Tris
as buffer, 839, 840
Tropocollagen, 838
Trousers-type
specimen, 421
Trypsin
-ETDA, 669

Tyrode, 839, 840

U

Ubellhode viscometer, 839
Ultrasonic
Rayleigh wave measurements,
175, 181
Ultraviolet
radiation, 551
-visible absorption, 97
Undercutting constant, 509
Uniaxial
deformation, 780
tension, 287
Un-notched strength, 468

V

Van der Waals
forces, 67, 204, 343,
535
interactions, 165, 802,
817
Van Laar
mixing energy, 632
Vertical displacement, 221
Vinyl
acetate grafts, 474
triethoxysilane, 759
Viscoelastic
losses, 203
material, 375
plastic theory, 296
shear behavior, 279
solid, 218
surface, 237
Viscoelasticity, 150
Viscoplasticity, 280
Viscosity coefficient, 282
Void, 154
Volume
contraction, 557
fraction, 577

W

Water
    absorption, 482, 557
    polymer solution, 716
Weak boundary layer, 372, 565
Weissenberg
    Rheogoniometer, 321, 499
Wenzel's relation, 246
Wettability, 474
Welhelmy method, 793, 848
WLF
    equation, 144, 439, 448,
      455, 499
    relationship, 333, 334
    shift factor, 217
Work
    of adhesion, 32, 46, 203,
      204, 217, 252, 303,
      425, 500, 510
    of adsorption, 500
    of fracture, 144, 468

X

X-ray
    beam (monochromatized),
      693
    diffraction (wide angle),
      775
    microanalyzer, 373
    photoelectron spectro-
      meter, 107

Y

Yield
    strength, 372
    stress, 149, 280
Young's modulus, 203, 282,
    374, 424, 429, 557, 577

Z

Zisman's
    critical surface tension, 53